T0155597

# Visual Quantum Mechanics

Bernd Thaller

# Visual Quantum Mechanics

## Selected Topics with Computer-Generated Animations of Quantum-Mechanical Phenomena

Bernd Thaller
Institute for Mathematics
University of Graz
A-8010 Graz
Austria
bernd.thaller@kfunigraz.ac.at

Additional material to this book can be downloaded from http://extras.springer.com

Library of Congress Cataloging-in-Publication Data
Visual quantum mechanics : selected topics with computer-generated animations of quantum-mechanical phenomena / Bernd Thaller.
p. cm.
Includes bibliographical references and index.
ISBN 978-1-4757-7428-3 ISBN 978-0-387-22770-2 (eBook)
DOI 10.1007/978-0-387-22770-2
1. Quantum theory. 2. Quantum theory—Computer simulation.
I. Title.
QC174.12.T45 2000
530.12'0113—dc21 99-42455

ISBN 978-1-4757-7428-3 Printed on acid-free paper.



9 8 7 6 5 4 3 2 (Corrected printing, 2002) SPIN 10876005

Typesetting: Pages created by the author using a Springer $\mathrm{T_EX}$ macro package.

www.springer-ny.com

# Preface

In the strange world of quantum mechanics the application of visualization techniques is particularly rewarding, for it allows us to depict phenomena that cannot be seen by any other means. *Visual Quantum Mechanics* relies heavily on visualization as a tool for mediating knowledge. The book comes with a CD-ROM containing about 320 digital movies in QuickTime™ format, which can be watched on every multimedia-capable computer. These computer-generated animations are used to introduce, motivate, and illustrate the concepts of quantum mechanics that are explained in the book. If a picture is worth a thousand words, then my hope is that each short animation (consisting of about a hundred frames) will be worth a hundred thousand words.

The collection of films on the CD-ROM is presented in an interactive environment that has been developed with the help of Macromedia Director™. This multimedia presentation can be used like an adventure game without special computer skills. I hope that this presentation format will attract the interest of a wider audience to the beautiful theory of quantum mechanics.

Usually, in my own courses, I first show a movie that clearly depicts some phenomenon and then I explain step-by-step what can be learned from the animation. The theory is further impressed on the students' memory by watching and discussing several related movies. Concepts presented in a visually appealing way are easier to remember. Moreover, the visualization should trigger the students' interest and provide some motivation for the effort to understand the theory behind it. By "watching" the solutions of the Schrödinger equation the student will hopefully develop a feeling for the behavior of quantum-mechanical systems that cannot be gained by conventional means.

The book itself is self-contained and can be read without using the software. This, however, is not recommended, because the phenomenological background for the theory is provided mainly by the movies, rather than the more traditional approach to motivating the theory using experimental results. The text is on an introductory level and requires little previous knowledge, but it is not elementary. When I considered how to provide the theoretical background for the animations, I found that only a more mathematical approach would lead the reader quickly to the level necessary to understand the more intricate details of the movies. So I took the opportunity to combine a vivid discussion of the basic principles with a more advanced presentation of some mathematical aspects of the formalism. Therefore, the book will certainly serve best as a companion in a theoretical physics course, while the material on the CD-ROM will be useful for a more general audience of science students.

The choice of topics and the organization of the text is in part due to purely practical considerations. The development of software parallel to writing a text is a time-consuming process. In order to speed up the publication I decided to split the text into two parts (hereafter called Book One and Book Two), with this first book containing selected topics. This enables me to adapt to the technological evolution that has taken place since this project started, and helps provide the individual volumes at an affordable price. The arrangement of the topics allows us to proceed from simple to more and more complicated animations. Book One mainly deals with spinless particles in one and two dimensions, with a special emphasis on exactly solvable problems. Several topics that are usually considered to belong to a basic course in quantum mechanics are postponed until Book Two. Book Two will include chapters about spherical symmetry in three dimensions, the hydrogen atom, scattering theory and resonances, periodic potentials, particles with spin, and relativistic problems (the Dirac equation).

Let me add a few remarks concerning the contents of Book One. The first two chapters serve as a preparation for different aspects of the course. The ideas behind the methods of visualizing wave functions are fully explained in Chapter 1. We describe a special color map of the complex plane that is implemented by *Mathematica* packages for plotting complex-valued functions. These packages have been created especially for this book. They are included on the CD-ROM and will, hopefully, be useful for the reader who is interested in advanced graphics programming using *Mathematica.*

Chapter 2 introduces some mathematical concepts needed for quantum mechanics. Fourier analysis is an essential tool for solving the Schrödinger equation and for extracting physical information from the wave functions. This chapter also presents concepts such as Hilbert spaces, linear operators, and distributions, which are all basic to the mathematical apparatus of quantum mechanics. In this way, the methods for solving the Schrödinger equation are already available when it is introduced in Chapter 3 and the student is better prepared to concentrate on conceptual problems. Certain more abstract topics have been included mainly for the sake of completeness. Initially, a beginner does not need to know all this "abstract nonsense," and the corresponding sections (marked as "special topics") may be skipped at first reading. Moreover, the symbol $\boxed{\Psi}$ has been used to designate some paragraphs intended for the mathematically interested reader.

Quantum mechanics starts with Chapter 3. We describe the free motion of approximately localized wave packets and put some emphasis on the statistical interpretation and the measurement process. The Schrödinger equation for particles in external fields is given in Chapter 4. This chapter on states and observables describes the heuristic rules for obtaining the correct quantum observables when performing the transition from classical to quantum mechanics. We proceed with the motion under the influence of boundary conditions (impenetrable walls) in Chapter 5. The particle in a box serves to illustrate the importance of eigenfunctions of the Hamiltonian and of the eigenfunction expansion. Once again we come back to interpretational difficulties in our discussion of the double-slit experiment.

Further mathematical results about unitary groups, canonical commutation relations, and symmetry transformations are provided in Chapter 6 which focuses on linear operators. Among the mathematically more sophisticated topics that usually do not appear in textbooks are the questions related to the domains of linear operators. I included these topics for several reasons. For example, solutions that are not in the domain of the Hamiltonian have strange temporal behavior and produce interesting effects when visualized in a movie. Some of these often surprising phenomena are perhaps not widely known even among professional scientists. Among these I would like to mention the strange behavior of the unit function in a Dirichlet box shown in the movie CD 4.11 (Chapter 5).

The remaining chapters deal with subjects of immediate physical importance: the harmonic oscillator in Chapter 7, constant electric and magnetic fields in Chapter 8, and some elements of scattering theory in Chapter 9. The exactly solvable quantum systems serve to underpin the theory by examples for which all results can be obtained explicitly. Therefore, these systems play a special role in this course although they are an exception in nature.

Many of the animations on the CD-ROM show wave packets in two dimensions. Hence the text pays more attention than usual to two-dimensional problems, and problems that can be reduced to two dimensions by exploiting their symmetry. For example, Chapter 8 presents the angular-momentum decomposition in two dimensions. The investigation of two-dimensional systems is not merely an exercise. Very good approximations to such systems do occur in nature. A good example is the surface states of electrons which can be depicted by a scanning tunneling microscope.

The experienced reader will notice that the emphasis in the treatment of exactly solvable systems has been shifted from a mere calculation of eigenvalues to an investigation of the dynamics of the system. The treatment of the harmonic oscillator or the constant magnetic field makes it very clear that in order to understand the motion of wave packets, much more is needed than just a derivation of the energy spectrum. Our presentation includes advanced topics such as coherent states, completeness of eigenfunctions, and Mehler's integral kernel of the time evolution. Some of these results certainly go beyond the scope of a basic course, but in view of the overwhelming number of elementary books on quantum mechanics the inclusion of these subjects is warranted. Indeed, a new book must also contain interesting topics which cannot easily be found elsewhere. Despite the presentation of advanced results, an effort has been made to keep the explanations on a level that can be understood by anyone with a little background in elementary calculus. Therefore I hope that the text will fill a gap between the classical texts (e.g., [39], [48], [49], [68]) and the mathematically advanced presentations (e.g., [4], [17], [62], [76]). For those who like a more intuitive approach it is recommended that first a book be read that tries to avoid technicalities as long as possible (e.g., [19] or [40]).

Most of the films on the CD-ROM were generated with the help of the computer algebra system *Mathematica*. While *Mathematica* has played an important role in the creation of this book, the reader is not required to have any knowledge of a computer algebra system. Alternate approaches which use symbolic mathematics packages on a computer to teach quantum mechanics can be found, for example, in the books [18] and [36], which are warmly recommended to readers familiar with both quantum mechanics and *Mathematica* or Maple. However, no interactive computer session can replace an hour of thinking just with the help of a pencil and a sheet of paper. Therefore, this text describes the mathematical and physical ideas of quantum mechanics in the conventional form. It puts no special emphasis on symbolic computation or computational physics. The computer is mainly used to provide quick and easy access to a large collection of animated illustrations, interactive pictures, and lots of supplementary material. The book teaches the concepts, and the CD-ROM engages the imagination. It is hoped that this combination will foster a deeper understanding of quantum mechanics than is usually achieved with more conventional methods.

While knowledge of *Mathematica* is not necessary to learn quantum mechanics with this text, there is a lot to find here for readers with some experience in *Mathematica*. The supplementary material on the CD-ROM includes many *Mathematica* notebooks which may be used for the reader's own computer experiments.

In many cases it is not possible to obtain explicit solutions of the Schrödinger equation. For the numerical treatment we used external C++ routines linked to *Mathematica* using the MathLink interface. This has been done to enhance computation speed. The simulations are very large and need a lot of computational power, but all of them can be managed on a modern personal computer. On the CD-ROM will be found all the necessary information as well as the software needed for the student to produce similar films on his/her own. The exploration of quantum-mechanical systems usually requires more than just a variation of initial conditions and/or potentials (although this is sometimes very instructive). The student will soon notice that a very detailed understanding of the system is needed in order to produce a useful film illustrating its typical behavior.

This book has a home page on the internet with URL

`http://www.kfunigraz.ac.at/imawww/vqm/`

As this site evolves, the reader will find more supplementary material, exercises and solutions, additional animations, links to other sites with quantum-mechanical visualizations, etc.

**Acknowledgments**

During the preparation of both the book and the software I have profited from many suggestions offered by students and colleagues. My thanks to M. Liebmann for his contributions to the software, and to K. Unterkofler for his critical remarks and for his hospitality in Millstatt, where part of this work was completed. This book would not have been written without my wife Sigrid, who not only showed patience and understanding when I spent 150% of my time with the book and only -50% with my family, but who also read the entire manuscript carefully, correcting many errors and misprints. My son Wolfgang deserves special thanks. Despite numerous projects of his own, he helped me a lot with his unparalleled computer skills. I am grateful to the people at Springer-Verlag, in particular to Steven Pisano for his professional guidance through the production process. Finally, a project preparation grant from Springer-Verlag is gratefully acknowledged.

Bernd Thaller

# Contents

*Chapter 1*

# Visualization of Wave Functions

**Chapter summary**: Although nobody can tell how a quantum-mechanical particle looks like, we can nevertheless visualize the complex-valued function (wavefunction) that describes the state of the particle. In this book complex-valued functions are visualized with the help of colors. By looking at Color Plate 3 and browsing through the section "Visualization" on the accompanying CD-ROM, you will quickly develop the necessary feeling for the relation between phases and colors. You need to study this chapter only if you want to understand the ideas behind this method of visualization in more detail and if you want to increase your familiarity with complex-valued functions. Here we derive the mathematical formulas describing the color map that associates a unique color to every complex number. This color map is defined with the help of the HLS color system (hue-lightness-saturation): The phase of a complex number is given by the hue and the absolute value is described by the lightness of the color (the saturation is always maximal). On the CD-ROM you will find the *Mathematica* packages `ArgColorPlot.m` and `ComplexPlot.m` which implement this color map on a computer. These packages have been used to create most of the color plates in this book and most of the movies on the CD-ROM. In this chapter you will also find a comparison of various other methods for visualizing complex-valued functions in one and more dimensions. Finally, we describe some ideas for a graphical representation of spinor wave functions.

## 1.1. Introduction

Many quantum-mechanical processes can be described by the Schrödinger equation, which is the basic dynamic law of nonrelativistic quantum mechanics. The solutions of the Schrödinger equation are called *wave functions* because of their oscillatory behavior in space and time. The accompanying CD-ROM contains many pictures and movies of wave functions.

Unfortunately, it is not at all straightforward to understand and interpret a graphical representation of a quantum phenomenon. Wave functions, like other objects of quantum theory, are idealized concepts from which statements about the physical reality can only be derived by means of certain interpretation rules. Therefore a picture of a wave function does not show the quantum system as it really looks like. In fact, the whole concept of "looking like something" cannot be used in the strange world of quantum mechanics. Most phenomena take place on length scales much smaller than the wavelength of light.

With the help of some mathematical procedures, a wave function allows us to determine the probability distributions of physical observables (like position, momentum, or spin). Thus, the wave function gives high-dimensional data at each point of space and time and it is a difficult task to visualize such an amount of information. Usually, it is not possible to show all that information in a single graph. One has to concentrate on particular aspects and to apply special techniques in order to display the information in a form that can be understood.

Mathematically speaking, a wave function is a complex-valued function of space and time; a spinor wave function even consists of several components. In this first chapter I describe some methods of visualizing such an object. In the following chapters you will learn how to extract the physically relevant information from the visualization.

For the visualization of high-dimensional data a color code can be very useful. Because the set of all colors forms a three-dimensional manifold (see Sect. 1.2.2), it is possible—at least in principle—to represent triples of data values using a color code. Unfortunately, the human visual system is not able to recognize colors with quantitative precision. But at least we can expect that an appropriately chosen color code helps to visualize the most important qualitative features of the data.

## 1.2. Visualization of Complex Numbers

As a first step, I want to discuss some possibilities to visualize complex values. It is my goal to associate a unique color to each complex number. You will learn about the various color systems in some detail because this subject is relevant for the actual implementation on a computer.

CD 1.1 and Color Plate 3 show an example of such a color map, designed mainly for on-screen use. Here the phase of the complex number determines the hue of the color, and the absolute value is represented by the lightness of the color. This color map will be now described in more detail.

### 1.2.1. The two-dimensional manifold of complex numbers

Any complex number $z$ is of the form

$$z = x + \mathrm{i}y, \qquad x = \operatorname{Re} z, \quad y = \operatorname{Im} z. \tag{1.1}$$

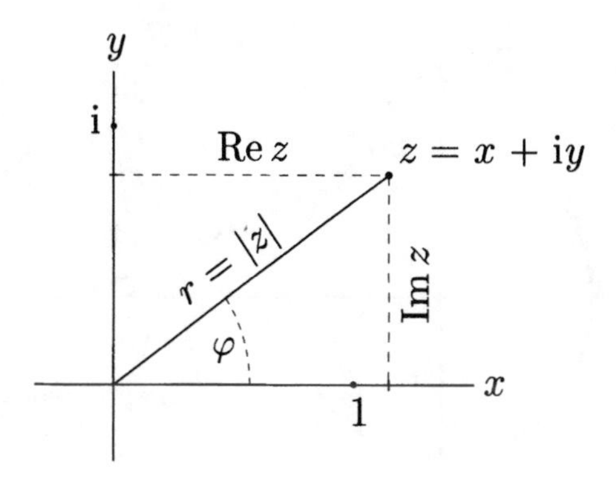


FIGURE 1.1. Graphical representation of a complex number $z$ in Cartesian and in polar coordinates.

Here i is the complex unit which is defined by the property $\mathrm{i}^2 = -1$. The values $x$ and $y$ are real numbers which are called the *real part* and the *imaginary part* of $z$, respectively. The field of all complex numbers is denoted by $\mathbb{C}$.

Thus, complex numbers $z \in \mathbb{C}$ can be represented by pairs $(x, y)$ of real numbers and visualized as points in the two-dimensional complex plane.

Using polar coordinates $(r, \varphi)$ in the complex plane gives another representation, the *polar form* of a complex number (see Fig. 1.1)

$$z = r\,\cos\varphi + \mathrm{i}\,r\,\sin\varphi = r\,\mathrm{e}^{\mathrm{i}\varphi}, \qquad r = |z|, \quad \varphi = \arg z. \tag{1.2}$$

Here we have used Euler's formula

$$\mathrm{e}^{\mathrm{i}\varphi} = \cos\varphi + \mathrm{i}\sin\varphi. \tag{1.3}$$

The non-negative real number $r$ is the *modulus* or *absolute value* of $z$ and the angle $\phi$ is called the *phase* or *argument* of $z$.

For $z = r\,\mathrm{e}^{\mathrm{i}\varphi} = x + \mathrm{i}y$ the *conjugate complex number* is $\overline{z} = r\,\mathrm{e}^{-\mathrm{i}\varphi} = x - \mathrm{i}y$.

One often adds the *complex infinity* $\infty$ to the complex numbers. This can be explained easily with the help of a stereographic projection.

**The stereographic projection**: You can interpret the complex plane as the $xy$-plane in the three-dimensional space $\mathbb{R}^3$. Consider a sphere of radius $R$ centered at the origin in $\mathbb{R}^3$. Draw the straight line which contains the point $(x, y, 0)$ (corresponding to the complex number $z = x + \mathrm{i}y$) and the north pole $(0, 0, R)$ of the sphere. Then the *stereographic projection* of $z$ is the intersection of that line with the surface of the sphere. Obviously, this gives a unique point on the sphere for each complex number $z$. Using polar coordinates $(\theta,\ \varphi)$ on the sphere, it is clear that the azimuthal angle $\varphi$ is

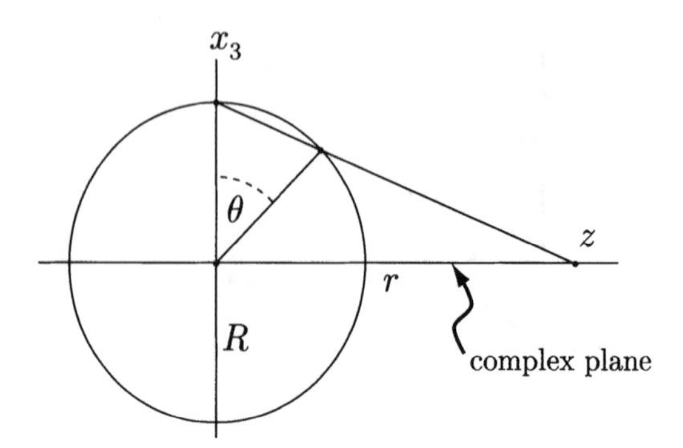


FIGURE 1.2. Stereographic projection of a complex number $z$ with $|z| = r$.

just the phase of $z = r\,\exp(\mathrm{i}\varphi)$,

$$\varphi = \arg z. \tag{1.4}$$

A little trigonometric exercise (see Fig. 1.2) shows that the polar angle $\theta$ is given by

$$\theta = \pi - 2\,\arctan\frac{r}{R}, \quad r = |z|. \tag{1.5}$$

In that way the circle with radius $R$ in $\mathbb{C}$ is mapped onto the equator of the sphere. A complex number $z = r\exp(\mathrm{i}\varphi)$ is mapped to the northern hemisphere if $r > R$, and to the southern hemisphere if $r < R$. The origin $z = 0$ is mapped onto the south pole of the sphere, $\theta = \pi$. Every point of the sphere—except the north pole—is the image of some complex number under the stereographic projection, and the correspondence is one-to-one. The north pole $\theta = 0$ of the sphere is interpreted as the image of a new element, called *complex infinity* and denoted by $\infty$. The complex infinity has an infinite absolute value and an undefined phase (like $z = 0$). Obviously, $\infty$ can be used to represent $\lim_{n\to\infty} z_n$ for all sequences $(z_n)$ that have no finite accumulation point.

With a stereographic projection, the whole set of complex numbers together with complex infinity can be mapped smoothly and in a one-to-one fashion onto a sphere. Because the sphere is a compact two-dimensional surface we can regard the set $\overline{\mathbb{C}} = \mathbb{C} \cup \{\infty\}$ as a compact two-dimensional manifold. It is called the *compactified complex plane.*

EXERCISE 1.1. *Check your familiarity with complex numbers. Express* $|z|$ *and* $\arg z$ *in terms of* $\operatorname{Re} z$ *and* $\operatorname{Im} z$, *and vice versa.*

EXERCISE 1.2. *Given two complex numbers* $z_1$ *and* $z_2$ *in polar form describe the absolute values and the phases of* $z_1 z_2$, $z_1/z_2$ *and* $z_1 + z_2$.

EXERCISE 1.3. *The stereographic projection is one-to-one and onto. Determine the inverse mapping from the sphere of radius $R$ onto the compactified complex plane $\overline{\mathbb{C}}$.*

### 1.2.2. The three-dimensional color manifold

For the purpose of visualization we want to associate a color to each complex number. Before doing so, let's have a short look at various methods of describing colors mathematically.

The set of all colors that can be represented in a computer is a compact, three-dimensional manifold. It can be described in many different ways. Perhaps the most common description is given by the RGB model (CD 1.2).

**The RGB color system**: In the RGB system the color manifold is defined as the three-dimensional unit cube $[0,1] \times [0,1] \times [0,1]$. The points in the cube have coordinates $(R, G, B)$ which describe the intensities of the primary colors red, green, and blue. The corners $(1,0,0)$, $(0,1,0)$, and $(0,0,1)$ (= red, green, and blue at maximal intensity) are regarded as basis elements from which all other colors $(R, G, B)$ can be obtained as linear combinations (additive mixing of colors). Of special importance are the complementary colors "yellow" $(1,1,0)$ (=red+green), "magenta" $(1,0,1)$, and "cyan" $(0,1,1)$, which are also corner points of the color cube. The two remaining corners are "black" $(0,0,0)$ and "white" $(1,1,1)$. All shades of gray are on the main diagonal from black to white. In *Mathematica*, the RGB colors are implemented by the color directive `RGBColor`.

In order to visualize a complex number by a color, we have to define a mapping from the two-dimensional complex plane into the three-dimensional color manifold. This can be done, of course, in an infinite number of ways. For our purposes we will define a mapping which is best described by another set of coordinates on the color manifold.

**The HSB and HLS color systems**: A measure for the distance between any two colors $C^{(1)} = (R^{(1)}, G^{(1)}, B^{(1)})$ and $C^{(2)} = (R^{(2)}, G^{(2)}, B^{(2)})$ in the color cube is given by the maximum metric

$$d(C^{(1)}, C^{(2)}) = \max\{|R^{(1)} - R^{(2)}|, |G^{(1)} - G^{(2)}|, |B^{(1)} - B^{(2)}|\}. \qquad (1.6)$$

The distance of a color $C = (R, G, B)$ from the black origin $O = (0,0,0)$ is called the *brightness* $b$ of $C$,

$$b(C) = d(C, O) = \max\{R, G, B\}. \qquad (1.7)$$

The *saturation* $s(C)$ is defined as the distance of $C$ from the gray point on the main diagonal which has the same brightness. Hence

$$s(C) = \max\{R, G, B\} - \min\{R, G, B\}. \qquad (1.8)$$

The possible values of the brightness $b$ range between 0 and 1. For each value of $b$, the saturation varies between 0 and the "maximal saturation at brightness $b$,"

$$s_{\max}^{b} = b. \tag{1.9}$$

The set of all the colors in the RGB cube with the same saturation and brightness is a closed polygonal curve $\Gamma_{s,b}$ of length $6s$ which is formed by edges of a cube with edge length s (see Color Plate 1a).

The hue $h(C)$ of a point $C$ is $\lambda/6s$, where $\lambda$ is the length of the part of $\Gamma_{s,b}$ between $C$ and the red corner (the corner of $\Gamma_{s,b}$ with maximal red component) in the positive direction (counter-clockwise, if viewed from the white corner). In that way $h = 0$ and $h = 1$ both give the red corner and it is most natural to define the hue as a cyclic variable modulo 1. Hence the pure colors at the corners of the RGB cube (red, yellow, green, cyan, blue, magenta) have the hue values $(0, 1/6, 1/3, 1/2, 2/3, 5/6) \pmod 1$.

For any color $C = (R, G, B)$ the *lightness* $l(C)$ is defined as the average of the maximal and the minimal component,

$$l(C) = \frac{\max\{R, G, B\} + \min\{R, G, B\}}{2} = b(C) - \frac{s(C)}{2}. \tag{1.10}$$

We have $0 \leq l \leq 1$ and, at a given lightness $l$, the brightness ranges in $l \leq b \leq \min\{1, 2l\}$. Lightness $l = 0$ denotes black, $l = 1$ (which implies $b = 1$, $s = 0$) is white. If we keep the lightness fixed, the saturation has values in the range $0 \leq s \leq s_{\max}^{l}$, where the maximal saturation at a given lightness $l$ is

$$s_{\max}^{l} = \begin{cases} 2l, & \text{if } l \leq 1/2, \\ 2(1-l), & \text{if } l \geq 1/2. \end{cases} \tag{1.11}$$

The set of color points which have the maximal saturation with respect to their lightness is just the surface of the RGB color cube.

In the *HSB color system* every color is characterized by the triple $(h, s, b)$ of hue, saturation, and brightness. We can interpret the color manifold as a cone in $\mathbb{R}^3$ with vertex at the origin (see Color Plate 1b and CD 1.3). The values $(2\pi h, s, b)$ are cylindrical coordinates where $b$ corresponds to the $z$-coordinate, $s$ specifies the radial distance from the axis of the cone, and $\varphi = 2\pi h$ gives the angle.

The coordinates $(h, l, s)$ describing the hue, lightness, and saturation of a color are used in the *HLS color system*. The color manifold in the HLS system can be interpreted as a double cone where the position of a color point $(h, l, s)$ is given by an angle $2\pi h$, the height $l$, and the radial distance $s$ from the axis (Color Plate 1c and CD 1.5).

In the HSB system one often redefines the saturation as $s' = s/b$ such that the maximal $s'$ at a given brightness $b$ is equal to 1. This provides a cylindrical color space, see CD 1.4. Likewise one renormalizes the saturation in the HLS system such that its values at a given *lightness* range between 0 and 1. In *Mathematica*, the HSB color system is implemented by `Hue`$[h, s', b]$. The standard package `Graphics`Colors`` adds the color directive `HLSColor`.

The movies CD 1.2–CD 1.5 present animated views of the color manifold as it appears in the various coordinate systems. See also Color Plate 1.

EXERCISE 1.4. *Try to invert the mapping between RGB and HLS coordinates. That is, find an expression for the red, green, and blue components of a color in terms of its hue, lightness, and saturation.*

### 1.2.3. A color code for complex numbers

This section finally describes the mapping from the compactified complex plane $\overline{\mathbb{C}}$ into the manifold of colors. This color map associates a color with each complex number in a unique way. Because $\overline{\mathbb{C}}$ is two-dimensional, there exists a unique correspondence between $\overline{\mathbb{C}}$ and the *surface* of the three-dimensional color manifold. (In fact, any mapping from $\overline{\mathbb{C}}$ to a two-dimensional (compact) submanifold of the color manifold could be used for the same purpose, but the colors on the surface of the color manifold have maximal saturation and thus can be distinguished most easily).

We are going to use a stereographic projection to obtain unique colors for complex numbers. As a first step we color the sphere by defining a mapping from the sphere to the surface of the color manifold. Each point in the complex plane will then receive the color of its stereographic image on the surface of the sphere.

CD 1.6 shows the surface of the color manifold represented as a sphere. In polar coordinates $(\phi, \theta)$ the angle $\phi$ gives the hue and $\theta$ gives the lightness of the color. See Color Plate 2. The animation in CD 1.7 explains the stereographic color map that projects colors from the surface of the colored sphere onto the complex plane.

**Color map of the sphere**: Every point $(\theta, \varphi)$ of the sphere (except the poles) will be colored with a hue given by $\varphi/(2\pi)$. The lightness of the color is defined to depend linearly on $\theta$,

$$l(\theta) = 1 - \frac{\theta}{\pi}, \qquad 0 \le \theta \le \pi. \tag{1.12}$$

We choose the maximal saturation corresponding to each value of the lightness, $s(\theta) = s_{\max}^{l(\theta)}$. In this way we have defined a homeomorphism (i.e., a mapping that is one-to-one, continuous, and has a continuous inverse) from the surface of the sphere onto the surface of the color manifold (see Color Plate 2 and CD 1.6). The north pole ($\theta = 0$, $z = \infty$) is white, the south pole ($\theta = \pi$, $z = 0$) is black. The equator ($\theta = \pi/2$, $|z| = R$) has lightness 1/2 and hence shows all colors with saturation 1.

EXERCISE 1.5. *Show that in the HSB system the mapping defined above can be described as follows: The southern hemisphere has a brightness that increases linearly in $\theta$ toward the equator, and a maximal saturation. The equator has maximal saturation and brightness. The northern hemisphere has maximal brightness with saturation decreasing linearly toward the north pole.*

**Color map of the complex plane**: The composition of the stereographic projection described in Section 1.2.1 with the color map of the sphere defines a coloring of the complex plane, which is shown in Color Plate 3. The color map is a homeomorphism from the compactified complex plane $\overline{\mathbb{C}}$ onto the surface of the color manifold. CD 1.7 illustrates this method of coloring the complex plane.

Color Plate 3 shows that each complex number (except $z = 0$, which is black, and $z = \infty$, which is white) is colored with a hue determined by its phase, $h = \varphi/(2\pi)$. Positive real values are red; negative real values are in cyan (green-blue). For any complex number $z$, the opposite $-z$ has the complementary hue. The additive elementary colors red, green, and blue, are at the angles $\varphi = 0$, $2\pi/3$, and $4\pi/3$, the subtractive elementary colors yellow, cyan, and magenta are at $\varphi = \pi/3$, $\pi$, and $5\pi/3$. The imaginary unit i has $\varphi = \pi/2$, and hence its hue $h = 1/4$ is between yellow and green.

EXERCISE 1.6. *How would the color map look if we used the brightness instead of the lightness in Eq.* (1.12)*?*

While the simple relations between the complex numbers and the HLS color system are easy to implement, they don't take into account the more subtle points of visual perception. Colors that have the same computer-defined lightness don't appear to have the same lightness on screen. In particular, yellow, magenta and cyan (the edges of the color cube) seem to be significantly brighter than their neighbors in the color circle, while blue appears to be rather dark. As a consequence, the colors with the same perceived lightness do not lie on a circle in the complex plane. Those nonlinear relationships between our mathematically defined lightness (and brightness) and the actually perceived lightness can only be dealt with in special color systems (e.g., CIE-Lab). Another drawback of our color map is that the colors with maximal saturation and brightness in RGB-based systems cannot be reproduced accurately in print. Thus, the color plates in this book look a little bit different from their counterparts on the CD-ROM.

## 1.3. Visualization of Complex-Valued Functions

A complex-valued function $\psi$ associates a complex number $\psi(x)$ to each value of an independent variable $x \in \mathbb{R}^n$. A color code such as the one explained above is very useful for the qualitative visualization of such an object—even in the one-dimensional case $n = 1$.

### 1.3.1. Complex-valued functions in one dimension

One of the simplest quantum systems is a single spinless particle in one space dimension. At a fixed time the particle is described by a complex-valued wave function $\psi$. This means that a complex number $\psi(x)$ is given at each point $x$. As an example of a complex-valued function we consider the one-dimensional "stationary plane wave" with wave number $k$,

$$\psi_k(x) = \exp(ikx), \qquad x \in \mathbb{R}. \tag{1.13}$$

The real number $k$ describes the wavelength $\lambda = 2\pi/k$. Using this example we illustrate several methods of visualizing complex-valued functions.

METHOD 1. *Real and imaginary part*: We can visualize a complex-valued function $\psi$ by separate plots of the real part and the imaginary part. For the function $\psi_k$ we have $\operatorname{Re}\psi_k(x) = \cos(kx)$ and $\operatorname{Im}\psi_k(x) = \sin(kx)$ (see Color Plate 4a). Later we will see that the splitting into real and imaginary parts does not have much physical meaning. It is more important to know the absolute value of the wave function.

METHOD 2. *Plot the graph*: One-dimensional wave functions can always be visualized using a three-dimensional plot. In three-dimensional space the plane orthogonal to the $x$-axis can be interpreted as the complex plane. At each point $x$ we may plot $\operatorname{Re}\psi(x)$ as the $y$-coordinate and $\operatorname{Im}\psi(x)$ as the $z$-coordinate. In this way the complex-valued function $\psi$ can be represented by a space curve. This space curve is called the *graph* of the function $\psi$. The orthogonal distance of the curve from the $x$-axis is just the absolute value $|\psi(x)|$. Color Plate 4b illustrates this method for the stationary plane wave $\psi_k$. This method of visualizing a complex-valued function has nevertheless some disadvantages. The plots are sometimes difficult to interpret, and the method cannot be generalized to higher dimensions.

METHOD 3. *Use a color code for the phase*: Color Plate 4c shows how a color can be used to visualize a complex-valued function $\psi(x)$ in one dimension. We plot the absolute value and fill the area between the $x$-axis and the graph with a color indicating the complex phase of the wave function at the point $x$. In this case we may use a simplified color map, because the absolute value is clearly displayed as the height of the graph. Hence we plot all colors at maximal saturation and brightness (i.e., with lightness 1/2). The hue $h$ at the point $x$ depends on the phase as discussed in Section 1.2.3, namely, $h(x) = \arg\psi(x)/(2\pi)$.

CD 1.8 shows several examples of one-dimensional complex-valued functions visualized using the methods described above.

EXERCISE 1.7. *Find the real and the imaginary parts of the function*

$$\phi(x) = \psi_2(x) + \psi_3(x), \tag{1.14}$$

*where $\psi_k$ are the plane waves defined above.*

EXERCISE 1.8. *Multiply the function $\phi$ defined in Exercise 1.7 by the phase factor $e^{i\pi/4}$. How does this affect the splitting into real and imaginary parts? How does this change the phase of the wave function?*

EXERCISE 1.9. *Draw a color picture of the functions* $\sin(x)$, $e^{ix}\sin(x)$, *and of other functions of your own choice. Check your results with the* Mathematica *notebook* `ArgColorPlot.m` *on the CD-ROM.*

EXERCISE 1.10. *A function $x \to \psi(x)$ is called periodic with period $\lambda$ if $\psi(x+\lambda) = \psi(x)$ holds for all $x$. The plane wave $\psi_k$ is obviously periodic. Is the sum $\psi_{k_1} + \psi_{k_2}$ of two plane waves again periodic?*

### 1.3.2. Higher-dimensional wave functions

Complex-valued functions of $x \in \mathbb{R}^2$ (i.e., functions of two variables) can again be visualized using several methods.

METHOD 4. *Real and imaginary part:* This is the same as the first method described in the previous section. If $\psi(x, y)$ is a complex-valued function of two variables, then the real-valued functions $\operatorname{Re}\psi$ and $\operatorname{Im}\psi$ can be visualized as three-dimensional surface plots. An example is shown in Fig 1.3 for the function $\psi(x, y) = (x + iy)^3 - 1$.

All the methods described here are presented in a sequence of movies on the CD-ROM. These examples show a time-dependent quantum-mechanical wave function that describes the propagation of a free quantum-mechanical particle in two dimensions. CD 1.12 shows the real part of this wave function. The other visualization methods are shown in CD 1.13–CD 1.16.

METHOD 5. *Plot of vector field:* A complex number $z$ can be interpreted as a two-dimensional vector with components $(\operatorname{Re} z, \operatorname{Im} z)$. Hence a function $\psi(x,y)$ may be regarded a vector field. Figure 1.4 visualizes the function $\psi(x,y) = (x+\mathrm{i}y)^3 - 1$ by plotting little arrows on a suitable grid of points. Of course, this method is not able to show very fine details of a function. See also CD 1.13.

METHOD 6. *Plot the graph:* The graph of a function $\psi(x,y)$ of two variables would have to be drawn in a four-dimensional space with coordinates $x$, $y$, $\operatorname{Re}\psi$, $\operatorname{Im}\psi$. Of course, this cannot be done easily on a sheet of paper. Hence this method does not work here.

METHOD 7. *Image of the coordinate lines:* Apart from giving separate surface plots of the real part and the imaginary part of the function, one could try to visualize how a grid of coordinate lines in $\mathbb{R}^2$ is mapped onto the complex plane. This is illustrated in Fig. 1.5 for the function $\psi(x,y) = (x+\mathrm{i}y)^3 - 1$. While this method is sometimes very instructive, the resulting plots are usually very difficult to interpret for functions with a complicated structure. See also CD 1.14. This method is implemented by the standard *Mathematica* package `Graphics`ComplexMap``.

METHOD 8. *Use a color map:* The method of using a color code for the visualization of a complex-valued function can be easily generalized to functions of two variables. An appropriately colored surface graphics or a density graphics can give a useful graphical representation of a complex-valued function. An example is given in Color Plate 5. Even in case of a

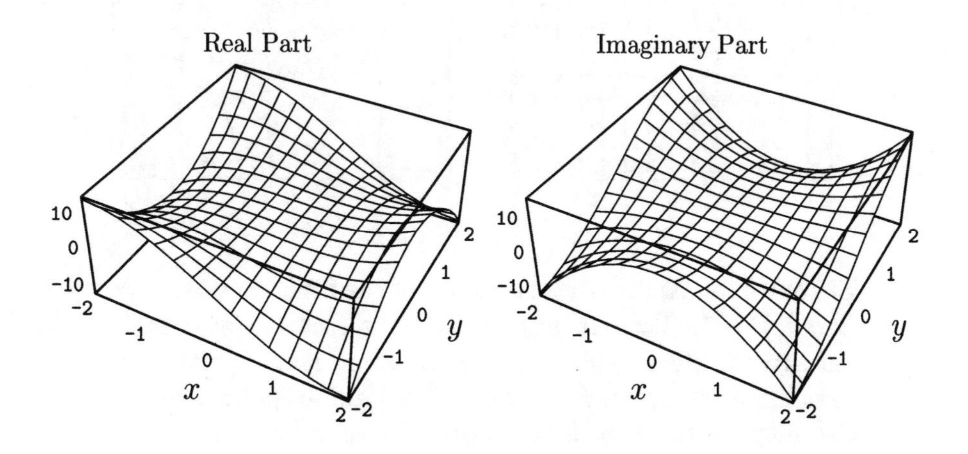


FIGURE 1.3. Visualization of the function $(x+\mathrm{i}y)^3 - 1$ by surface plots of the real and the imaginary part.

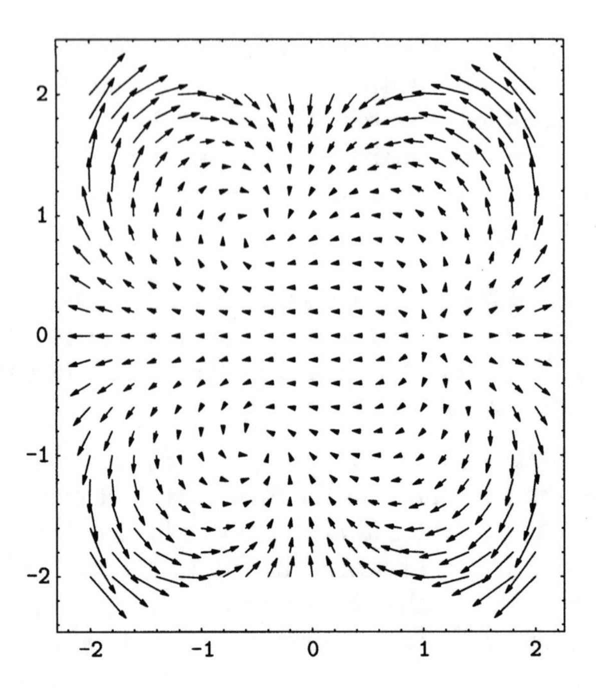


FIGURE 1.4. The function $(x + iy)^3 - 1$ as a vector field.

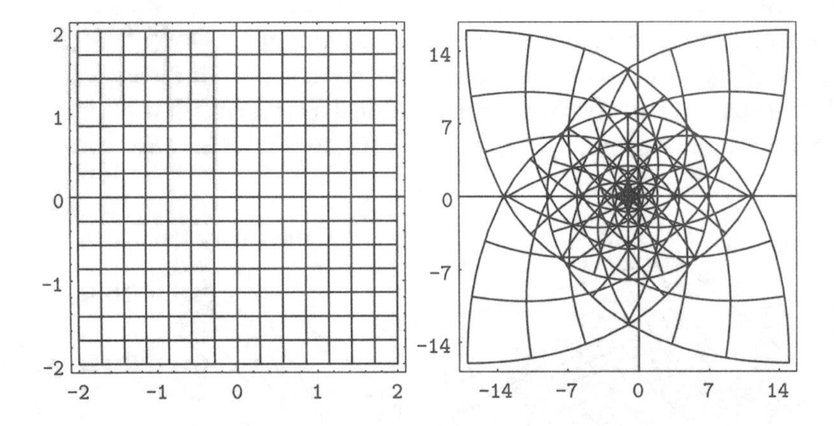


FIGURE 1.5. Another visualization of the function $\psi(x, y) = (x + iy)^3 - 1$. The left graphics shows a rectangular coordinate grid in the $x$-$y$-plane. These lines are mapped onto the complex plane by the function $\psi$. The right graphics displays the image of the coordinate grid in the complex plane.

surface graphics we find it useful to indicate the absolute value also by the lightness of the color, for example, by using the color map shown in Color

Plate 3. For comparison, we show in Color Plate 6 a density plot of the function $\psi(x,y) = (x+\mathrm{i}y)^3 - 1$ which has been used for illustrating the other methods. Now we see immediately that the function has three zeros of first order on the unit circle.

As you can see from the examples, already in two dimensions the color map described in Section 1.2.3 becomes an indispensable tool for the visualization of complex-valued functions. Most visualizations in this book or on the CD-ROM use Method 8.

But this method can be generalized even to three dimensions. For complex-valued functions depending on three variables one may use isosurfaces to represent the absolute value of the function. This surface can be colored according to the phase of the wave function.

CD 1.9 shows the graphical representations of complex functions on $\mathbb{R}^2$ using the stereographic color map. As an example, the Riemann zeta function is discussed in more detail. CD 1.10 contains many more examples of analytic functions and some special functions of mathematical physics. CD 1.11 is an animation showing the dependence of the Jacobi function $sn(z|n)$, $z \in \mathbb{C}$, on the parameter $n$.

CD 1.17 is an example of a wave function in three dimensions. An isosurface of the absolute value is colored according to the phase. The example shows a highly excited state of the hydrogen atom; see also Color Plate 7. Many more visualizations of three-dimensional wave functions will be presented in Book Two.

## 1.4. Special Topic: Wave Functions with an Inner Structure

Elementary particles usually have an inner structure described by wave functions with several components. The simplest case of a two-component wave function occurs for particles with spin-1/2 which will be treated among other things in the second volume of this title. In this section we describe a possible method of visualizing such a "spinor wave function."

A spin-1/2 wave function is a function of a space variable $x$ with values in the vector space $\mathbb{C}^2$ of pairs of complex numbers,

$$\mathbb{C}^2 = \left\{ z = \begin{pmatrix} z_1 \\ z_2 \end{pmatrix} \mid z_i \in \mathbb{C} \right\}. \tag{1.15}$$

For two vectors $y$ and $z$ in $\mathbb{C}^2$ a *scalar product* is defined by

$$\langle y, z \rangle = \overline{y_1} z_1 + \overline{y_2} z_2, \tag{1.16}$$

and the absolute value of $z \in \mathbb{C}^2$ is given by

$$|z| = \sqrt{\langle z, z \rangle} = \sqrt{|z_1|^2 + |z_2|^2}. \tag{1.17}$$

A spinor wave function is a mapping

$$x \to \psi(x) = \begin{pmatrix} \psi_1(x) \\ \psi_2(x) \end{pmatrix} \tag{1.18}$$

which consists of four independent real-valued functions, the real and imaginary parts of the components $\psi_1$ and $\psi_2$. In order to visualize such a high-dimensional object, we introduce the *Pauli matrices*

$$\sigma_0 = \begin{pmatrix} 1 & 0 \\ 0 & 1 \end{pmatrix}, \quad \sigma_1 = \begin{pmatrix} 0 & 1 \\ 1 & 0 \end{pmatrix}, \quad \sigma_2 = \begin{pmatrix} 0 & -\mathrm{i} \\ \mathrm{i} & 0 \end{pmatrix}, \quad \sigma_3 = \begin{pmatrix} 1 & 0 \\ 0 & -1 \end{pmatrix}, \tag{1.19}$$

and define the four real-valued functions

$$V_i(x) = \langle \psi(x), \sigma_i\, \psi(x) \rangle, \qquad i = 0, 1, 2, 3. \tag{1.20}$$

We note that $V_0$ depends on the other functions because

$$V_0(x) = \sqrt{V_1(x)^2 + V_2(x)^2 + V_3(x)^2} = |\psi(x)|^2. \tag{1.21}$$

With the three functions $(V_1(x), V_2(x), V_3(x))$ we form a vector field $\vec{V}(x)$ that can be visualized in a three-dimensional graphic by arrows attached to a grid of $x$-values or by flux lines. In Book Two, we discuss how this vector field describes a "local spin density" (the integral of $\vec{V}(x)$ over $x$ gives twice the expectation value of the spin). Hence this method of visualization displays physically interesting information. By comparison, a visualization that just plots the real and imaginary parts of both components of $\psi$ is not very instructive.

We finally note that the spinor wave function $\psi(x)$ is not represented uniquely by the vector field $\vec{V}(x)$: Multiplication of $\psi(x)$ with a phase factor

$$\psi(x) \to \mathrm{e}^{\mathrm{i}\theta(x)}\psi(x) = \begin{pmatrix} \mathrm{e}^{\mathrm{i}\theta(x)}\psi_1(x) \\ \mathrm{e}^{\mathrm{i}\theta(x)}\psi_2(x) \end{pmatrix} \tag{1.22}$$

would not change any of the functions $V_i(x)$.

*Chapter 2*

# Fourier Analysis

**Chapter summary**: Fourier analysis is of utmost importance in many areas of mathematics, physics, and engineering. In quantum mechanics, the Fourier transform is an essential tool for the solution and the interpretation of the Schrödinger equation. It will help you to understand how a wave function can describe simultaneously the localization properties and the momentum distribution of a particle.

In this chapter we collect many results from Fourier analysis which will be used frequently in later chapters. In passing, you will be introduced to the most important mathematical concepts of quantum mechanics, such as Hilbert spaces and linear operators. Moreover, you will learn that the famous uncertainty relation is just a property of the Fourier transformation. If you need some more motivation, you may read Chapter 3, Sections 3.1 and 3.2 first.

This chapter starts by describing the Fourier series of a complex-valued periodic function. The Fourier series describes the given function as an infinite linear combination of stationary plane waves, each characterized by an amplitude and a wave number. In order to understand in which sense the Fourier sum converges, we need to introduce the concept of a Hilbert space.

As the period of the complex-valued function goes to infinity, the Fourier series becomes a Fourier integral which represents the function as a "continuous superposition" of stationary plane waves. The spectrum of wave numbers is described by a function on "$k$-space." This is the space of all possible wave numbers, which in the context of quantum mechanics is called the momentum space. It is a very important observation that the original function and the function describing the continuous spectrum of wave numbers depend on each other in a very symmetrical way. This relationship—the Fourier transformation—can be described as a linear operator acting in the Hilbert space of square integrable functions.

The properties of the Fourier transformation make it a very useful tool in quantum mechanics. For example, the derivative of a function corresponds via the Fourier transformation to a simple multiplication by $k$ in momentum space. This fact will be exploited in Chapter 3 to solve the free Schrödinger equation with arbitrary initial conditions.

While this chapter contains some material that is indispensable for a thorough description of quantum mechanics, there are some mathematically more elaborate sections that may be skipped at first reading. These sections are labeled "special topics."

## 2.1. Fourier Series of Complex-Valued Functions

Fourier analysis is the art of writing arbitrary wave functions as superpositions of trigonometric functions. As a first step we consider periodic functions and the associated Fourier series.

### 2.1.1. Basic definitions

Given a real number $L > 0$, we define the numbers

$$k_n^{(L)} = n\,\frac{\pi}{L}, \qquad n = 0, \pm 1, \pm 2, \dots, \tag{2.1}$$

and the associated functions

$$u_n^{(L)}(x) = \frac{1}{\sqrt{2L}}\, e^{ik_n^{(L)}x} \quad \text{for all } x \text{ and } n. \tag{2.2}$$

In view of the quantum-mechanical applications, we call $u_n^{(L)}$ a *stationary plane wave* with *wave number* $k_n$. An example of a stationary plane wave is shown in Color Plate 4. The function $u_n^{(L)}$ is a complex-valued trigonometric function. The real and imaginary parts are given by cosine and sine functions, respectively. If you remember Euler's formula (1.3), you will notice immediately that

$$u_n^{(L)}(x) = \frac{1}{\sqrt{2L}}\left(\cos(k_n^{(L)}x) + i\,\sin(k_n^{(L)}x)\right) \quad \text{for all } x \text{ and } n. \tag{2.3}$$

The choice of the normalization factor $(2L)^{-1/2}$ will be explained later.

Each of the functions $u_n^{(L)}$ is periodic with period $2L$, that is,

$$u_n^{(L)}(x + 2L) = u_n^{(L)}(x) \quad \text{for all } x \text{ and } n. \tag{2.4}$$

Because of the periodicity, it is sufficient to restrict the consideration to an interval of length $2L$, say, the interval $[-L, L]$.

Obviously, any finite sum (*superposition* or *linear combination*) of the form

$$\psi(x) = \sum_{n=-N}^{N} c_n\, u_n^{(L)}(x), \quad \text{all } x \in [-L, L], \tag{2.5}$$

(with arbitrary complex numbers $c_n$) is a smooth function on $[-L, L]$ with the property $\psi(-L) = \psi(L)$. It can, of course, also be considered a periodic function on $\mathbb{R}$.

The expression (2.5) is called a *trigonometric sum* or *Fourier sum*. For $n \geq 0$ we call

$$c_n\, u_n^{(L)}(x) + c_{-n}\, u_{-n}^{(L)}(x) \tag{2.6}$$

the *summand of order n*.

CD 2.1 is an interactive demonstration showing how new functions can be built by adding trigonometric functions. Summands of increasing order can be added step-by-step in order to generate Fourier sums that approximate Gaussian functions.

EXERCISE 2.1. *Consider the important special case where* $c_n$ *and* $c_{-n}$ *are complex conjugate numbers, say,* $c_{\pm n} = a_n \pm \mathrm{i} b_n$. *Show that the summand of order* $n$ *and hence the trigonometric sum* (2.5) *is real. Moreover, the summand of order* $n$ *can be written as*

$$c_n\, u_n^{(L)}(x) + c_{-n}\, u_{-n}^{(L)}(x) = d_n \cos\bigl(k_n^{(L)} x + \phi_n\bigr), \tag{2.7}$$

*with*

$$d_n = \sqrt{\frac{2(a_n^2 + b_n^2)}{L}}, \qquad \phi_n = \arccos\Bigl(\frac{a_n}{\sqrt{a_n^2 + b_n^2}}\Bigr). \tag{2.8}$$

EXERCISE 2.2. *Consider the superposition of two stationary plane waves*

$$f(x) = \exp(\mathrm{i}k_1 x) + \exp(\mathrm{i}k_2 x), \quad k_1, k_2 \in \mathbb{R}. \tag{2.9}$$

*Find the period of the absolute value* $|f(x)|$. *Under what condition on* $k_1$ *and* $k_2$ *is* $f(x)$ *a periodic function?*

### 2.1.2. Fourier expansion of square-integrable functions

The set of functions that can be generated by superpositions of plane waves is huge. It is a fundamental mathematical result that every function that is square-integrable on $[-L, L]$ can be approximated by a superposition of plane waves. This result is quoted in the box below.

Let me first give you the definition of square-integrability. This definition is very important for quantum mechanics, because all wave functions with a physical interpretation have to be square-integrable.

DEFINITION 2.1. A (complex-valued) function $\psi$ is called *square-integrable* on the interval $[a, b]$ if

$$\int_a^b |\psi(x)|^2\, dx < \infty. \tag{2.10}$$

The set of all square-integrable functions forms the *Hilbert space* $L^2([a, b])$. You will learn more about Hilbert spaces soon.

**Fourier series of a square-integrable function**:

Let $\psi$ be any square-integrable function on the interval $[-L, L]$. Then

$$\psi(x) = \sum_{n=-\infty}^{\infty} c_n \, u_n^{(L)}(x), \quad c_n = \int_{-L}^{L} \overline{u_n^{(L)}(x)} \, \psi(x) \, dx. \tag{2.11}$$

This infinite trigonometric sum is called a *Fourier series*. The coefficients (*amplitudes*) $c_n$ are square-summable, that is,

$$\sum_{n=-\infty}^{\infty} |c_n|^2 = \int_a^b |\psi(x)|^2 \, dx < \infty. \tag{2.12}$$

Let me define a function $\hat{\psi}$ by

$$\hat{\psi}(k) = \frac{1}{\sqrt{2\pi}} \int_{-L}^{L} e^{-ikx} \, \psi(x) \, dx, \qquad \text{for } k \in \mathbb{R}. \tag{2.13}$$

Now the Fourier series of $\psi$ can be written as

$$\psi(x) = \frac{1}{\sqrt{2\pi}} \sum_{n=-\infty}^{\infty} \frac{\pi}{L} \, \hat{\psi}(k_n^{(L)}) \, e^{ik_n^{(L)}x}. \tag{2.14}$$

This way of writing the Fourier series has cosmetic reasons which will become apparent soon when I make the transition to Fourier integrals.

CD 2.2 shows again the approximation of real and complex Gaussian functions by finite trigonometric sums on the interval $[-2\pi, 2\pi]$. Now the approximation is visualized by showing the spectrum of the Fourier amplitudes (as in Color Plate 8) together with the corresponding partial sum. (See also Fig. 2.1).

### 2.1.3. The convergence of the Fourier series

The mathematical interpretation of the infinite sums contains a more subtle point. The convergence of the Fourier series has to be understood as a *convergence in the mean*, that is,

$$\lim_{m \to \infty} \int_{-L}^{L} \Big| \sum_{n=-m}^{m} c_n \, u_n^{(L)}(x) - \psi(x) \Big|^2 \, dx = 0. \tag{2.15}$$

The convergence in the mean does not imply that the sum converges for a fixed value of $x$, that is, in a "pointwise sense." Indeed, this observation is important if the function has discontinuous jumps.

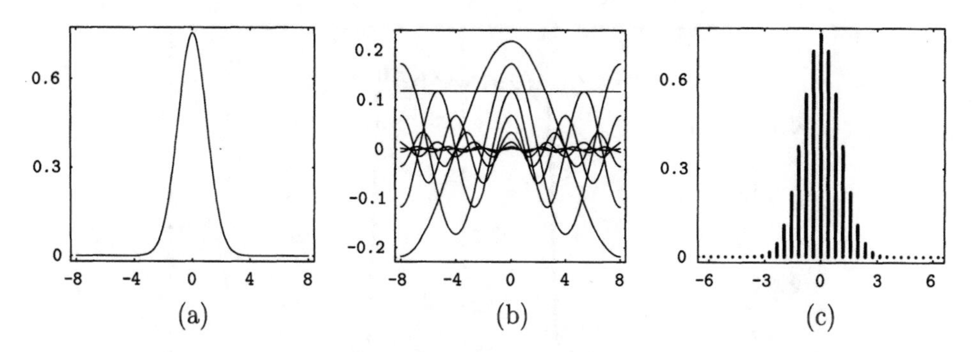


FIGURE 2.1. Approximation by a finite Fourier sum. (a) Symmetric Gaussian function $\exp(-x^2/2)$. (b) The summands in the Fourier-cosine expansion Eq. (2.20) up to order $n = 8$. (c) The coefficients $\hat{\psi}(k_n^{(L)})$, visualized as a bar graph. Vertical lines of length $|\hat{\psi}(k_n^{(L)})|$ are drawn at each $k_n^{(L)}$. For arbitrary functions, we could use a color to indicate the phase of the complex coefficients $\hat{\psi}(k_n^{(L)})$. Several examples are given in Color Plate 8.

A square-integrable function need not be continuous. For example, the *characteristic function* of the interval $[-1, 1]$,

$$\chi_{[-1,1]}(x) = \begin{cases} 1, & \text{for } -1 \leq x \leq 1, \\ 0, & \text{elsewhere,} \end{cases} \tag{2.16}$$

has discontinuous jumps at $x = \pm 1$, but it is certainly square-integrable on any interval $[-L, L]$.

EXERCISE 2.3. *Assuming $L > 1$, find the Fourier series of the characteristic function* (2.16).

If you take only finitely many summands of the Fourier series of a square-integrable function, you get an approximation by a finite Fourier sum. Any finite Fourier sum is a smooth function with the property $\psi_N(-L) = \psi_N(L)$. You can see in Fig. 2.2 how such an approximation of a discontinuous function looks like. Near the discontinuities the Fourier sum shows rapid oscillations. The amplitude of the oscillations near the discontinuities does not become smaller by adding more and more terms, and the Fourier series does not converge in a pointwise sense to the correct value at these points (the *Gibbs' phenomenon*). Instead, the Fourier series converges only in the more moderate sense of Eq. (2.15).

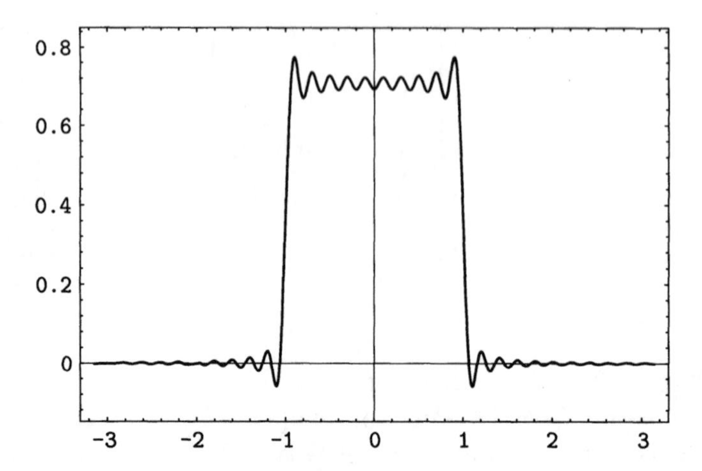


FIGURE 2.2. Approximation of a step function by a finite Fourier sum. Increasing the order of the approximation does not reduce the amplitude of the oscillation near the discontinuities of the function.

CD 2.3 and CD 2.4 show the approximation of discontinuous functions by finite Fourier sums illustrating the Gibbs phenomenon (see also Fig. 2.2). It can be seen that the Fourier series converges very slowly and oscillates near the discontinuities.

A square-integrable function $\psi$ on the interval $[-L, L]$ may be regarded as a periodic function on $\mathbb{R}$ with period $2L$. If $\psi$ is continuous on the interval, but has different boundary values $\psi(-L) \neq \psi(L)$, then $\psi$ has in fact a discontinuity at $\pm L$, and you can expect the Fourier approximations to oscillate as in Fig. 2.2 near the borders of the interval $[-L, L]$.

If $\psi$ itself is a smooth function, then there are stronger results on the convergence of the Fourier series. For example: Let $\psi$ be continuously differentiable on $[-L, L]$, with $\psi(-L) = \psi(L)$. Then the infinite sum in Eq. (2.14) converges even pointwise and uniformly in $x$, that is

$$\max_{x:-L\leq x\leq L} \Bigl\{ \sum_{n=-N}^{N} c_n\, u_n^{(L)}(x) - \psi(x) \Bigr\} \to 0, \quad \text{as } N \to \infty. \tag{2.17}$$

EXERCISE 2.4. *Verify the following assertion: With respect to the* scalar product (*see Sect. 2.2.2 below*)

$$\langle f, g \rangle = \int_{-L}^{L} \overline{f(x)}\, g(x)\, dx \tag{2.18}$$

*the functions* $u_n^{(L)}$ *have the property*

$$\langle u_n^{(L)}, u_m^{(L)} \rangle = \delta_{nm} = \begin{cases} 1, & \textit{for } n = m, \\ 0, & \textit{for } n \neq m. \end{cases} \tag{2.19}$$

*We say that the set* $\{u_n^{(L)} \mid n = 0, \pm 1, \pm 2, \dots\}$ *is an* orthonormal set. *Equation* (2.19) *looks so nice because of our choice of the normalization constant in the definition of the stationary plane waves, see Eq.* (2.2).

EXERCISE 2.5. *If a function* $\psi$ *is real-valued and symmetric,* $\psi(x) = \psi(-x)$*, show that Eq.* (2.14) *can be written as*

$$\psi(x) = \sum_{n=0}^{\infty} a_n \cos(k_n^{(L)} x). \tag{2.20}$$

*with real coefficients* $a_n$ *(see also Fig. 2.1). What happens for antisymmetric functions, that is,* $\psi(x) = -\psi(-x)$*?*

## 2.2. The Hilbert Space of Square-Integrable Functions

The set $L^2([a, b])$ of all square-integrable functions on an interval (see Definition 2.1) has the structure of a Hilbert space. In the following, the most important concepts of Hilbert space theory are explained as far as they are needed for Fourier analysis. In many respects the functions in a Hilbert space can be treated like ordinary vectors. For example, we can define linear combinations and scalar products of functions. This is not merely an exercise in abstract mathematics, but will be useful for understanding quantum mechanics. In the common interpretation of quantum mechanics wave functions appear as elements of a suitable Hilbert space. Thus, Hilbert spaces are a central element of the modern mathematical apparatus of quantum mechanics. They will be encountered very often later in this course.

### 2.2.1. Linear structure

The set $L^2([a, b])$ of square-integrable functions has the structure of a linear space. For the following it is useful to bear in mind the analogy with the $n$-dimensional complex space $\mathbb{C}^n$ which is a more elementary example of a linear space. The vectors $\vec{v} \in \mathbb{C}^n$ are $n$-tuples

$$\vec{v} = (v_1, \dots, v_n), \quad \text{with } v_i \in \mathbb{C}. \tag{2.21}$$

These $n$-tuples can be added and multiplied by scalars. For $a \in \mathbb{C}$, $b \in \mathbb{C}$, $\vec{v} \in \mathbb{C}^n$ and $\vec{w} \in \mathbb{C}^n$ we can define the *linear combination* $a\vec{v} + b\vec{w}$ by

$$(a\vec{v} + b\vec{w})_i = a\, v_i + b\, w_i, \quad \text{for } i = 1, \dots, n. \tag{2.22}$$

Linear combinations of functions $\psi$ and $\phi$ in $L^2([a,b])$ are defined in a pointwise sense:

$$(a\psi + b\phi)(x) = a\,\psi(x) + b\,\phi(x), \quad \text{for all } x \in [a,b]. \tag{2.23}$$

If $\psi$ and $\phi$ are square-integrable, then the linear combination $a\psi + b\phi$ is again a square-integrable function.

### 2.2.2. Norm and scalar product

The linear spaces $L^2([a,b])$ and $\mathbb{C}^n$ have much more in common than the possibility of forming linear combinations. For example, the *length* (or *norm*) of a vector in $\mathbb{C}^n$ is defined by

$$\|\vec{v}\| = \left( \sum_{i=1}^{n} |v_i|^2 \right)^{1/2}. \tag{2.24}$$

In the Hilbert space $L^2([a,b])$ this corresponds to the norm of the function $\psi$, which is defined by

$$\|\psi\| = \left( \int_a^b |\psi(x)|^2 \, dx \right)^{1/2}. \tag{2.25}$$

Hence we can also define the *distance* $d$ between two functions $\psi$ and $\phi$ by

$$d(\phi, \psi) = \|\phi - \psi\|. \tag{2.26}$$

Likewise, the scalar product of two vectors $\vec{v}$ and $\vec{w}$,

$$\vec{v} \cdot \vec{w} = \sum_{i=1}^{n} \overline{v_i}\, w_i, \tag{2.27}$$

has an analog for functions $\psi$ and $\phi$:

$$\langle \psi, \phi \rangle = \int_a^b \overline{\psi(x)}\, \phi(x)\, dx. \tag{2.28}$$

As for vectors, we have

$$\|\psi\|^2 = \langle \psi, \psi \rangle. \tag{2.29}$$

Notice that the scalar product has been defined to be antilinear in the first, and linear in the second argument:

$$\langle a\psi_1 + b\psi_2, \phi \rangle = \overline{a}\, \langle \psi_1, \phi \rangle + \overline{b}\, \langle \psi_2, \phi \rangle, \tag{2.30}$$

$$\langle \psi, a\phi_1 + b\phi_2 \rangle = a\, \langle \psi, \phi_1 \rangle + b\, \langle \psi, \phi_2 \rangle. \tag{2.31}$$

Moreover,

$$\langle \psi, \phi \rangle = \overline{\langle \phi, \psi \rangle}. \tag{2.32}$$

We also state without proof the important *Cauchy–Schwarz inequality*:

$$|\langle \psi, \phi \rangle| \leq \|\psi\| \, \|\phi\|. \tag{2.33}$$

Equality holds if and only if $\psi = \alpha\phi$ with some $\alpha \in \mathbb{C}$.

### 2.2.3. Other Hilbert spaces

So far, we have only considered the Hilbert space $L^2([a,b])$ of square-integrable functions over a finite interval $[a,b]$. But in a completely analogous way we can also define Hilbert spaces that are defined on some other set. Among the most important examples are the Hilbert spaces $L^2([a,\infty))$, and $L^2(\mathbb{R})$ of square-integrable functions on an infinite interval, and the Hilbert space $L^2(\mathbb{R}^n)$ of square-integrable functions on $\mathbb{R}^n$. All we have to do is to define the norm and the scalar product by taking the integrals in (2.25) and (2.28) over the respective domain of definition. For example, a vector in the Hilbert space $L^2(\mathbb{R}^n)$ is a function

$$\psi : \mathbb{R}^n \to \mathbb{C}$$

for which the norm

$$\|\psi\| = \left( \int_{\mathbb{R}^n} |\psi(x)|^2 \, d^n x \right)^{1/2} \tag{2.34}$$

is finite.

Ψ Abstractly, a Hilbert space is defined as a vector space that is equipped with a scalar product, and that is complete with respect to the norm induced by the scalar product. The completeness of the Hilbert space means that every sequence $(\psi_n)$ with the property (*Cauchy sequence*)

$$\|\psi_n - \psi_m\| \to 0, \quad \text{for } n, m \to \infty, \tag{2.35}$$

has a limit $\psi = \lim \psi_n$, that is,

$$\lim_{n \to \infty} \|\psi_n - \psi\| = 0. \tag{2.36}$$

For example, the set of real numbers is complete, the set of rational numbers is not. Because the vector space $\mathbb{C}^n$ with the previously defined scalar product is complete, it is a Hilbert space.

EXAMPLE 2.2.1. **The Hilbert space $l^2$.** The Hilbert space of square-summable sequences is an infinite-dimensional analog of $\mathbb{C}^n$. A sequence $c = (c_i) = (c_1, c_2, c_3, \dots)$ of complex numbers is *square-summable* if

$$\sum_i |c_i|^2 < \infty. \tag{2.37}$$

The set $l^2$ of all square-summable sequences forms a vector space if the addition and multiplication by a scalar $a \in \mathbb{C}$ is defined by

$$(c_i) + (d_i) = (c_i + d_i), \qquad a\,(c_i) = (ac_i). \tag{2.38}$$

With the scalar product

$$\langle (c_i)\,,\,(d_i)\rangle = \sum_i \overline{c_i}\, d_i \tag{2.39}$$

the vector space $l^2$ becomes a Hilbert space.

### 2.2.4. Orthogonality

Two functions $\psi$ and $\phi$ in a Hilbert space are *orthogonal* if $\langle \psi,\phi\rangle = 0$. They are called *linearly independent* if $a\psi + b\phi = 0$ implies $a = b = 0$. If two functions are orthogonal, then they are also linearly independent. A set of functions $\{\phi_i\}$ is called an *orthonormal set* if $\langle \phi_i,\phi_j\rangle = \delta_{ij}$. By Exercise 2.4 the set $\{u_n^{(L)}\}$ is orthonormal in $L^2([-L, L])$.

An orthonormal set $\{\phi_i\}$ in a Hilbert space $\mathfrak{H}$ is a *basis*, if and only if the *completeness property* holds. The completeness property means that every vector $\psi \in \mathfrak{H}$ can be written as a (possibly infinite) linear combination of the basis vectors in the form

$$\psi = \sum_i c_i\, \phi_i, \qquad c_i = \langle \phi_i,\psi\rangle. \tag{2.40}$$

The form of the coefficients $c_i$ is an immediate consequence of the orthonormality of the basis:

$$\langle \phi_i,\psi\rangle = \langle \phi_i, \sum_j c_j\, \phi_j\rangle = \sum_j c_j\, \langle \phi_i,\phi_j\rangle = \sum_j c_j\, \delta_{ij} = c_i. \tag{2.41}$$

The infinite sum is assumed to converge with respect to the norm of the Hilbert space, that is,

$$\lim_{N\to\infty} \|\sum_{i\le N} c_i\, \phi_i - \psi\| = 0. \tag{2.42}$$

With the help of the distance defined in Eq. (2.26) this can be written as

$$d\Bigl(\sum_{i\le N} c_i\, \phi_i\,,\psi\Bigr) \to 0, \qquad \text{as } N \to \infty. \tag{2.43}$$

In the finite-dimensional Hilbert space $\mathbb{C}^n$ the set $\{e_i \mid i = 1, \dots, n\}$ is a basis, where $e_i = (0, \dots, 1, \dots, 0)$ is the vector with 1 as the $i$th component, all other components being zero.

EXERCISE 2.6. *Assuming that the set $\{\phi_i \mid i = 1, 2, \dots\}$ is a basis, prove that*

$$\|\psi\|^2 = \sum_{i=1}^{\infty} |c_i|^2, \tag{2.44}$$

*where $c_i = \langle \phi_i, \psi \rangle$ are the coefficients in the expansion of $\psi$. Thus, the sequence $(c_i)$ belongs to the Hilbert space $l^2$.*

$\boxed{\Psi}$ With respect to a fixed orthonormal basis $\{\phi_i\}$, every vector $\psi$ in the Hilbert space can be represented uniquely by the square-summable sequence $(c_i)$ of expansion coefficients. This establishes an isomorphism between the given Hilbert space and the Hilbert space $l^2$ of sequences. It can be shown that every (separable) Hilbert space has a (countable) basis in the sense defined above.

### 2.2.5. Fourier series

The importance of the concepts introduced above for the treatment of Fourier series is quite obvious. For example, the theorem on the Fourier series of a square-integrable function can simply be rephrased as follows.

> The orthonormal set $\{u_n^{(L)} \mid n = 0, \pm 1, \dots\}$ is a basis in $L^2([-L, L])$.

This can be seen by noting that Eq. (2.11) is equivalent to

$$\psi = \sum_n \langle u_n^{(L)}, \psi \rangle \, u_n^{(L)}. \tag{2.45}$$

You should compare the condition (2.42) for the convergence of this sum with the condition (2.15) for the convergence of the Fourier series.

The orthonormality of the set of plane waves $u_n^{(L)}$ is the subject of Exercise 2.4. The completeness property of an orthonormal set is usually rather difficult to prove. For the set $\{u_n^{(L)}\}$ the interested reader will find the details in almost any book on Fourier series.

## 2.3. The Fourier Transformation

### 2.3.1. From the Fourier series to the Fourier integral

If the length of the periodicity interval tends to infinity, it finally fills the whole real axis and the periodicity vanishes. It is interesting to investigate the behavior of the Fourier series in this limit because this will lead us to the study of nonperiodic functions.

Let $\psi$ be a smooth function which vanishes outside an interval $[-L_0, L_0]$. For any $L \geq L_0$ and all $x \in [-L, L]$ the Fourier series of $\psi$ is given as in Eq. (2.14) by

$$\psi(x) = \frac{1}{\sqrt{2\pi}} \sum_{n=-\infty}^{\infty} \hat{\psi}(k_n^{(L)})\, \mathrm{e}^{\mathrm{i}k_n^{(L)}x}\, \frac{\pi}{L} \tag{2.46}$$

$$= \frac{1}{\sqrt{2\pi}} \sum_{n=-\infty}^{\infty} \hat{\psi}(k_n^{(L)})\, \mathrm{e}^{\mathrm{i}k_n^{(L)}x}\, (k_{n+1}^{(L)} - k_n^{(L)}), \tag{2.47}$$

where

$$k_n^{(L)} = \frac{n\pi}{L}, \qquad k_{n+1}^{(L)} - k_n^{(L)} = \frac{\pi}{L} \equiv \Delta^{(L)}k. \tag{2.48}$$

CD 2.6 shows the spectrum of Fourier amplitudes of a Gaussian function depending on the length $L$ of the interval. We see the transition from the Fourier series to the Fourier transform. This transition is achieved by letting the length $L$ tend to infinity. See also Color Plate 9.

Equation (2.47) can be interpreted as the approximation of an integral by a Riemann sum:

$$\sum_{n=-\infty}^{\infty} \hat{\psi}(k_n^{(L)})\, \mathrm{e}^{\mathrm{i}k_n^{(L)}x} \Delta^{(L)}k \approx \int_{-\infty}^{\infty} \hat{\psi}(k)\, \mathrm{e}^{\mathrm{i}kx}\, dk. \tag{2.49}$$

Because $\psi(x) = 0$ for $|x| \geq L_0$ we can replace the borders of the integral in Eq. (2.13) by $\pm\infty$ and finally obtain the very symmetrically looking relations

$$\psi(x) = \frac{1}{\sqrt{2\pi}} \int_{-\infty}^{\infty} \hat{\psi}(k)\, \mathrm{e}^{\mathrm{i}kx}\, dk, \tag{2.50}$$

$$\hat{\psi}(k) = \frac{1}{\sqrt{2\pi}} \int_{-\infty}^{\infty} \psi(x)\, \mathrm{e}^{-\mathrm{i}kx}\, dx. \tag{2.51}$$

The function $\hat{\psi}$ is called the *Fourier transform* of $\psi$. The mapping $\mathcal{F} : \psi \mapsto \hat{\psi}$ is called *Fourier transformation*. We will describe its properties in the next section.

$\boxed{\Psi}$ The Fourier transform can be defined for all functions $\psi$ that are integrable (in the sense of Lebesgue). The existence of $\int \psi(x)\, dx$ is equivalent to the existence of the integral $\int |\psi(x)|\, dx$ and hence also to the existence of $\int \mathrm{e}^{\mathrm{i}kx}\, \psi(x)\, dx$.

### 2.3.2. Fourier transformation in $n$ dimensions

The Fourier transform can easily be generalized to functions of several variables. For example, take some integrable function $\psi(x_1, x_2)$ on $\mathbb{R}^2$, and do a Fourier transformation first with respect to the variable $x_1$, and then with respect to $x_2$. You will obtain

$$\hat{\psi}(k_1, k_2) = \frac{1}{\sqrt{2\pi}} \int_{-\infty}^{\infty} \left\{ \frac{1}{\sqrt{2\pi}} \int_{-\infty}^{\infty} \psi(x_1, x_2)\, \mathrm{e}^{-\mathrm{i}k_1 x_1}\, dx_1 \right\} \mathrm{e}^{-\mathrm{i}k_2 x_2}\, dx_2 \tag{2.52}$$

$$= \frac{1}{2\pi} \int_{\mathbb{R}^2} \psi(x_1, x_2)\, \mathrm{e}^{-\mathrm{i}(k_1 x_1 + k_2 x_2)}\, dx_1\, dx_2. \tag{2.53}$$

The double integral in Eq. (2.53) is well defined and independent of the order of integration whenever $\psi$ is an integrable function on $\mathbb{R}^2$. This suggests the following generalization of the Fourier transformation to functions of several variables:

Let $\psi(\mathbf{x})$, $\mathbf{x} \in \mathbb{R}^n$, be an integrable function. The Fourier transformation $\mathcal{F}$ maps $\psi$ onto the function $\hat{\psi}$ defined by

$$\hat{\psi}(\mathbf{k}) = (\mathcal{F}\psi)(\mathbf{k}) = \frac{1}{(2\pi)^{n/2}} \int_{\mathbb{R}^n} \mathrm{e}^{-\mathrm{i}\mathbf{k}\cdot\mathbf{x}}\, \psi(\mathbf{x})\, d^n x. \tag{2.54}$$

The space $\mathbb{R}^n$ formed by the independent variables $\mathbf{k} = (k_1, \ldots, k_n) \in \mathbb{R}^n$ is usually called the *Fourier space* or—in view of the application to quantum mechanics—the *momentum space*. This helps to distinguish it from the *position space* formed by the variables $\mathbf{x} \in \mathbb{R}^n$. Correspondingly, one calls $\psi$ the function in position space and $\hat{\psi}$ the function in momentum space.

Moreover, for an integrable function $\varphi(\mathbf{k})$, $\mathbf{k} \in \mathbb{R}^n$, we define another function $\check{\varphi}$ by

$$\check{\varphi}(\mathbf{x}) = \frac{1}{(2\pi)^{n/2}} \int_{\mathbb{R}^n} \mathrm{e}^{\mathrm{i}\mathbf{k}\cdot\mathbf{x}}\, \varphi(\mathbf{k})\, d^n k. \tag{2.55}$$

In view of Eq. (2.50) the mapping $\varphi \mapsto \check{\varphi}$ is called the *inverse Fourier transformation* and is denoted by $\mathcal{F}^{-1}$. Indeed, even for this more general situation one can prove the relation

$$\psi = \mathcal{F}^{-1}(\mathcal{F}\psi) \tag{2.56}$$

for every integrable function $\psi$ for which $\mathcal{F}\psi$ is also integrable. This is the famous Fourier inversion theorem.

**Fourier inversion theorem:**

If $\psi$ is integrable on $\mathbb{R}^n$ and if the Fourier transform of $\psi$,

$$\hat{\psi}(\mathbf{k}) = \frac{1}{(2\pi)^{n/2}} \int_{\mathbb{R}^n} \mathrm{e}^{-\mathrm{i}\mathbf{k}\cdot\mathbf{x}}\, \psi(\mathbf{x})\, d^n x, \tag{2.57}$$

is also integrable (as a function of $\mathbf{k} \in \mathbb{R}^n$), then $\psi$ has the representation

$$\psi(\mathbf{x}) = \frac{1}{(2\pi)^{n/2}} \int_{\mathbb{R}^n} \mathrm{e}^{\mathrm{i}\mathbf{k}\cdot\mathbf{x}}\, \hat{\psi}(\mathbf{k})\, d^n k. \tag{2.58}$$

A function and its Fourier transform depend on each other in a very symmetric way. Apart from the sign in the exponent the inverse transformation is just a Fourier transformation from momentum space to position space. More precisely, we have

$$(\mathcal{F}^{-1}\varphi)(\mathbf{x}) = (\mathcal{F}\varphi)(-\mathbf{x}). \tag{2.59}$$

EXERCISE 2.7. *Prove that the composition $\mathcal{F}^2$ of two Fourier transforms amounts to a space reflection, that is,*

$$(\mathcal{F}^2\psi)(\mathbf{x}) = \psi(-x). \tag{2.60}$$

*Therefore, $(\mathcal{F}^4\psi)(\mathbf{x}) = \psi(x)$, or*

$$\mathcal{F}^4 = \mathbf{1}. \tag{2.61}$$

## 2.4. Basic Properties of the Fourier Transform

The following results describe the range of the Fourier transform. You can find the proofs in any book about Fourier analysis.

### 2.4.1. Riemann–Lebesgue lemma

The Fourier transform $\hat{\psi}$ of an integrable function $\psi$ is a continuous function with the following properties:

1. $\hat{\psi}$ is bounded,

$$\sup_{k\in\mathbb{R}^n} |\hat{\psi}(k)| \leq \int_{\mathbb{R}^n} |\psi(x)|\, d^n x. \tag{2.62}$$

2. $\hat{\psi}$ vanishes at infinity,

$$\hat{\psi}(k) \to 0, \quad \text{as } |k| \to \infty. \tag{2.63}$$

The same result applies to the inverse Fourier transform of a $\phi$ which is integrable with respect to $\mathbf{k}$.

The Riemann–Lebesgue lemma does not state how fast the Fourier transform vanishes at infinity. Hence the Fourier transform of an integrable function need not be integrable. Indeed, if $\psi$ is discontinuous at some point $\mathbf{x}_0$, then the Fourier transform $\hat{\psi}$ is not integrable in momentum space. (If it were, $\psi$ would be the inverse Fourier transform of $\hat{\psi}$ and hence a continuous function). See CD 2.10 and CD 2.11.

### 2.4.2. Fourier–Plancherel theorem

If $\psi$ is both integrable and square-integrable, then the Fourier transform $\hat{\psi}$ is also square-integrable, and

$$\int_{\mathbb{R}^n} |\psi(\mathbf{x})|^2 \, d^n x = \int_{\mathbb{R}^n} |\hat{\psi}(\mathbf{k})|^2 \, d^n k. \tag{2.64}$$

EXERCISE 2.8. *By giving examples, show that there exist functions $\psi$ that are integrable on $\mathbb{R}$ but not square-integrable. Similarly, find a square-integrable function $\psi \in L^2(\mathbb{R})$ that is not integrable.*

EXERCISE 2.9. *Prove: A function that is square-integrable over a finite interval $[a, b]$ is also integrable on $[a, b]$.*

The Fourier–Plancherel theorem states that the Fourier transform $\mathcal{F}$ is continuous with respect to the norm in the Hilbert space $L^2(\mathbb{R}^n)$. If $\psi$ and $\phi$ are close together, then also $\hat{\psi}$ and $\hat{\phi}$ are close together. Here "close together" means that $\|\psi - \phi\|$ is small. The Fourier–Plancherel theorem implies that $\hat{\psi}$ and $\hat{\phi}$ are close together in the same sense because $\|\hat{\psi} - \hat{\phi}\| = \|\psi - \phi\|$.

## 2.5. Linear Operators

### 2.5.1. Basic definitions

The Fourier transformation $\mathcal{F}$ is a mapping from a set of integrable functions $\psi$ in a Hilbert space to functions $\hat{\psi}$ belonging to the same Hilbert space. The mapping $\mathcal{F}$ is an example of a linear operator. You will learn in Chapter 4 that in the quantum-mechanical formalism the linear operators play an essential role (all physical observables are represented by linear operators). Thus, it is important familiarize yourself with this concept at an early stage.

Mathematically, linear operators are defined as mappings on a vector space that are compatible with the linear structure of the vector space. It is typical for infinite-dimensional function spaces that linear operators often can only be defined on a suitable subspace.

DEFINITION 2.2. A linear operator $T$ in a Hilbert space $\mathfrak{H}$ is a transformation mapping vectors $\psi$ from a *domain* $\mathfrak{D}(T)$, which is a linear subspace of $\mathfrak{H}$, to other vectors $T\psi$ in such a way that

$$T(a\psi + b\phi) = a\,T\psi + b\,T\phi \tag{2.65}$$

for all $\psi, \phi \in \mathfrak{D}(T)$ and all scalars $a, b \in \mathbb{C}$.

EXERCISE 2.10. *Verify that the set $\mathfrak{D}$ of integrable and square-integrable functions is a linear subspace of $L^2(\mathbb{R}^n)$. Show that the Fourier transform $\mathcal{F} : \psi \to \hat{\psi}$ is a linear operator defined on the domain $\mathfrak{D}$, that is, prove that Eq.* (2.65) *holds for $\mathcal{F}$.*

Two linear operators $S$ and $T$ can be multiplied. The product is simply defined as the composition,

$$(ST)\psi \equiv S \circ T\,\psi = S(T\psi). \tag{2.66}$$

The domain of the product $ST$ consists of those elements $\psi$ in the domain of $T$, for which $T\psi$ is in the domain of $S$.

$$\mathfrak{D}(ST) = \{\psi \in \mathfrak{D}(T) \mid T\psi \in \mathfrak{D}(S)\}. \tag{2.67}$$

**The commutator of two linear operators:**

The product of two operators is not commutative. This means that in general $ST$ is different from $TS$. The degree of noncommutativity is described by the *commutator*

$$[S, T] = ST - TS. \tag{2.68}$$

The commutator is again a linear operator which is defined on the intersection of the domains of $ST$ and of $TS$, that is, on the linear subspace

$$\mathfrak{D}([S, T]) = \mathfrak{D}(ST) \cap \mathfrak{D}(TS). \tag{2.69}$$

EXERCISE 2.11. *The commutator is linear in both arguments, that is, for any complex numbers $\alpha_1$ and $\alpha_2$,*

$$[\alpha_1 S_1 + \alpha_2 S_2, T] = \alpha_1 [S_1, T] + \alpha_2 [S_2, T], \tag{2.70}$$

*and similar for the second argument. Moreover, the commutator has the following algebraic properties which are often useful for practical calculations:*

$$[S,T] = [-T,S], \qquad (\textit{antisymmetry}) \tag{2.71}$$

$$[R,[S,T]] + [S,[T,R]] + [T,[R,S]] = 0, \qquad (\textit{Jacobi identity}) \tag{2.72}$$

$$[R,ST] = S\,[R,T] + [R,S]\,T. \tag{2.73}$$

*Verify these identities by a formal calculation using the definition of the commutator.*

### 2.5.2. Boundedness

A linear operator $T$ is *bounded* if there is a constant $c > 0$ such that

$$\|T\psi\| \le c\,\|\psi\|, \qquad \text{for all } \psi \in \mathfrak{D}(T). \tag{2.74}$$

The smallest such constant is called the *norm* of $T$ and is denoted by $\|T\|$,

$$\|T\| = \sup_{\psi\in\mathfrak{H}} \frac{\|T\psi\|}{\|\psi\|}. \tag{2.75}$$

If the supremum above does not exist, then the linear operator is called *unbounded*.

$\boxed{\Psi}$ **Unboundedness criterion**: For an unbounded operator $T$ there exists a sequence $\psi_n$ in the domain $\mathfrak{D}(T)$, such that $\psi_n$ converges to zero while $T\psi_n$ diverges.

PROOF. If $T$ is not bounded, then for every $n$ there is a $\phi_n$ such that $\|T\phi_n\| > n\,\|\phi_n\|$. Setting

$$\psi_n = \frac{1}{\sqrt{n}}\,\frac{\phi_n}{\|\phi_n\|},$$

we find that $\|\psi_n\| = 1/\sqrt{n} \to 0$, and $\|T\psi_n\| > \sqrt{n} \to \infty$. □

The Fourier transformation $\mathcal{F}$ is an example of a bounded linear operator. It is defined on the domain of functions that are integrable and square-integrable at the same time. The Fourier–Plancherel relation states that

$$\|\mathcal{F}\| = 1. \tag{2.76}$$

### 2.5.3. Special topic: Continuity

A linear operator is *continuous* if it preserves the convergence of sequences.

DEFINITION 2.3. A linear operator $T$ is *continuous* if $\lim T\psi_n$ exists whenever $\lim \psi_n$ exists.

Here we state the basic theorem about continuous linear operators

THEOREM 2.1. *Each of the following statements is completely equivalent to the continuity of the linear operator* $T$.

1. $T$ *is bounded.*
2. $T$ *is continuous at* 0, *that is,* $\lim T\psi_n = 0$, *whenever* $\lim \psi_n = 0$.

Let us sketch the proof of this theorem.

(1) The continuity of $T$ implies the boundedness: If $T$ were unbounded, then the unboundedness criterion implies the existence of a convergent sequence that is mapped to a divergent one (which clearly contradicts the definition of continuity).

(2) Boundedness implies continuity at 0 because

$$\|T\psi_n\| \leq \|T\| \, \|\psi_n\|.$$

Hence, if the sequence $(\psi_n)$ converges to 0, so does the sequence $(T\psi_n)$.

(3) The continuity at zero implies the continuity at any other point $\psi \in \mathfrak{D}(A)$: If $(\psi_n)$ is an arbitrary convergent sequence, $\lim \psi_n = \psi \in \mathfrak{D}(A)$, then the sequence $(\psi_n - \psi)$ converges to 0 and hence, because of the continuity at 0, the sequence $T\psi_n - T\psi$ also converges to 0. Hence $T\psi_n$ converges.

In the definition of the continuity it is assumed that $\psi_n \in \mathfrak{D}(T)$ for all $n$. Let us now assume that the limit $\psi$ of the convergent sequence $\{\psi_n\}$ is not in the domain of $T$. In this case we can nevertheless *define* an action of $T$ on $\psi$ in a completely natural way:

$$T\psi = \lim_{n \to \infty} T\psi_n.$$

In this way one can always extend a continuous operator $T$ to an eventually larger domain which contains all limit points of sequences in the original domain. This process preserves the norm of the operator. If—as it is usually the case—the original domain is "dense" in $\mathfrak{H}$, then the extended domain is the whole Hilbert space.

DEFINITION 2.4. A set $A$ is called *dense* in $\mathfrak{H}$ if every vector $\psi \in \mathfrak{H}$ is the limit of a convergent sequence of vectors $\psi_n \in A$.

$\boxed{\Psi}$ By the definition above, the subset $A$ is dense if every vector in the Hilbert space can be approximated from within the subset. We can reformulate this property as follows: Given $\psi \in \mathfrak{H}$ and $\varepsilon > 0$, there is always a $\phi \in A$ within the distance $\varepsilon$ from $\psi$. A useful criterion for a linear subspace to be dense is the following: A linear subspace $A$ of a Hilbert space is dense if and only if the only vector that is orthogonal to all vectors in $A$ is the zero-vector: $A$ is dense if and only if $\langle \psi, \phi \rangle = 0$ for all $\psi \in A$ implies $\phi = 0$.

Let $A$ be a subset of a Hilbert space. If we add all limit points of $A$ to $A$, we obtain the *closure* of $A$, sometimes denoted by $\overline{A}$. A set which already contains all its limit points is called a *closed set*. A set $A$ is dense in $\mathfrak{H}$ if and only if $\overline{A} = \mathfrak{H}$.

Let $T$ be a linear operator that is defined on a domain $\mathfrak{D}(T)$. The linear operator obtained from $T$ by extending its action to the closure of $\mathfrak{D}(T)$ is called the *closure* of $T$. If the original domain is dense in $\mathfrak{H}$, then the closure of $T$ is defined everywhere in $\mathfrak{H}$.

The process of extending a densely defined linear operator to the whole Hilbert space will now be demonstrated for the Fourier transformation $\mathcal{F}$.

### 2.5.4. Special topic: Extension of the Fourier transform

We can extend the definition of the Fourier transform $\mathcal{F}$ to all square-integrable functions in the Hilbert space $L^2(\mathbb{R}^n)$. Let $\psi$ be a square-integrable function. As you know, $\psi$ need not be integrable and hence the Fourier transform of $\psi$ is perhaps ill-defined. However, we can approximate $\psi$ by integrable functions. Define

$$\psi_n(\mathbf{x}) = \begin{cases} \psi(\mathbf{x}), & \text{if } |\mathbf{x}| \leq n, \\ 0, & \text{if } |\mathbf{x}| > n. \end{cases} \tag{2.77}$$

Then $\psi_n$ is integrable and you can see that

$$\|\psi - \psi_n\|^2 = \int_{|\mathbf{x}|>n} |\psi(\mathbf{x})|^2 \, d^n x \to 0, \quad \text{as } n \to \infty. \tag{2.78}$$

Thus, an arbitrary square-integrable function is the limit of a sequence of integrable functions. This proves that the set of functions in $L^2(\mathbb{R}^n)$ that are integrable is dense in $L^2(\mathbb{R}^n)$.

Each $\psi_n$ is in the domain of $\mathcal{F}$, and the sequence $(\psi_n)$ converges in $\mathfrak{H}$. The continuity of $\mathcal{F}$ implies the convergence of the sequence $(\mathcal{F}\psi_n)$ and thus it makes sense to define the Fourier transform of $\psi$ to be the limit of the sequence $\mathcal{F}\psi_n$.

**Fourier transform in $L^2$:**

The Fourier transform of an $L^2$-function $\psi$ is defined as

$$\mathcal{F}\psi = \lim_{n\to\infty} \hat{\psi}_n, \tag{2.79}$$

where $\psi_n$ is any sequence of integrable functions converging to $\psi$, that is,

$$\psi = \lim_{n\to\infty} \psi_n. \tag{2.80}$$

## 2.6. Further Results About the Fourier Transformation

The following discussion quotes some elementary properties that make the Fourier transform a very important tool in analysis. For the proofs, we refer again to the books dedicated to Fourier analysis.

### 2.6.1. Translation, phase shift, scaling transformation

For a function $\psi$ on $\mathbb{R}^n$ one can define the following operations ($a$, $b \in \mathbb{R}$, $\lambda > 0$):

$$(\tau_{\mathbf{a}}\psi)(\mathbf{x}) = \psi(\mathbf{x}-\mathbf{a}), \qquad \text{translation by } \mathbf{a} \in \mathbb{R}^n, \tag{2.81}$$

$$(\mu_{\mathbf{b}}\psi)(\mathbf{x}) = \mathrm{e}^{\mathrm{i}\mathbf{b}\cdot\mathbf{x}}\psi(\mathbf{x}), \qquad \text{phase shift}, \tag{2.82}$$

$$(\delta_\lambda\psi)(\mathbf{x}) = \lambda^{-n/2}\,\psi(\mathbf{x}/\lambda), \qquad \text{scaling transformation.} \tag{2.83}$$

For each $a$, $b \in \mathbb{R}$, $\lambda > 0$, the operations $\tau_{\mathbf{a}}$, $\mu_{\mathbf{b}}$, and $\delta_\lambda$ are linear operators defined on the Hilbert space of square-integrable functions. The action of these operators on a given function is depicted in Fig. 2.3.

Using a substitution of variables in the Fourier integral it is easy to see that

$$(\mathcal{F}\tau_{\mathbf{a}}\psi)(\mathbf{k}) = (\mu_{-\mathbf{a}}\mathcal{F}\psi)(\mathbf{k}) = \mathrm{e}^{-\mathrm{i}\mathbf{k}\cdot\mathbf{a}}\,\hat{\psi}(\mathbf{k}), \tag{2.84}$$

$$(\mathcal{F}\mu_{\mathbf{b}}\psi)(\mathbf{k}) = (\tau_{\mathbf{b}}\mathcal{F}\psi)(\mathbf{k}) = \hat{\psi}(\mathbf{k}-\mathbf{b}), \tag{2.85}$$

$$(\mathcal{F}\delta_\lambda\psi)(\mathbf{k}) = (\delta_{1/\lambda}\mathcal{F}\psi)(\mathbf{k}) = \lambda^{n/2}\,\hat{\psi}(\lambda\mathbf{k}). \tag{2.86}$$

**Translations in position and momentum space:**
The translation $\tau_{\mathbf{a}}$ in position space corresponds to a phase shift $\mathrm{e}^{-\mathrm{i}\mathbf{k}\cdot\mathbf{a}}$ in momentum space. The translation $\mu_{\mathbf{b}}$ in momentum space corresponds to a phase shift $\mathrm{e}^{\mathrm{i}\mathbf{x}\cdot\mathbf{b}}$ in position space.

The relation between phase shift and translation of the Fourier transform is illustrated in Color Plate 10. The behavior of scaled functions under the Fourier transform is related to the well-known uncertainty relation (see Section 2.8.1 below). Equation (2.86) states that a scaling transformation in position space corresponds to the inverse scaling transformation in momentum space. Narrowing a function in position space makes it wide in momentum space.

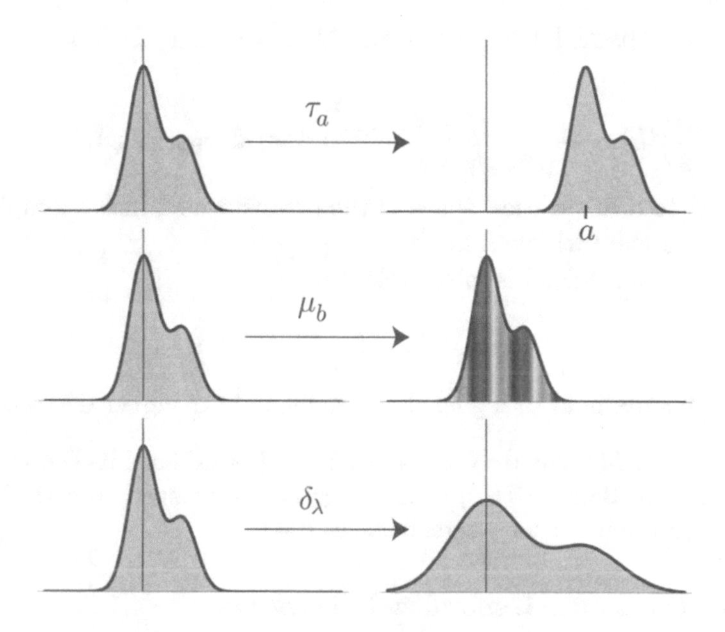


FIGURE 2.3. Action of the translation $\tau_a$, the phase shift $\mu_b$, and the scaling transformation $\delta_\lambda$ (for $\lambda > 1$). The translation shifts the wave packet $\psi$ by $a$; the phase shift adds color stripes (with wavelength proportional to $1/|b|$); and the scaling transformation widens ($\lambda > 1$) or narrows ($\lambda < 1$) the wave packet.

CD 2.7, CD 2.8, and CD 2.9 illustrate the behavior of a Gaussian function and of its Fourier transform under translations, phase shifts, and scaling transformations.

EXERCISE 2.12. *Show that the operators $\tau_{\mathbf{a}}$, $\mu_{\mathbf{b}}$, and $\delta_\lambda$ are defined everywhere in $L^2(\mathbb{R}^n)$ and that they are bounded with norm 1.*

EXERCISE 2.13. *Show that the operators $\tau_{\mathbf{a}}$, and $\mu_{\mathbf{b}}$ do not commute, that is, the composition of these linear mappings depends on the order of the operations. To this end, calculate*

$$\tau_{\mathbf{a}}\,\mu_{\mathbf{b}}\,\psi - \mu_{\mathbf{b}}\,\tau_{\mathbf{a}}\,\psi \tag{2.87}$$

*for arbitrary $\psi \in L^2$. See CD 2.8. This result will be discussed further in Section 6.8.*

### 2.6.2. Derivative, multiplication

Let $\psi$ be an integrable function on $\mathbb{R}$. Assume that $\psi$ is differentiable and vanishes, as $|x| \to \infty$, such that the derivative $d\psi(x)/dx$ is also integrable.

Then (using a partial integration and the fact that the boundary terms vanish)

$$\left(\mathcal{F}\frac{d\psi}{dx}\right)(k) = -\frac{1}{\sqrt{2\pi}}\int_{\mathbb{R}}\left(\frac{d}{dx}e^{-ikx}\right)\psi(x)\,dx = ik(\mathcal{F}\psi)(k). \tag{2.88}$$

A differentiation in position space is thus transformed into a simple multiplication with the variable $k$ in Fourier space.

If the function $x\psi(x)$ is integrable, then

$$(\mathcal{F}x\psi)(k) = i\frac{d}{dk}(\mathcal{F}\psi)(k). \tag{2.89}$$

Analogous results hold in higher dimensions and for higher derivatives.

CD 2.10 shows the function $f(x) = a/(1+bx^2)$ and its Fourier transform. Because $xf(x)$ is not integrable with respect to $x$, the Fourier transform is not differentiable at $k = 0$.

### 2.6.3. Special topic: Generalized derivative

With the Fourier transform you can do fantastic things. For example, it is possible to define in a mathematically clean way the derivative of functions that are not differentiable in the sense of ordinary calculus. The expression

$$\psi' = \mathcal{F}^{-1}(ik\hat{\psi}) \tag{2.90}$$

is called the "*generalized derivative*" of $\psi$. Let me explain this concept with an example.

The function (see Fig. 2.4)

$$\psi = e^{-|x|} \tag{2.91}$$

is integrable and its Fourier transform is

$$\hat{\psi}(k) = \sqrt{\frac{2}{\pi}}\frac{1}{1+k^2}. \tag{2.92}$$

As $|k| \to \infty$, this function decreases so slowly that $ik\hat{\psi}(k)$ is not integrable. This, of course, reflects the fact that $\psi$ is not differentiable at $x = 0$. But since $ik\hat{\psi}(k)$ is square-integrable, we can define its inverse Fourier transform in the $L^2$-sense as in Eq. (2.79). Thus, the inverse Fourier transform is the limit (with respect to the metric in $L^2$) of the sequence

$$\phi_n(x) = \frac{1}{\sqrt{2\pi}}\int_{-n}^{n} e^{ikx}\, ik\hat{\psi}(k)\,dk. \tag{2.93}$$

It can be shown that the $L^2$-limit of this sequence is given by

$$\phi(x) = -e^{-|x|}\frac{x}{|x|}. \tag{2.94}$$

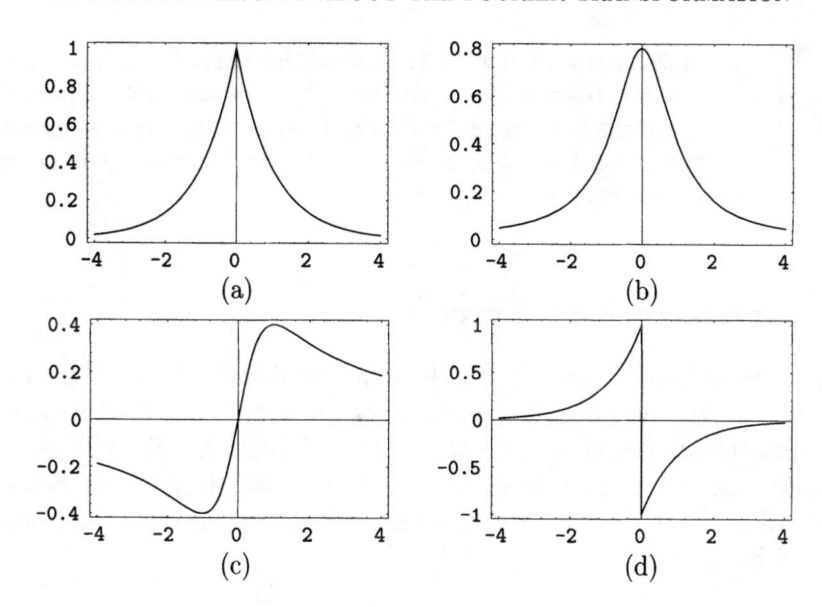


FIGURE 2.4. The Fourier transform allows us to define a generalized derivative. (a) This function $\psi(x)$ is not differentiable at the origin. (b) The Fourier transform $\hat{\psi}(k)$ of (a). (c) The function $k\hat{\psi}(k)$ is not integrable because it decays too slowly, as $|k| \to \infty$. (d) Because $\mathrm{i}k\hat{\psi}(k)$ is square-integrable, the inverse Fourier transform of this function can still be defined and gives the derivative of $\psi$ at all points $x$ where $\psi$ is differentiable.

At all points where $\psi$ is differentiable, $\phi$ obviously coincides with the derivative of $\psi$.

The Fourier transform even allows us to define functions of the linear operator $d/dx$ by

$$f\Big(\frac{d}{dx}\Big) = \mathcal{F}^{-1} f(\mathrm{i}k)\,\hat{\psi}. \tag{2.95}$$

This is, for example, important in relativistic quantum mechanics where one defines a relativistic energy operator by $(c^2p^2 - m^2c^4)^{1/2}$ with $p = -\mathrm{i}d/dx$.

**Generalized derivative of order** $s$: Using the Fourier transform one can generalize the notion of the derivative to noninteger orders. The CD 2.18 shows the application of $d^s/dx^s$ (for noninteger $s$) on the Gaussian function $\exp(-x^2/2)$.

CD 2.20 deals with another function of the derivative operator $d/dx$. This is the *resolvent* $(-d^2/dx^2 - z)^{-1}$. The animation shows the action of the resolvent on a Gaussian function and the dependence on the complex number $z$. In CD 2.20 the resolvent operator acts on a step function.

## 2.7. Gaussian Functions

Gaussian functions are smooth functions that vanish rapidly in all directions, as $|x| \to \infty$. In quantum mechanics, they are very well suited to describe the prototypical case of a fairly well localized particle. We will therefore frequently use Gaussian wave packets as initial conditions for the Schrödinger equation. In this section we derive some formulas for Gaussian functions that will be used later.

### 2.7.1. The Fourier transform of a Gaussian function

A Gaussian function on $\mathbb{R}^n$ is an exponential function of the form

$$f(\mathbf{x}) = \exp\left(-\sum_{i=1}^{n} \alpha_i x_i^2\right), \quad \alpha_i > 0. \tag{2.96}$$

Because $f$ can be written as a product of Gaussian functions on $\mathbb{R}$, it follows that the Fourier transform of $f$ is a product of the Fourier transforms of one-dimensional Gaussians. Thus, we only need to calculate the Fourier transform of $\exp(-\alpha x^2)$ (see Fig. 2.5).

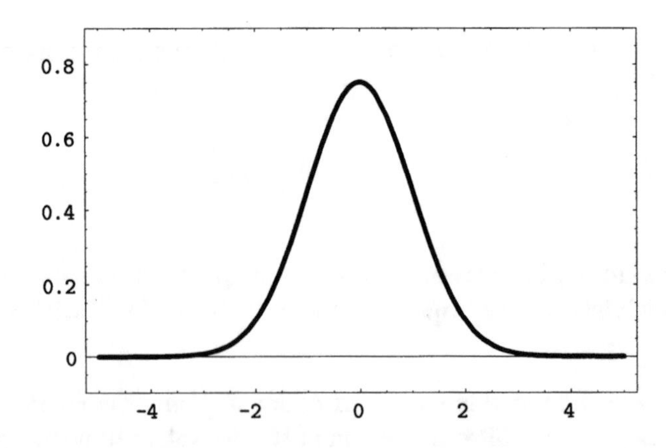


FIGURE 2.5. Plot of the Gaussian function $(1/\pi)^{1/4} \exp(-x^2/2)$.

**Fourier transform of a Gaussian:**

The Fourier transform of the Gaussian function $f(x) = \exp(-\alpha x^2/2)$, $\alpha > 0$, is given by

$$\hat{f}(k) = \frac{1}{\sqrt{\alpha}} \exp\Bigl(-\frac{k^2}{2\alpha}\Bigr). \tag{2.97}$$

PROOF. Obviously, the Gaussian function $\mathrm{e}^{-x^2/2}$ on $\mathbb{R}$ is the unique solution of the differential equation

$$\frac{d}{dx} f(x) + x\, f(x) = 0, \quad \text{with } f(0) = 1. \tag{2.98}$$

We now perform a Fourier transformation on both sides of this equation. If $f$ is a Gaussian function, then $f$, its derivatives, and $xf$ vanish rapidly, as $|x| \to \infty$, and are thus integrable functions. We can therefore use Eqs. (2.88) and (2.89) in order to obtain

$$\mathrm{i}k\, \hat{f}(k) + \mathrm{i}\frac{d}{dk} \hat{f}(k) = 0. \tag{2.99}$$

This is the same differential equation as in (2.98) above. In order to determine $\hat{f}$ uniquely, we need the initial condition

$$\hat{f}(0) = \frac{1}{\sqrt{2\pi}} \int_{\mathbb{R}} \mathrm{e}^{-x^2/2}\, dx. \tag{2.100}$$

The value of the integral is easily determined with the help of *Mathematica*. Using the following trick, it is easy to do the calculation by oneself. Setting $I = \int_{\mathbb{R}} \mathrm{e}^{-x^2/2}\, dx$, we calculate $I^2$, which can be written as

$$I^2 = \int_{\mathbb{R}^2} \mathrm{e}^{-(x^2+y^2)/2}\, dx\, dy = 2\pi \int_0^\infty \mathrm{e}^{-r^2/2}\, r\, dr \tag{2.101}$$

$$= 2\pi \int_0^\infty \mathrm{e}^{-s} ds = -2\pi \mathrm{e}^{-s}\Big|_0^\infty = 2\pi. \tag{2.102}$$

Hence we find $I = \sqrt{2\pi}$. Hence $\hat{f}(0) = I/\sqrt{2\pi} = 1$. Therefore, $\hat{f}$ satisfies the same differential equation as $f$, with the same initial condition. We conclude that $\hat{f}(k) = f(k) = \exp(-k^2/2)$.

This proves Eq. (2.97) for $\alpha = 1$. The general case now follows easily with a scaling transformation according to Eq. (2.86). □

Gaussian functions and their translates in $x$- and $k$-space are investigated in CD 2.7–CD 2.9. The movies CD 2.10–CD 2.17 show various other examples of the Fourier transformation. These examples include oscillating and discontinuous functions and functions with a nonintegrable Fourier transform.

### 2.7.2. Special topic: A dense set of Gaussian functions

In a certain sense the set of Gaussian functions is a very large subset of the Hilbert space. Here you will learn that Gaussian functions and their translates span the whole Hilbert space $L^2(\mathbb{R})$. This should justify the predominant use of Gaussian initial functions in the movies of quantum-mechanical wave functions.

THEOREM 2.2. *Define the functions*

$$G_q(x) = \left(\frac{1}{\pi}\right)^{1/4} \exp\left(-\frac{x^2}{2} + \mathrm{i}qx\right). \tag{2.103}$$

*Then the linear span (the set of all linear combinations) of the set* $\{G_q | q \in R\}$ *is a dense subspace of the Hilbert space* $L^2(\mathbb{R})$. *In other words, every vector in the Hilbert space can be approximated by a finite linear combination of the functions* $G_q$.

PROOF. You can use the criterion cited after Definition 2.4 to convince yourself that the set of all $G_q$ is dense. You only have to show that the only vector which is orthogonal to all the functions $G_q$ is the zero-vector. Consider the scalar product

$$\langle G_q, \psi \rangle = \int_{-\infty}^{\infty} \mathrm{e}^{-\mathrm{i}qx}\, G_0(x)\, \psi(x)\, dx = \sqrt{2\pi}\, \mathcal{F}(G_0\psi)(q). \tag{2.104}$$

Hence if $\psi$ is orthogonal to all $G_q$, we have

$$\langle G_q, \psi \rangle = 0 \quad \Leftrightarrow \quad \mathcal{F}(G_0\psi) = 0 \quad \Leftrightarrow \quad G_0\psi = 0 \quad \Leftrightarrow \quad \psi = 0. \tag{2.105}$$

The last step follows because $G_0(x) \neq 0$ for all $x$. This proves that only $\psi = 0$ is orthogonal to the set $\{G_q | q \in R\}$. This set is not a linear subspace, because the linear combination of two Gaussians $G_{q_1}$ and $G_{q_2}$ is not of the same form. But the set of all possible linear combinations of the functions $G_q$ is a linear subspace that fulfills the density criterion. □

In Fourier space, the functions $\hat{G}_q = \mathcal{F}G_q$ are just shifted Gaussian functions. Hence the set of all linear combinations of

$$\hat{G}_q(k) = \left(\frac{1}{\pi}\right)^{1/4} \exp\left(-\frac{(k-q)^2}{2}\right), \quad q \in \mathbb{R}, \tag{2.106}$$

is also dense in the Hilbert space of square-integrable functions.

## 2.8. Inequalities

In this section you will read about some inequalities related to Fourier analysis. Here you will find a proof of the famous uncertainty relation that has been of great importance in the development of quantum mechanics and its interpretation. But there are other inequalities sometimes even more useful in the mathematical theory of quantum mechanics.

### 2.8.1. The uncertainty relation

The uncertainty relation is an inequality that relates the width of a function $\psi$ in position space with the width of the Fourier transform $\hat{\psi}$ in momentum space. A measure for the spread around $\mathbf{x} = 0$ is, for example, the integral

$$\int_{\mathbb{R}^n} x^2 \, |\psi(\mathbf{x})|^2 \, d^n x = \|\mathbf{x}\psi\|^2. \tag{2.107}$$

Obviously, this integral is small if the function $|\psi(\mathbf{x})|^2$ is well localized around the origin in $\mathbb{R}^n$. Conversely, the integral will be large if a significant part of the function $\psi$ is located at large values of $\mathbf{x}$. In Fourier space, consider the expression

$$\int_{\mathbb{R}^n} k^2 \, |\hat{\psi}(\mathbf{k})|^2 \, d^n x = \|\nabla\psi\|^2. \tag{2.108}$$

Here we used the Fourier–Plancherel relation Eq. (2.64). Of course we had to assume that (each component of) $\mathbf{k}\hat{\psi}(\mathbf{k})$ is square-integrable.

**Heisenberg's uncertainty relation:**

For all square-integrable functions $\psi$ for which $|\mathbf{x}| \, \psi(\mathbf{x})$ and $|\mathbf{k}| \, \hat{\psi}(\mathbf{k})$ are also square-integrable, we have the inequality

$$\|\nabla\psi\| \, \|\mathbf{x}\psi\| \geq \frac{n}{2} \, \|\psi\|^2. \tag{2.109}$$

Equality holds if and only if $\psi$ is a Gaussian function of the form

$$\psi(\mathbf{x}) = N \exp\Bigl(-\alpha \frac{\mathbf{x}^2}{2}\Bigr) \tag{2.110}$$

with some $\alpha > 0$.

PROOF. The following proof holds for differentiable functions that vanish, as $|\mathbf{x}| \to \infty$. With the Cauchy–Schwarz inequality (2.33) you can see that

$$\|\nabla\psi\|^2 \|\mathbf{x}\psi\|^2 \geq |\langle \nabla\psi, \mathbf{x}\psi \rangle|^2. \tag{2.111}$$

Equality holds if and only if $\nabla\psi = \alpha\mathbf{x}\psi$ with some $\alpha \in \mathbb{C}$. This differential equation already implies that in this case $\psi$ must be a Gaussian function as in Eq. (2.110), with $\operatorname{Re}\alpha > 0$, in order to be square-integrable.

For arbitrary $\psi$ you can perform the following partial integration

$$\begin{aligned}\langle\nabla\psi, \mathbf{x}\psi\rangle &= \int \sum_{i=1}^{n} \frac{\partial}{\partial x_i}\overline{\psi(\mathbf{x})}\, x_i\psi(\mathbf{x})\, d^n x \\ &= -\int \sum_{i=1}^{n} \overline{\psi(\mathbf{x})} \frac{\partial}{\partial x_i}\, x_i\psi(\mathbf{x})\, d^n x = -\langle\psi, \nabla\cdot\mathbf{x}\psi\rangle.\end{aligned} \tag{2.112}$$

Notice that the integrated terms $\overline{\psi(\mathbf{x})}\, x_i\psi(\mathbf{x})$ vanish at the boundaries $\pm\infty$. Now

$$\nabla\cdot\mathbf{x}\psi = n\psi + \mathbf{x}\cdot\nabla\psi, \tag{2.113}$$

and hence

$$\langle\psi, \nabla\cdot\mathbf{x}\,\psi\rangle = \langle\mathbf{x}\psi, \nabla\psi\rangle + n\|\psi\|^2 \tag{2.114}$$

Put this together with Eq. (2.112) and you will find

$$2\operatorname{Re}\langle\nabla\psi, \mathbf{x}\psi\rangle = \langle\mathbf{x}\psi, \nabla\psi\rangle + n\|\psi\|^2, \tag{2.115}$$

and hence

$$|\langle\nabla\psi, \mathbf{x}\psi\rangle|^2 = (\operatorname{Re}\langle\nabla\psi, \mathbf{x}\psi\rangle)^2 + (\operatorname{Im}\langle\nabla\psi, \mathbf{x}\psi\rangle)^2 \geq \frac{n^2}{4}\,\|\psi\|^2. \tag{2.116}$$

Equality holds if $\operatorname{Im}\langle\nabla\psi, \mathbf{x}\psi\rangle = 0$. For the Gaussian function above this means $\alpha \in \mathbb{R}$, hence $\alpha > 0$. □

One of the crucial steps in the calculation above was to observe that

$$[\nabla, \mathbf{x}]\psi \equiv \nabla\cdot\mathbf{x}\psi - \mathbf{x}\cdot\nabla\psi = n\psi, \tag{2.117}$$

see Eq. (2.113). The expression $[\nabla, \mathbf{x}]$ is the commutator of $\nabla$ and $\mathbf{x}$; see Section 2.5.

Because there is nothing special about the origin in $\mathbb{R}^n$, you can do the same calculation with respect to other points $\mathbf{x}_0$, resp. $\mathbf{k}_0$. This amounts to replacing $\mathbf{x}$ by $\mathbf{x}-\mathbf{x}_0$, and $\mathbf{k}$ by $\mathbf{k}-\mathbf{k}_0$ (or $\nabla$ by $\nabla + i\mathbf{k}_0$) in the calculations above. Note in particular that $[\nabla + i\mathbf{k}_0, \mathbf{x}-\mathbf{x}_0] = n$. As a result one obtains the inequality

$$\|(-i\nabla - \mathbf{k}_0)\psi\|\,\|(\mathbf{x}-\mathbf{x}_0)\psi\| \geq \frac{n}{2}\,\|\psi\|^2. \tag{2.118}$$

Equality holds for a Gaussian that is translated in position space by $\mathbf{x}_0$ and in momentum space by $\mathbf{k}_0$. The translation in momentum space, of course, amounts to a phase shift in position space. Hence the Gaussian functions that are optimal with respect to the above inequality are given by

$$\psi(\mathbf{x}) = N \exp\Bigl(\mathrm{i}\mathbf{k}_0 \cdot \mathbf{x} - \alpha \frac{(\mathbf{x} - \mathbf{x}_0)^2}{2}\Bigr), \qquad \alpha > 0. \tag{2.119}$$

Finally, we note that for a given function $\psi$ the expressions $\|(\mathbf{x} - \mathbf{x}_0)\psi\|$ and $\|(-\mathrm{i}\nabla - \mathbf{k}_0)\psi\|$ can be minimized separately by choosing $\mathbf{x}_0 = \langle \mathbf{x} \rangle_\psi$ and $\mathbf{k}_0 = \langle \mathbf{k} \rangle_\psi$, where

$$\langle \mathbf{x} \rangle_\psi \equiv \langle \psi, \mathbf{x}\psi \rangle = \int_{\mathbb{R}^n} \mathbf{x}\, |\psi(\mathbf{x})|^2 \, d^n x, \tag{2.120}$$

and

$$\langle \mathbf{k} \rangle_\psi \equiv \langle \hat{\psi}, \mathbf{k}\hat{\psi} \rangle = \int_{\mathbb{R}^n} \mathbf{k}\, |\hat{\psi}(\mathbf{k})|^2 \, d^n k. \tag{2.121}$$

In quantum mechanics, these quantities are called the *expectation values* of position and momentum. The expressions

$$\Delta x \equiv \|(\mathbf{x} - \langle \mathbf{x} \rangle_\psi)\psi\| = \sqrt{\langle (x - \langle x \rangle_\psi)^2 \rangle_\psi} \tag{2.122}$$

and

$$\Delta k \equiv \|(-\mathrm{i}\nabla - \langle \mathbf{k} \rangle_\psi)\psi\| = \sqrt{\langle (k - \langle k \rangle_\psi)^2 \rangle_\psi} \tag{2.123}$$

are called the *uncertainties* in position and momentum. In terms of these quantities the uncertainty relation has the following form:

**Uncertainty relation:**

With $\Delta x$ and $\Delta k$ defined as above for $\psi$ with $\|\psi\| = 1$, the uncertainty relation reads

$$\Delta x \, \Delta k \geq \frac{n}{2}. \tag{2.124}$$

Equality holds for the Gaussian

$$\psi(\mathbf{x}) = N \exp\Bigl(\mathrm{i}\langle \mathbf{k} \rangle_\psi \cdot \mathbf{x} - \alpha \frac{(\mathbf{x} - \langle \mathbf{x} \rangle_\psi)^2}{2}\Bigr), \qquad \alpha > 0. \tag{2.125}$$

In quantum mechanics, it is always assumed that $\psi$ satisfies the normalization condition $\|\psi\| = 1$. This allows us to consider the function $\mathbf{x} \to |\psi(\mathbf{x})|^2$ to be a probability distribution. The same is true for $\mathbf{k} \to |\hat{\psi}(\mathbf{k})|^2$ because of the Fourier–Plancherel relation. In that context the numbers $\langle \mathbf{x} \rangle_\psi$ and $\langle \mathbf{k} \rangle_\psi$ are called the *mean values* of $\mathbf{x}$ and $\mathbf{k}$. The uncertainties $\Delta x$ and $\Delta k$ are called *standard deviations* in the language of probability theory.

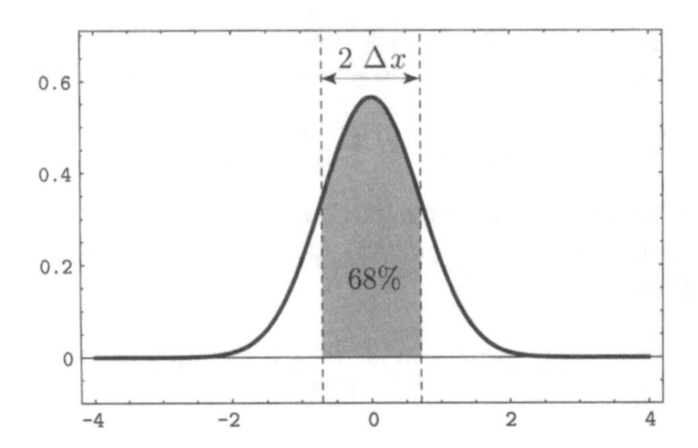


FIGURE 2.6. This plot shows the square of the Gaussian function $\psi(x) = (1/\pi)^{1/4} \exp(-x^2/2)$. The expectation value of the position is $\langle x \rangle_\psi = 0$. The vertical lines are at $\langle x \rangle_\psi \pm \Delta x$. The shaded area is about 68% of the whole area below the curve. About 95% of the total area is between the points $\langle x \rangle_\psi \pm 2\Delta x$.

EXERCISE 2.14. *Calculate the expectation values of position and momentum and the uncertainties $\Delta x$ and $\Delta k$ for the Gaussian function in Eq.* (2.119).

EXERCISE 2.15. *Show that the normalization constant $N$ in Eq.* (2.125) is given by

$$N = \left(\frac{\alpha}{\pi}\right)^{n/4}, \tag{2.126}$$

where $n$ is the space dimension.

### 2.8.2. Special topic: Sobolev and Hardy inequalities

For normalized functions in three dimensions, the uncertainty relation can be written as

$$\|\nabla \psi\| \geq \frac{3}{2} \frac{1}{\|\mathbf{x}\, \psi\|}, \quad \text{with } \|\psi\| = 1. \tag{2.127}$$

Here, for the sake of completeness, we mention some other inequalities which, like the uncertainty relation, give an estimate from below of the norm of the gradient of $\psi$.

The movies CD 2.13, CD 2.15, and CD 2.16 illustrate that the spread in position space is not always related to the spread in momentum space. In these cases the following inequalities often give quantitative better estimates than the uncertainty relation.

A first example is *Hardy's inequality* which is also sometimes called the *uncertainty principle lemma*,

$$\|\nabla\psi\| \geq \frac{1}{2}\,\|\frac{1}{|\mathbf{x}|}\,\psi\| \tag{2.128}$$

(here $\psi$ need not be normalized).

Another "uncertainty relation" is expressed by the following *Sobolev inequality*,

$$\|\nabla\psi\| \geq \sqrt{3}\left(\frac{\pi}{2}\right)^{2/3}\|\psi\|_6. \tag{2.129}$$

Here the $q$-norm of $\psi$ is defined by

$$\|\psi\|_q = \left(\int_{\mathbb{R}^3} |\psi(\mathbf{x})|^q\, d^3x\right)^{1/q}. \tag{2.130}$$

Equality holds in Sobolev's inequality (2.129) precisely for the functions

$$\psi_m(\mathbf{x}) = \frac{1}{\sqrt{x^2+m^2}}, \quad \text{with } m>0 \text{ arbitrary}, \tag{2.131}$$

as well as for all functions obtained from $\psi_m$ by translations. (Sobolev's inequality—unlike the uncertainty relation (2.127)—is insensitive to translations of the function $\psi$).

## 2.9. Special Topic: Dirac Delta Distribution

A formal manipulation with the formulas (2.57) and (2.58) gives

$$\psi(\mathbf{y}) = \frac{1}{(2\pi)^{n/2}}\int_{\mathbb{R}^n} e^{i\mathbf{k}\cdot\mathbf{y}}\left(\frac{1}{(2\pi)^{n/2}}\int_{\mathbb{R}^n} e^{-i\mathbf{k}\cdot\mathbf{x}}\,\psi(\mathbf{x})\,d^nx\right)d^nk \tag{2.132}$$

$$= \int\left(\frac{1}{(2\pi)^n}\int e^{-i\mathbf{k}\cdot(\mathbf{x}-\mathbf{y})}\,d^nk\right)\psi(\mathbf{x})\,d^nx. \tag{2.133}$$

This calculation is formal and mathematically not justified because we have interchanged the order of integration where nonintegrable functions are involved. The manipulation above nevertheless may serve as a useful reminder of some important mathematical results. We define the *Dirac delta function*

$$\delta(\mathbf{x}) = \frac{1}{(2\pi)^n}\int e^{-i\mathbf{k}\cdot\mathbf{x}}\,d^nk. \tag{2.134}$$

It looks like the Fourier transform of the constant function $1/(2\pi)^{n/2}$ (which is, of course, not integrable). From Eq. (2.133) we see that the delta function must have the formal property

$$\psi(\mathbf{y}) = \int \delta(\mathbf{x} - \mathbf{y})\, \psi(\mathbf{x})\, d^n x. \tag{2.135}$$

The expression $\delta(\mathbf{x})$ is a very strange object. If $\psi$ is an arbitrary positive function that vanishes at 0, we have $\int \delta(\mathbf{x})\, \psi(\mathbf{x})\, d^n x = 0$. Hence we must have $\delta(\mathbf{x}) = 0$ for all $\mathbf{x}$ except for $\mathbf{x} = 0$. Moreover, for functions $\psi$ that are equal to 1 in a neighborhood of $\mathbf{x} = 0$, we find that $\psi(0) = 1 = \int \delta(\mathbf{x})\, d^n x$, that is, the area below the "graph" of $\delta$ must be 1. Obviously, there exists no (locally integrable) function with these properties. (By the usual definition of the (Lebesgue) integral, the integral yields zero for any function that vanishes almost everywhere, that is, except on a set of measure zero).

As long as the delta-function is only used "under the integral sign", it can be given a rigorous meaning "in the sense of distributions". For sufficiently smooth functions $\psi$, the operation described in Eq. (2.135) amounts to assigning the *number* $\psi(\mathbf{y})$ (the value of $\psi$ at the point $\mathbf{y}$) to the *function* $\psi$. The mapping $\psi \to \psi(\mathbf{y})$ is a continuous linear mapping from a suitable function space into the field of complex numbers $\mathbb{C}$. Such a mapping is called a *linear functional* or a *distribution*. Of particular importance are the *Dirac delta distributions*

$$\delta(\psi) = \psi(0), \quad \text{and} \quad \delta_{\mathbf{y}}(\psi) = \psi(\mathbf{y}). \tag{2.136}$$

Usually, distributions are applied to functions $\psi$ that are infinitely differentiable and go to zero, as $|\mathbf{x}| \to \infty$ (faster than $|\mathbf{x}|^{-n}$ for arbitrary $n$). These functions are called *test functions*.

Any distribution $t$ can be differentiated using the rule

$$t'(\psi) = t(-\psi'). \tag{2.137}$$

The Fourier transform of a distribution is defined as

$$\hat{t}(\psi) = t(\hat{\psi}). \tag{2.138}$$

A distribution is said to be *regular* if there exists a (locally integrable) function $t(\cdot)$ such that

$$t(\psi) = \int t(\mathbf{x})\, \psi(\mathbf{x})\, d^n x. \tag{2.139}$$

If a distribution is regular, then the linear mapping $t$ determines the function $t(\mathbf{x})$ uniquely (almost everywhere), and vice versa. Hence one usually does not distinguish between the regular distribution and the function describing it.

Any square-integrable function $\psi$ can also be interpreted as a regular distribution. If the distributional derivative of $\psi$ is again a regular distribution $\psi'$, and if $\psi'$ is square-integrable, then the distributional derivative $\psi'$ coincides with the generalized derivative defined in (2.90) with the help of a Fourier transform.

The Dirac-delta-distribution is not regular because according to our discussion above, $\delta(\mathbf{x})$ is not well defined as a function. Nevertheless, one often writes $\int \delta(\mathbf{x})\, \psi(\mathbf{x})\, d^n x$ instead of $\delta(\psi)$ because this is, after all, a convenient and well-established notation.

According to the definition (2.138), the Fourier transform of $\delta$ is a regular distribution,

$$\hat{\delta}(\psi) = \delta(\hat{\psi}) = \hat{\psi}(0) = \int \frac{1}{(2\pi)^{n/2}}\, \psi(\mathbf{x})\, d^n x. \tag{2.140}$$

That is, $\hat{\delta}$ is the regular distribution given by the constant function $1/(2\pi)^{n/2}$. The distributional Fourier transform of the constant function 1 is given by

$$\hat{1}(\psi) = 1(\hat{\psi}) = \int \hat{\psi}(\mathbf{k})\, d^n k = (2\pi)^{n/2} \psi(0) = (2\pi)^{n/2}\, \delta(\psi). \tag{2.141}$$

This result gives a rigorous meaning to Eq. (2.134).

*Chapter 3*

# Free Particles

**Chapter summary**: We start our exposition of quantum mechanics with a "derivation" of the Schrödinger equation for free particles. This is just a first step. In realistic situations, particles interact with force fields and other particles and can only be detected through their interaction with some measurement device. Nevertheless, a good understanding of the free motion is important, for example, for the asymptotic description of interacting particles in scattering experiments.

The (time-dependent) free Schrödinger equation will be obtained as the differential equation for de Broglie's plane waves. The plane waves have been introduced to describe the wavelike behavior of a beam of particles, thereby relating the wave number to the momentum and the frequency to the energy of the particles. However, the property of being localized in some region—which is typical for particles—cannot be described by plane waves. Therefore, we exploit the linearity of the Schrödinger equation to form (continuous) superpositions of plane waves. This process can be described as a Fourier transform and leads to wave packets corresponding to fairly localized phenomena. In this way, the problem of solving the Schrödinger equation with an arbitrary initial function can be reduced to the calculation of a Fourier integral.

Unlike classical particles, wave packets can neither have a sharp position nor a sharp momentum. The extension of the wave packets in position and momentum space can be described by the uncertainties $\Delta x$ and $\Delta k$, which satisfy Heisenberg's uncertainty relation.

The wave function is usually interpreted statistically. The square of the absolute value of the complex-valued wave packet describes a position probability distribution. The Fourier transform of the wave packet is related in the same way to the distribution of momenta. According to the statistical interpretation, the predictions of quantum mechanics concern the probability distributions of measurement results. This is rather a theory of statistical ensembles than of individual particles. The status of individual systems within quantum theory depends on the interpretation of the measuring process. In this context we mention the paradox with Schrödinger's cat which occurs whenever quantum systems are in a superposition of rather distinct wave functions.

The chapter ends with a description of the asymptotic form of the free time evolution and a discussion of the energy representation. Both will be important for the formulation of scattering theory in Chapter 9.

# 3.1. The Free Schrödinger Equation

## 3.1.1. Particles and waves

According to the classical physical theories the basic entities of the physical world are either particles moving through space as well-localized chunks of matter, or fields that are spread out in space and propagate as waves. This picture had to be revised when the availability of new experimental methods allowed physicists to investigate phenomena on very small scales, which are normally beyond the reach of our senses. At the beginning of the development of modern quantum mechanics in the 1920s it was clear that many physical phenomena exhibit an inherent wave–particle dualism. For example, the explanation of the photoelectric effect (see below) seemed to indicate that the propagation of light, which has been considered a classical wave phenomenon, also has the characteristics of a beam of particles (photons). On the other hand, under certain circumstances material particles such as electrons showed wave properties.

Speaking of wave phenomena, it is perhaps a good time to review some basic definitions.

**Wave phenomena**: In the simplest case a wave phenomenon is described by a plane wave. It describes some quantity that varies periodically in space and time and is characterized by a wavelength $\lambda$ and an oscillation period $T$. We define the wave number $k$ and frequency $\omega$, respectively, as

$$k = \frac{2\pi}{\lambda} \quad \text{and} \quad \omega = \frac{2\pi}{T}. \tag{3.1}$$

The *plane wave* with wave number $k$ is the complex-valued function

$$u(x,t) = \exp(\mathrm{i}kx - \mathrm{i}\omega t) \ . \tag{3.2}$$

The stationary plane wave in Color Plate 4 is a snapshot of $u_k$ (with $k = 1$) at $t = 0$. The frequency $\omega$ may depend on $k$ and $\omega = \omega(k)$ is called the *dispersion relation* of the wave phenomenon. The function

$$\phi(x,t) = kx - \omega t \tag{3.3}$$

is called the *phase* of the plane wave. A point where the phase has a fixed value moves with the *phase velocity*

$$v = \frac{\lambda}{T} = \frac{\omega}{k}. \tag{3.4}$$

For wave phenomena in higher dimensions we define the wave vector $\mathbf{k}$, which has the magnitude $2\pi/\lambda$ and points into the direction of wave propagation. The plane wave is the function

$$u(\mathbf{x},t) = \exp(\mathrm{i}\mathbf{k}\cdot\mathbf{x} - \mathrm{i}\omega t). \tag{3.5}$$

For a fixed time $t$, the points $\mathbf{x}$ with a fixed value of the phase are characterized by $\mathbf{k} \cdot \mathbf{x} = \text{const}$. These points form a plane that is orthogonal to the wave vector $\mathbf{k}$. The phase planes move with the phase velocity $\omega/|\mathbf{k}|$ in the direction of $\mathbf{k}$.

CD 1.18 visualizes the time dependence of a plane wave propagating in two space dimensions. In this case the phase planes appear as lines of a certain color.

In order to describe the photoelectric effect, Albert Einstein introduced in the year 1905 the relations

$$E = \hbar\omega, \qquad \mathbf{p} = \hbar\mathbf{k}, \tag{3.6}$$

between the energy $E$ of the photons in the beam and the frequency $\omega$ of the light wave (resp. between momentum $\mathbf{p}$ of photons and wave vector $\mathbf{k}$). The constant $\hbar$ is known as Planck's constant. It has the value

$$\hbar = 1.0546 \cdot 10^{-34} \text{ Joule} \cdot \text{sec}. \tag{3.7}$$

The physical dimension energy $\times$ time is called *action*. It can also be interpreted as momentum $\times$ length or as angular momentum $\times$ angle.

**Photoelectric effect**: If light shines on a metal, electrons are set free. The energy of the released electrons depends linearly on the frequency of the light wave (not on the intensity, as one might have expected). Einstein described the photo effect by assuming that the electrons are knocked out by light particles (*photons*). During this scattering process, the energy of the photons is transferred to the electrons. If the energy of the photons is large enough, the electrons are emitted with a kinetic energy $E_{\text{kin}} = E - W$, where $W$ is the work needed to release an electron and $E$ is the energy received from a photon. Because of the observed linear relationship between the kinetic energy of the electrons and the frequency $\omega$ of the beam of light, Einstein concluded that the energy of a photon in the beam must be proportional to the frequency, that is, $E = \hbar\omega$. Because photons move with the velocity of light, they must have the rest mass $m = 0$. For particles with vanishing rest mass the relativistic relation between energy and momentum becomes

$$E = \sqrt{c^2p^2 + m^2c^4} = cp \quad (c = \text{velocity of light}) \tag{3.8}$$

and hence the photons must have the momentum $p = E/c = \hbar\omega/c = \hbar k$.

**Matter waves**: In 1924, Louis de Broglie postulated a wave phenomenon to be associated with a beam of particles. He assumed that a beam of electrons can be described by a complex plane wave function

$$u(x,t) = e^{i\mathbf{k}\cdot\mathbf{x} - i\omega t}, \tag{3.9}$$

where again the wave vector $\mathbf{k}$ and the frequency $\omega$ are determined by the momentum and energy of the particles in the beam as in Eq. (3.6). These relations make a connection between wave and particle properties and therefore have no obvious interpretation. An experimental verification of de Broglie's assumptions can be seen in the experiments of Davisson and Germer in 1927, where a beam of electrons scattered from a crystal showed the interference pattern of a wave phenomenon.

**Scattering from a crystal**: Light waves sent through a crystal are scattered by the atoms in the crystal. Every atom becomes the origin of a scattered wave which is sent into all directions. But because the atoms are arranged in a regular grid, there are only a few directions where all scattered waves are in phase and get amplified by constructive interference. The intensity of the scattered wave therefore shows sharp maxima in certain directions. The condition for a maximum at an angle $\theta$ according to Bragg is $2d\sin\theta = n\lambda$, where $n$ is an arbitrary integer, $d$ is the lattice constant of the crystal, and $\lambda$ is the wavelength of the incident light. In this way, the german physicist Max von Laue was able to prove the wave nature of x-rays. If the wavelength is known, the interference pattern of the scattered wave can be used to analyze the structure of the crystal with the help of the Bragg condition. Essentially the same interference phenomena can be observed for the scattering of electrons at thin metal foils. If the lattice constant of the material is known, the Bragg condition allows one to determine the wavelength in a beam of electrons. The experimentally measured intensity maxima are obtained if one assumes that electrons with energy $E = p^2/2m$ have the wavelength $\lambda = 2\pi\hbar/p$.

### 3.1.2. The Schrödinger equation

The equations (3.6) and (3.9) are a common starting point for motivating the Schrödinger equation. The nonrelativistic relation between energy and momentum of free particles with mass $m$,

$$E = \frac{\mathbf{p}^2}{2m}, \tag{3.10}$$

corresponds via Eq. (3.6) to the dispersion relation

$$\omega = \hbar\mathbf{k}^2/2m. \tag{3.11}$$

Hence it is easy to verify that the function $u$ must be a solution of the following partial differential equation:

$$i\hbar\frac{\partial}{\partial t}\psi(\mathbf{x},t) = -\frac{\hbar^2}{2m}\,\Delta\,\psi(\mathbf{x},t). \tag{3.12}$$

Here $\Delta$ is the *Laplace operator*

$$\Delta = \nabla \cdot \nabla = \frac{\partial^2}{\partial x_1^2} + \cdots + \frac{\partial^2}{\partial x_n^2}, \tag{3.13}$$

and $n$ is the dimension of the configuration space (also called *position space*). Usually we have $n = 1$, 2, or 3. Eq. (3.12) is the famous Schrödinger equation for free particles which was introduced by the Austrian physicist Erwin Schrödinger in 1926.

EXERCISE 3.1. *Replace Eq.* (3.11) *by the dispersion relation for light and use the ansatz* $v(x,t) = \cos(\mathbf{k}\cdot\mathbf{x} - \omega t)$ *to obtain a partial differential equation for light waves.*

EXERCISE 3.2. *It is not so good to use the real-valued function* $v(x,t) = \cos(\mathbf{k}\cdot\mathbf{x} - \omega t)$ *instead of Eq.* (3.9) *in the description of electrons. Assuming the relation Eq.* (3.11), *show that the wave equation for* $v$ *would depend on* $\mathbf{k}$ *and hence on the momentum of the electrons to be described.*

EXERCISE 3.3. *Write down a relativistic wave equation assuming that the particles have nonvanishing rest mass* $m$ *and satisfy the relativistic energy-momentum relation*

$$E = \sqrt{c^2 p^2 + m^2 c^4} \tag{3.14}$$

*instead of Eq.* (3.10).

### 3.1.3. Scaling the unit of length

For the purpose of theoretical considerations and in order to standardize image scales, we apply a little trick that allows us to get rid of the physical constants $\hbar$ and $m$ in the free Schrödinger equation. This simplification can be achieved by a suitable scaling transformation. We simply change the length unit such that 1 meter corresponds to $\sqrt{m/\hbar}$ new length units. The position vector with respect to the new length unit will be denoted temporarily by $\tilde{\mathbf{x}}$. Obviously, we have

$$\mathbf{x} = \tilde{\mathbf{x}}\sqrt{\hbar/m}. \tag{3.15}$$

Inserting this into $\psi(\mathbf{x},t)$, we can define a new function $\tilde{\psi}$, which depends on $\tilde{\mathbf{x}}$ by

$$\tilde{\psi}(\tilde{\mathbf{x}},t) = \psi(\tilde{\mathbf{x}}\sqrt{\hbar/m}, t). \tag{3.16}$$

The new function satisfies the differential equation

$$\mathrm{i}\frac{\partial}{\partial t}\tilde{\psi}(\tilde{\mathbf{x}},t) = -\frac{1}{2}\Delta_{\tilde{\mathbf{x}}}\tilde{\psi}(\tilde{\mathbf{x}},t), \tag{3.17}$$

where $\Delta_{\tilde{\mathbf{x}}}$ is the Laplacian with respect to the new coordinates. Because we could have used the new units from the very beginning, we will omit the "$\sim$" from now on and consider the Schrödinger equation in the form

$$\mathrm{i}\,\frac{\partial}{\partial t}\,\psi(\mathbf{x},t) = -\frac{1}{2}\,\Delta\,\psi(\mathbf{x},t). \tag{3.18}$$

EXERCISE 3.4. *Verify that the function* $\tilde{\psi}$ *defined in Eq.* (3.16) *indeed satisfies the differential equation* (3.17), *whenever* $\psi$ *is a solution of* (3.12).

EXERCISE 3.5. *For* $\lambda \in \mathbb{C}$ *define the scaling transformation* $U_\lambda$ *by*

$$(U_\lambda\,\psi)(\mathbf{x}) = (\lambda)^{n/2}\psi(\lambda\mathbf{x}). \tag{3.19}$$

*Calculate* $U_\lambda \nabla\psi$, $\nabla U_\lambda \psi$, $U_\lambda \Delta\psi$, $\Delta U_\lambda \psi$.

### 3.1.4. Plane waves

It is clear from our considerations in Section 3.1.1 that a set of smooth solutions of the Schrödinger equation Eq. (3.12) is given by the plane waves

$$\exp\Bigl(\frac{\mathrm{i}}{\hbar}\mathbf{p}\cdot\mathbf{x} - \frac{\mathrm{i}}{\hbar}\,\frac{\mathbf{p}^2}{2m}\,t\Bigr), \quad \text{all } \mathbf{p} \in \mathbb{R}^n. \tag{3.20}$$

These solutions describe a beam of particles with momentum $\mathbf{p}$. The transition to the scaled units can be performed by substituting $\mathbf{x}\,\sqrt{\hbar/m}$ for $\mathbf{x}$. This gives the plane wave solutions of Eq. (3.18),

$$u_{\mathbf{k}}(\mathbf{x},t) = \exp\Bigl(\mathrm{i}\mathbf{k}\cdot\mathbf{x} - \mathrm{i}\,\frac{k^2}{2}\,t\Bigr), \quad \text{with } \mathbf{k} = \mathbf{p}/\sqrt{\hbar m}. \tag{3.21}$$

The wave vector $\mathbf{k}$ describes the momentum of the plane wave in the new units. Hence the vector space $\mathbb{R}^n$ of all possible $\mathbf{k}$ is called the *momentum space*.

EXERCISE 3.6. *The phase velocity* $c$ *of light waves in a vacuum is independent of the wavelength. For massive particles the phase velocity depends on the wavelength. Calculate the phase velocity of the plane wave* $u_{\mathbf{k}}$ *with wave vector* $\mathbf{k}$.

EXERCISE 3.7. *How would the momentum of a classical mechanical particle change under a scaling transformation* $\mathbf{x} \to \lambda\mathbf{x}$? *Show that after the coordinate transformation* $\mathbf{x} \to \mathbf{x}\sqrt{\hbar/m}$ *the momentum of the particle is given by* $\hbar\mathbf{k}$.

Plane waves are often used to model a beam of particles. Obviously a plane wave has the same absolute value all over space and time. A localized particle cannot be described in this way.

CD 3.1 visualizes plane waves using different methods. Also shown is the motion of a sum of three plane waves with different momenta. If the momenta are commensurable, the resulting motion is periodic in both space and time. (Two real numbers $a$ and $b$ are called commensurable if $a/b$ is a rational number). The sum of plane waves with incommensurable momenta is a nonperiodic function.

It should be clear that such a complicated thing as the interaction of the particles in the beam cannot be described in this simple way. Therefore, the description in terms of plane waves will only be good for beams of low intensity where the particles are separated far enough so that they may be considered independent physical objects.

## 3.2. Wave Packets

In Chapter 2, we constructed almost arbitrary functions as superpositions of stationary plane waves $\exp(ikx)$. Here we are going to do the same with the time-dependent plane waves. The linearity of the Schrödinger equation guarantees that superpositions of the plane waves are again solutions. Hence the analysis of solutions of the Schrödinger equation essentially amounts to a time-dependent Fourier analysis.

### 3.2.1. Superpositions of plane waves

The wavelike behavior of particles manifests itself in the experimentally observed interference patterns. Similar to the case of optical interference, the interference patterns produced by particles can be described by a linear superposition of two or more plane waves. This leads to the *principle of superposition*: For any two wave functions $\psi_1(x,t)$ and $\psi_2(x,t)$ the sum $\psi_1(x,t) + \psi_2(x,t)$ again describes a possible physical situation. Indeed, the Schrödinger equation

$$\mathrm{i}\frac{\partial}{\partial t}\psi(\mathbf{x},t) = -\frac{1}{2}\Delta\,\psi(\mathbf{x},t) \tag{3.22}$$

is in accordance with the principle of superposition because it is linear in $\psi$ (i.e., it does not contain powers of $\psi$ or products of $\psi$ with its derivatives). Hence, because the derivative of a sum is the sum of the derivatives, the sum of two solutions is again a solution. More precisely, if $\psi_1$ and $\psi_2$ are any two solutions of the Schrödinger equation, then also any superposition or linear combination $\psi = a\psi_1 + b\psi_2$, where $a \in \mathbb{C}$ and $b \in \mathbb{C}$, is a solution. This observation can be easily generalized to superpositions of an arbitrary number of solutions:

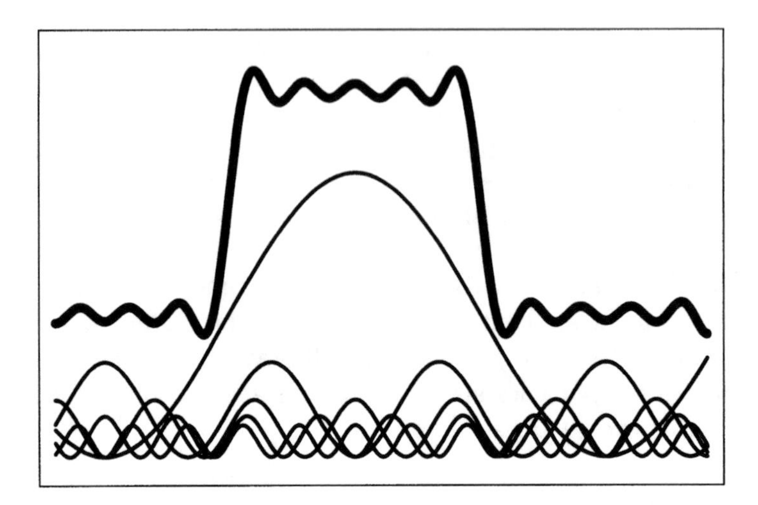

FIGURE 3.1. A linear combination (superposition) of sine-functions generates a new wave form (see also CD 3.2). Similarly, superpositions of the complex-valued plane waves give new solutions of the Schrödinger equation.

**Linearity of the Schrödinger equation:**

Any finite linear combination

$$\psi = \sum_{i=1}^{n} a_i\, \psi_i \qquad \text{with } a_i \in \mathbb{C} \tag{3.23}$$

of solutions $\psi_i$ is again a solution of the Schrödinger equation.

In particular, we can obtain new waveforms by combining plane waves $u_{\mathbf{k}}$ with different momenta. For example, any time-dependent Fourier series

$$\psi(\mathbf{x}, t) = \sum_i a_i\, u_{\mathbf{k}_i}(\mathbf{x}, t) \tag{3.24}$$

is a solution. The coefficient $a_i$ is the *amplitude* of the plane wave with wave number $\mathbf{k}_i$.

CD 3.2 shows how the superposition of several plane waves can generate solutions of the Schrödinger equation that—at least in certain regions—look more localized than single plane waves. See also Fig. 3.1. Because each of the constituent plane waves has a different phase velocity, the superposition pattern changes with time. Note, however, that any finite linear combination of periodic functions is a quasi-periodic function. One cannot describe a truly localized, aperiodic phenomenon in this way.

### 3.2.2. Continuous superposition

A finite superposition of plane waves is not sufficient if one wants to obtain a localized function. It turns out that one has to use a superposition of uncountably many plane waves with different momenta $\mathbf{k}$. Such a continuous superposition is the integral

$$\psi(\mathbf{x}, t) = \int_{\mathbb{R}^n} \phi(\mathbf{k})\, u_{\mathbf{k}}(\mathbf{x}, t)\, d^n k. \tag{3.25}$$

Here the plane wave $u_{\mathbf{k}}$ is multiplied by the scalar factor $\phi(\mathbf{k})$. Like the $a_i$'s in the discrete linear combination in Eq. (3.24) the coefficients $\phi(\mathbf{k})$ may be complex numbers. Hence we will allow the amplitude function $\phi$ to be a complex-valued function of the real variable $\mathbf{k}$.

Now a natural question arises: Assume that a certain function $\psi$ is given. How can one find an amplitude function $\phi$ such that $\psi$ can be written as a continuous superposition of plane waves as in Eq (3.25)? Put differently: How can one determine the distribution of momenta within $\psi$?

### 3.2.3. Fourier transformation

For simplicity we first consider the situation at time $t = 0$. The plane wave with momentum $\mathbf{k}$ at time $t = 0$ is the function $\exp(\mathrm{i}\mathbf{k} \cdot \mathbf{x})$. The continuous summation of plane waves with different momenta at $t = 0$ is given by the following integral:

$$\psi(\mathbf{x}, t)\Big|_{t=0} = \psi_0(\mathbf{x}) = \int_{\mathbb{R}^n} \mathrm{e}^{\mathrm{i}\mathbf{k}\cdot\mathbf{x}}\, \phi(\mathbf{k})\, d^n k. \tag{3.26}$$

From our considerations in Chapter 2 we know how to determine the amplitude function $\phi$ for a given function $\psi_0$.

Recall that any integrable function $\psi_0$ with an integrable Fourier transform $\hat{\psi}_0$ can be written as

$$\psi_0(\mathbf{x}) = \frac{1}{(2\pi)^{n/2}} \int_{\mathbb{R}^n} \mathrm{e}^{\mathrm{i}\mathbf{k}\cdot\mathbf{x}}\, \hat{\psi}_0(\mathbf{k})\, d^n k, \tag{3.27}$$

where

$$\hat{\psi}_0(\mathbf{k}) = \frac{1}{(2\pi)^{n/2}} \int_{\mathbb{R}^n} \mathrm{e}^{-\mathrm{i}\mathbf{k}\cdot\mathbf{x}}\, \psi_0(\mathbf{x})\, d^n x. \tag{3.28}$$

If you compare Eq. (3.26) with Eq. (3.27), you can see that the continuous superposition of plane waves can be interpreted as an inverse Fourier transform. The amplitude function $\phi$ is just the Fourier transform of the wave function.

**Momentum distribution of a wave function:**

Any wave function $\psi_0$ at time $t = 0$ can be understood as a continuous superposition of plane waves $\exp(i\mathbf{k} \cdot \mathbf{x})$, where the plane wave with momentum $\mathbf{k}$ has the amplitude

$$\phi(\mathbf{k}) = \frac{1}{(2\pi)^{n/2}} \hat{\psi}_0(\mathbf{x}). \tag{3.29}$$

Hence the momentum distribution in $\psi$ is given by its Fourier transform $\hat{\psi}$.

A localized wave packet $\psi$ consists of plain waves $\exp(i\mathbf{k}\cdot\mathbf{x})$ with various momenta. The contribution of a particular momentum $\mathbf{k}$ is not in any way localized in position space because the plane wave has everywhere the same absolute value. We cannot point at particular parts of a wave packet and say: "This part has momentum $\mathbf{k}_1$ and that part has momentum $\mathbf{k}_2$". The part with momentum $\mathbf{k}$ in a wave packet extends to every point of space-time.

### 3.2.4. Example: Gaussian amplitude function

In one dimension the Gaussian amplitude function

$$\phi(k) = \frac{1}{\sqrt{2\pi}} \Big(\frac{1}{\alpha\pi}\Big)^{1/4} \exp\Big(-\frac{k^2}{2\alpha}\Big), \qquad \operatorname{Re}\alpha > 0, \tag{3.30}$$

belongs to the wave packet

$$\psi_0(x) = \Big(\frac{\alpha}{\pi}\Big)^{1/4} \exp\Big(-\alpha\frac{x^2}{2}\Big). \tag{3.31}$$

The calculation leading to the result (3.31) is shown in Section 2.7. Here we want to emphasize the following points:

1. The function $\psi$ defined in Eq. (3.31) is reasonably localized inside the interval $(-\sqrt{1/2\alpha}, \sqrt{1/2\alpha})$ and decreases very rapidly outside. Nevertheless, a physical system described by $\psi$ does not have a sharp (i.e., pointlike) position.
2. $\psi$ has no sharp momentum either because many plane waves with different momenta $\mathbf{k}$ are necessary to build $\psi$. The momentum distribution of the constituent plane waves is given by Eq. (3.30).
3. By choosing $\alpha$ very large, we see that $\psi$ becomes very well localized in position space, but less localized in momentum space. The main contributions to the momenta come from the interval $(-\sqrt{\alpha/2}, \sqrt{\alpha/2})$ in momentum space.

CD 3.3.1 shows a Gaussian wave packet corresponding to a particle at rest. Here "at rest" means that the average momentum of the particle is zero, as it is the case for the amplitude function (3.30). Because the Gaussian consists of plane waves with many different momenta, the wave packet spreads with time in all directions.

## 3.3. The Free Time Evolution

### 3.3.1. Solution of the Schrödinger equation

Having described the function $\psi_0$ as the inverse Fourier transform of a function $\hat{\psi}_0$, we can easily determine a solution $\psi(\mathbf{x}, t)$ of the Schrödinger equation which at time $t = 0$ is equal to $\psi_0(\mathbf{x})$. We can do this because we know the time evolution of the plane wave functions $\exp(\mathrm{i}\mathbf{k}\cdot\mathbf{x})$ whose superposition forms the initial function $\psi_0$. We only have to insert this time dependence into Eq. (3.26). Then we obtain the formula

$$\psi(\mathbf{x}, t) = \frac{1}{(2\pi)^{n/2}} \int_{\mathbb{R}^n} e^{\mathrm{i}\mathbf{k}\cdot\mathbf{x} - \mathrm{i}k^2 t/2} \, \hat{\psi}_0(\mathbf{k}) \, d^n k. \tag{3.32}$$

We can verify in a formal way that this function is indeed a solution by inserting it into the Schrödinger equation and then interchanging the order of differentiation and integration. (In order to make this argument rigorous, the function $\hat{\psi}_0$ has to be sufficiently well behaved. The mathematical condition allowing the interchange of differentiation and integration is that $\hat{\psi}_0$ and $k^2\hat{\psi}_0$ be integrable).

Eq. (3.32) just states that the solution $\psi(\mathbf{x}, t)$ of the Schrödinger equation at time $t$ is the inverse Fourier transform of the function

$$\hat{\psi}(\mathbf{k}, t) = \exp(-\mathrm{i}k^2 t/2) \, \hat{\psi}_0(\mathbf{k}). \tag{3.33}$$

Hence we have solved the *initial-value problem*: Given a suitable initial function $\psi_0$, determine a solution $\psi(\mathbf{x}, t)$ of the Schrödinger equation such that $\psi(\mathbf{x}, 0) = \psi_0(\mathbf{x})$. The procedure can be summarized as follows:

**Solution of the initial-value problem for free particles:**

1. Determine the Fourier transform $\hat{\psi}_0$ of the initial function $\psi_0$.
2. Determine the inverse Fourier transform of $\exp(-\mathrm{i}k^2 t/2) \, \hat{\psi}_0(\mathbf{k})$.

For a few initial functions this procedure can indeed be carried through by an explicit calculation. We give an example in the next section.

CD 3.7 shows the time evolution of a wave packet in momentum space. The time dependence is described by the phase factor $\exp(-\mathrm{i}k^2t/2)$. The absolute value of the function does not change with time. This reflects the fact that for the free time evolution the momentum is a conserved quantity.

$\boxed{\Psi}$ The procedure above works for every initial function $\psi_0$ in the Hilbert space $L^2$, because the Fourier transform can be defined for all square-integrable functions (see Section 2.5.4). Hence we can define a time evolution $\psi(\mathbf{x}, t)$ even if the initial function $\psi_0(\mathbf{x})$ is not differentiable.

CD 3.13 shows a solution of the Schrödinger equation with a non-differentiable initial function.

### 3.3.2. Example: Gaussian function

We calculate the time evolution of the initial function

$$\psi_0(x) := \mathrm{e}^{-\alpha x^2/2}\,\mathrm{e}^{\mathrm{i}px}, \quad x \in \mathbb{R}, \tag{3.34}$$

for arbitrary $\alpha > 0$ and $p \in \mathbb{R}$. The result for $p = 2$ is shown in Color Plate 11.

**First Step**: The Fourier transform of this function is given by

$$\hat{\psi}_0(k) = \frac{1}{\sqrt{\alpha}} \exp\Bigl(-\frac{(k-p)^2}{2\alpha}\Bigr). \tag{3.35}$$

Note that the momenta $k$ in $\psi_0$ are distributed around the average momentum $p$. If you compare this with the example in Section 3.2.4 you will see that the Fourier transform has just been shifted by $p$ to the right. The shift in momentum space is caused by the phase factor $\mathrm{e}^{\mathrm{i}px}$ in position space.

**Second Step**: The solution at time $t$ of the initial-value problem is given by

$$\psi(x,t) = \frac{1}{\sqrt{\alpha}}\frac{1}{\sqrt{2\pi}} \int \mathrm{e}^{\mathrm{i}kx}\mathrm{e}^{-\mathrm{i}k^2t/2}\mathrm{e}^{-(k-p)^2/2\alpha}\,dk \tag{3.36}$$

$$= \frac{1}{\sqrt{\alpha}}\mathrm{e}^{-\frac{p^2}{2\alpha}}\frac{1}{\sqrt{2\pi}} \int \mathrm{e}^{-(\mathrm{i}t+1/\alpha)k^2/2+k(\mathrm{i}x+p/\alpha)}\,dk. \tag{3.37}$$

The exponent in the last integral can be written as

$$-\frac{\mathrm{i}t+1/\alpha}{2}q(k)^2 + \frac{(\mathrm{i}x+p/\alpha)^2}{2(\mathrm{i}t+1/\alpha)} \tag{3.38}$$

with

$$q(k) = k - \frac{\mathrm{i}x+p/\alpha}{\mathrm{i}t+1/\alpha} \tag{3.39}$$

from which we obtain

$$\psi(x,t) = \frac{1}{\sqrt{\alpha}} \exp\left(-\frac{p^2}{2\alpha} + \frac{(\mathrm{i}x + p/\alpha)^2}{2(\mathrm{i}t + 1/\alpha)}\right) \frac{1}{\sqrt{2\pi}} \int \mathrm{e}^{-(\mathrm{i}t+1/\alpha)q(k)^2/2}\, dk. \tag{3.40}$$

We find

$$\int \mathrm{e}^{-(\mathrm{i}t+1/\alpha)q(k)^2/2}\, dk = \int \mathrm{e}^{-(\mathrm{i}t+1/\alpha)k^2/2}\, dk = \frac{\sqrt{2\pi}}{\sqrt{\mathrm{i}t + 1/\alpha}} \tag{3.41}$$

and hence

$$\psi(x,t) = \frac{1}{\sqrt{1+\mathrm{i}\alpha t}} \exp\Bigl(-\frac{\alpha x^2 - 2\mathrm{i}xp + \mathrm{i}p^2 t}{2(1+\mathrm{i}\alpha t)}\Bigr). \tag{3.42}$$

CD 3.3.2 and CD 3.4 show Gaussian wave packets with various average momenta. The complex phase of the wave function is determined by the imaginary part of the exponent in Eq. (3.42). The local wavelength is shorter in front of the maximum at $pt$ and longer behind the maximum (see also Color Plate 11). This can be easily explained because the Gaussian consists of components with many different momenta moving with different velocities. The parts with higher momenta (shorter wavelength) are faster and are therefore found in front of the average position. The slower parts with longer wavelength clearly accumulate behind the maximum, which moves precisely with the average velocity corresponding to the momentum distribution of the Gaussian.

**Discussion**: A little calculation shows that the real part of the exponent in Eq. (3.42) is given by

$$-\frac{\alpha}{2(1+\alpha^2 t^2)}\,(x - pt)^2. \tag{3.43}$$

Hence the absolute value of the Gaussian wave function at time $t$ is again a Gaussian function,

$$|\psi(x,t)| = (1+\alpha^2 t^2)^{(-1/4)} \exp\Bigl(-\frac{\alpha(t)}{2}\,(x-pt)^2\Bigr), \tag{3.44}$$

where

$$\alpha(t) = \frac{\alpha}{1+\alpha^2 t^2}. \tag{3.45}$$

The Gaussian function $|\psi(x,t)|$ is centered around $x(t) = pt$ which is just the position at time $t$ of a free classical particle with mass 1 and momentum $p$ and initial condition $x(0) = 0$. The width of the Gaussian at time $t$ is given by $1/\sqrt{\alpha(t)}$ and increases roughly like $t$. This is called the *spreading of the wave packet.* At the same time the maximum value of the wave packet decreases like $1/\sqrt{t}$.

The spreading is a consequence of the distribution of momenta in the wave packet. For the slow wave packet in CD 3.3 the momentum uncertainty $\Delta k$ cannot be neglected in comparison with the average momentum $\langle k \rangle$. Faster wave packets have a lower relative uncertainty $\Delta k / \langle k \rangle$ of the momentum. As a consequence, very fast wave packets can move over large distances without notable spreading and thus behave in a more classical way (see CD 3.4).

The decay rate of the wave function depends on the dimension of the configuration space. In $n$ dimensions, the modulus of the wave function decreases like $t^{-(n/2)}$.

CD 3.5 and CD 3.6 show the motion of Gaussian wave packets with various average momenta in two space dimensions. CD 3.7 shows the motion of a Gaussian wave packet in Fourier space. The free motion in phase space is visualized by CD 3.8.

### 3.3.3. Conservation of the norm

The spreading of the wave packet $\psi$ and the simultaneous decreasing of its absolute value leads to the conservation of the norm

$$\|\psi(t)\|^2 = \int_{\mathbb{R}^n} \|\psi(x,t)\|^2 \, d^n x.$$

**Time-invariance of the norm:**

For any square-integrable solution $\psi(x,t)$ of the Schrödinger equation the norm is independent of time,

$$\|\psi(t)\|^2 = \|\psi(0)\|^2. \tag{3.46}$$

PROOF. In momentum space, the time evolution is described by the phase factor $\exp(-ik^2t/2)$, which has absolute value 1. Hence the norm of the wave function in momentum space,

$$\int |\hat{\psi}(k,t)|^2 \, d^n k = \int |\hat{\psi}(k,0)|^2 \, d^n k, \tag{3.47}$$

is independent of time. By the Fourier–Plancherel relation (2.64), the same is true for the wave function in position space. □

### 3.3.4. The propagator

Here we derive a formula that allows us to calculate the free time evolution of an arbitrary initial state. Let us assume that the initial state is given by a wave function $\psi_0(\mathbf{x})$. We know already that the wave function at time $t$ is given by

$$\psi(\mathbf{x},t) = \frac{1}{(2\pi)^{n/2}} \int \mathrm{e}^{\mathrm{i}\mathbf{k}\cdot\mathbf{x}-\mathrm{i}k^2t/2}\,\hat{\psi}_0(\mathbf{k})\,d^nk \tag{3.48}$$

$$= \frac{1}{(2\pi)^{n/2}} \int \mathrm{e}^{\mathrm{i}\mathbf{k}\cdot\mathbf{x}-\mathrm{i}k^2t/2} \left\{ \frac{1}{(2\pi)^{n/2}} \int \mathrm{e}^{-\mathrm{i}\mathbf{k}\cdot\mathbf{y}}\,\psi_0(\mathbf{y})\,d^ny \right\} d^nk \tag{3.49}$$

If we formally interchange the order of the $k$ and $y$ integration, we obtain

$$\psi(\mathbf{x},t) = \int K(\mathbf{x}-\mathbf{y},t)\,\psi(\mathbf{y})\,d^ny \tag{3.50}$$

where we introduced the symbol

$$K(\mathbf{x}-\mathbf{y},t) = \frac{1}{(2\pi)^n} \int \mathrm{e}^{\mathrm{i}\mathbf{k}\cdot(\mathbf{x}-\mathbf{y})-\mathrm{i}k^2t/2}\,d^nk \tag{3.51}$$

Hence $K(\mathbf{x},t)$ is the inverse Fourier transformation of the function

$$\frac{1}{(2\pi)^{n/2}} \exp\left(-\mathrm{i}k^2t/2\right). \tag{3.52}$$

This function is not integrable. If we still apply Eq. (2.97) (with complex $\alpha = 1/\mathrm{i}t$) we obtain formally

$$K(\mathbf{x}-\mathbf{y},t) = \frac{1}{(2\pi\mathrm{i}t)^{n/2}} \exp\left\{ \mathrm{i}\frac{(\mathbf{x}-\mathbf{y})^2}{2t} \right\}. \tag{3.53}$$

Although the steps leading to this result are mathematically not justified, the final formula is nevertheless correct and can be proved rigorously by an approximation argument.

## 3.4. The Physical Meaning of a Wave Function

### 3.4.1. Interpretation of the wave function

What does a wave function describe? What is its physical meaning? The discussion so far relates the wave function $\psi$ to the position smeared over some region and the Fourier transform $\hat{\psi}$ to the contributions of the various momenta.

As we have seen, wave packets tend to spread over larger and larger regions of space during their time evolution. But no matter how large the region occupied by the wave function, it has never been observed that the mass or charge of a single particle is actually spread over that region. On the contrary, elementary particles like electrons are always detected as point-like quantities. If many detectors are distributed over some region, we can observe that the particle is found in one and only one of the detectors and that it arrives there as a whole. The connection between the detection of a particle and the wave function is a statistical one. If one repeats the same experiment over and over again, the particle will be found most frequently in regions where the wave function is large. For the regions where the wave function is small, the probability to detect the particle is very small.

Thus, a wave packet does not describe a matter wave, that is, a continuous distribution of mass or charge over some region of space. Instead, the spreading of a wave function describes an uncertainty in the precise position at which the particle will be detected. Most physicists would agree on the following *statistical interpretation*, which was suggested by Max Born in 1926.

**Statistical interpretation of the wave function:**

Let us assume that a particle is described by a wave function $\psi(\mathbf{x})$ satisfying the normalization condition

$$\int_{\mathbb{R}^n} |\psi(\mathbf{x})|^2 \, d^n x = 1 \tag{3.54}$$

Then the expression

$$p(B) = \int_B |\psi(\mathbf{x})|^2 \, d^n x \tag{3.55}$$

is the probability of finding the particle in the region $B$ of the configuration space $\mathbb{R}^n$. Similarly,

$$\int_G |\hat{\psi}(\mathbf{k})|^2 \, d^n k \tag{3.56}$$

is the probability that the momentum of the particle is found in the subset $G$ of the momentum space $\mathbb{R}^n$.

The wave function contains information about the position and—via Fourier transformation—about the momentum at the same time. Note that the position probability $|\psi(x)|^2$ alone does not tell you how the Fourier transformed wave packet looks. See Figure 3.2. In order to describe the dynamical state of the particle, we need the information contained in the phase of the wave function. For example, the wave packets shown in the movies CD 3.3 and CD 3.4 all have the same position distribution at $t = 0$, but nevertheless move with different velocities. Hence it is necessary to visualize the complex-values of a wave function and not just the associated

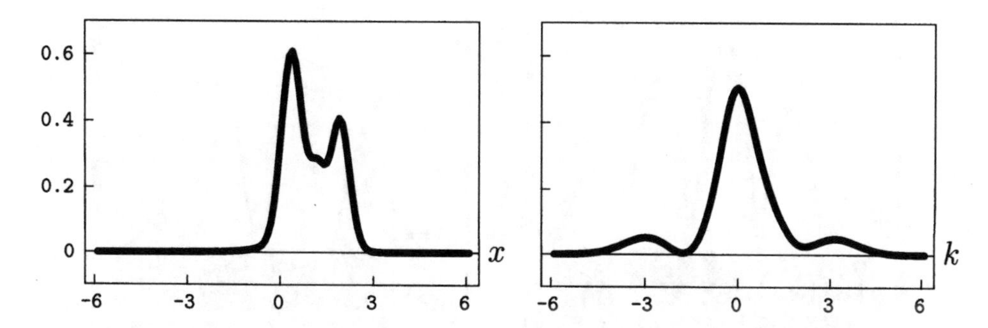


FIGURE 3.2. Every wave function contains information about both the position and the momentum of a particle. Here we see the position distribution $|\psi(x)|^2$ (left) and the momentum distribution $|\hat{\psi}(k)|^2$ (right) of some wave function $\psi$.

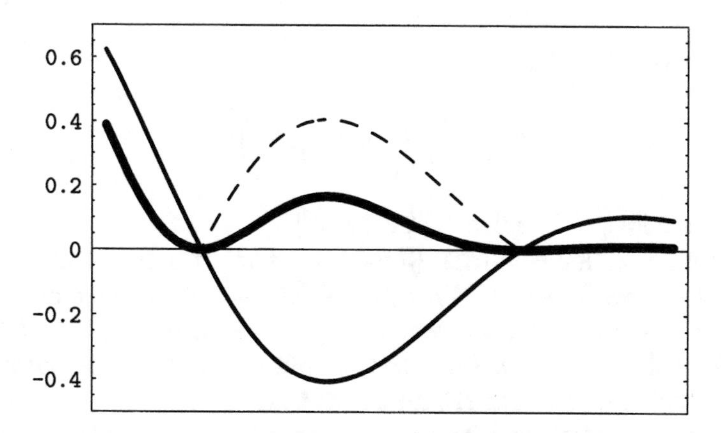


FIGURE 3.3. Comparison of a (real-valued) wave function (thin line) with its absolute value (dashed line), and with the square of the absolute value (thick solid line). It is only the square of the absolute value that has immediate physical significance as a position probability density.

position probability density. The visualizations in the book and on the CD-ROM usually show the absolute value of the wave function with a color code for the phase. You should always remember that it is the square of the absolute value that has a physical significance as a position probability density. With a little experience, it is not difficult to imagine the position probability density when looking at an image of the absolute value, see Figs. 3.3 and 3.4.

In order to apply the statistical interpretation to a square-integrable wave function, it has to be normalized first. The normalization procedure

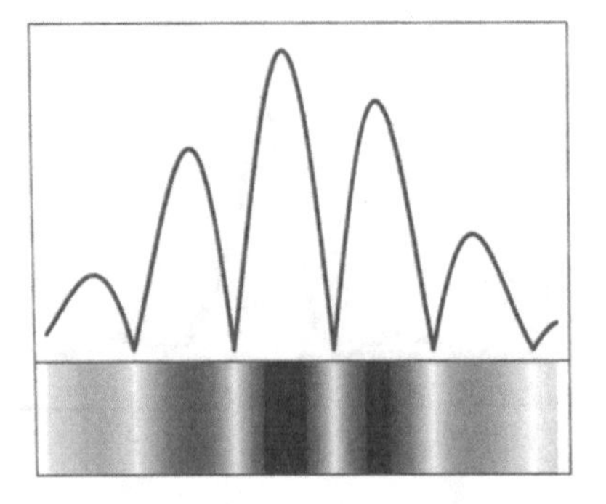
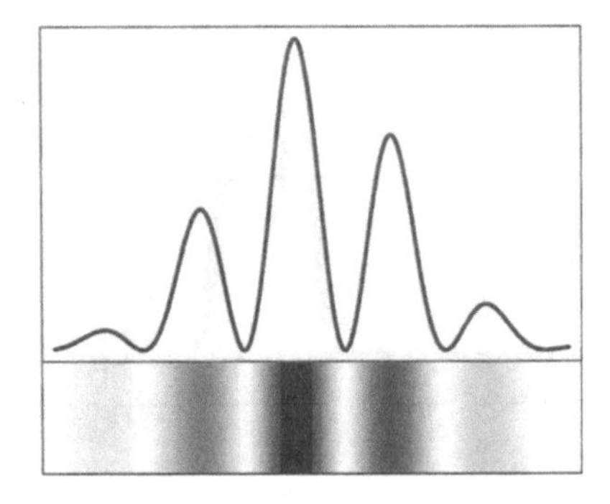

FIGURE 3.4. The left picture shows the absolute value of the wave function; the right picture displays the associated position probability density. The lower part in each picture visualizes the function as a density graph.

runs as follows. Calculate

$$\|\psi\| = \left( \int_{\mathbb{R}^n} |\psi(x)| \, d^n x \right)^{1/2} \tag{3.57}$$

and whenever $\|\psi\| \neq 1$ replace $\psi$ by $\psi / \|\psi\|$.

The normalization condition (3.54) means that the probability of finding the particle *somewhere* is 1. This probability is independent of time if the wave function depends on time according to the Schrödinger equation. The Fourier–Plancherel relation Eq. (2.64) implies that the same normalization condition holds for the momentum distribution.

How does a particle get from one region in space to another? Quantum-mechanically, this is described as a change in the position probabilities associated to these regions. The mechanism behind this change is the motion of a wave packet according to the Schrödinger equation. But the interpretation rule says nothing about how the particle actually moves. There is no such thing as a classical trajectory.

**Wave–particle dualism**: It is an experimental fact that the wavelike behavior of particles can only be observed by recording the statistical distribution of position measurements. Whenever the particle is detected, then it is detected as a whole and it appears as a pointlike particle. There is no need to assume that the particle itself is spread over large regions of space. The statistical interpretation, however, is minimal in the sense that it makes no statement about the nature of a quantum object. It does not exclude the possibility that a quantum particle is wavelike and that the statistical distribution of sharp positions is a result of the observation procedure.

### 3.4.2. Elementary measurements

According to the statistical interpretation, it makes no sense to speak of the wave function of an individual particle. A quantum-mechanical experiment can only verify the probability distributions predicted by the formalism. This can only be done by repeating some measurement many times under identical conditions. A quantum-mechanical experiment thus consists of many single experiments (*elementary measurements*). In order to determine, for example, the position probability for a region $B$, one has to perform many position measurements. In each elementary experiment, a particle is prepared in the same initial state and after the time $t$ a position measurement is carried out in $B$ to see if the particle has arrived there. The probability is determined by counting the number $n(B)$ of experiments in which the particle is detected in $B$ and dividing this number by the total number $n$ of experiments, that is,

$$p(B) = \lim_{n\to\infty} \frac{n(B)}{n}. \tag{3.58}$$

We want to stress the following: An elementary experiment usually does not yield a reproducible result (except in the case where the probability for some result is 1). The quantum-mechanical formalism makes no predictions about the result of a single measurement. The outcome of a single experiment is therefore not the subject of quantum mechanics. Only probability distributions can be predicted and can be checked by repeating the same state preparation and the same measurement many times. According to the statistical interpretation, quantum mechanics is a theory of statistical ensembles and not a theory of individuals. The wave function does not describe an individual particle, but represents a conceptual infinite set of identically prepared particles. If we speak of the wave function of a single particle, we mean, in fact, the ensemble of all single particles that have undergone some state preparation procedure.

It is necessary to make a clear distinction between the *preparation* of a state and the measurement which finally *determines* the value of an observable. The preparation—generally achieved by interaction with a suitable apparatus—has to guarantee that repeated measurements are always performed under the same conditions. Due to unavoidable experimental errors, it will not always be possible to repeat the preparation with perfect accuracy. In general, the best we can say is that after the preparation the system has a random state in a certain subset of all possible states. A state that can actually be realized will thus be a *mixed state*, that is, a statistical mixture of wave functions. This is the subject of statistical quantum mechanics. Here we only deal with *pure states* that can be described by a single wave function.

The statistical interpretation has been the source of many discussions which have not been finished up to now. Some physicists have argued that the theory cannot be complete physically, because it makes no statements about *individual* systems. Does quantum mechanics already describe the objective reality in the most complete possible way? Is it just our lacking knowledge that lets us make predictions of limited accuracy? Is it possible to speak of the value of a physical observable prior to measurement? However, the search for hidden variables, which should provide a more complete description of individual systems, has not yet produced testable consequences. The unpredictability appears as a fundamental property of nature. Therefore, most physicists stick with the statistical interpretation, which has proved to be perfectly consistent with experimental tests.

### 3.4.3. Expectation value

Let $\psi(\mathbf{x})$ be a normalized wave function. Because the function $|\psi(\mathbf{x})|^2$ describes a position probability density, we can calculate the *mean value* (or *expectation value*) of the results of many position measurements. It is given by

$$\langle \mathbf{x} \rangle_\psi = \int_{\mathbb{R}^n} \mathbf{x}\, |\psi(\mathbf{x})|^2 \, d^n x. \tag{3.59}$$

Similarly, the expectation value of the momentum $\mathbf{p}$ in a state $\psi$ follows from the interpretation of the Fourier transform $\hat{\psi}$ as a momentum probability density,

$$\langle \mathbf{p} \rangle_\psi = \int_{\mathbb{R}^n} \mathbf{k}\, |\hat{\psi}(\mathbf{k})|^2 \, d^n k. \tag{3.60}$$

The results of many measurements will be scattered around the mean values. The width of the distribution of the measured values are the uncertainties $\Delta x$ and $\Delta k$ introduced earlier. As we know from Section 2.8.1, they satisfy Heisenberg's uncertainty relation

$$\Delta x \, \Delta k \geq \frac{n}{2}. \tag{3.61}$$

The uncertainty relation states that a sharp position distribution corresponds to a wide momentum distribution, and vice versa. There are no states where both position and momentum have sharp distributions.

In principle, there is no limit to the accuracy of either the position or the momentum in a wave function. However, there exists no wave function with a perfectly sharp momentum. Plane waves do have a sharp momentum, but they cannot be normalized. Hence the plane waves have no probabilistic interpretation and are not admissible as quantum-mechanical wave functions. Hence it is not possible to prepare a particle such that a momentum measurement gives some value with certainty. Similarly, there is no wave function with a perfectly sharp position. Such a wave function would have to be like a Dirac-delta function (see Section 2.9). But the delta function is not square-integrable and hence cannot be normalized.

### 3.4.4. The measurement process

The measurement of a physical observable involves the interaction of a quantum system with a measuring apparatus. This interaction can be quite complicated and might change the state in an uncontrolled way or even destroy the system under consideration. For example, the position of a photon can be measured by a photographic plate. But after the absorption the photon is gone and it is not reasonable to speak of its wave function any longer. We call this type of experiment a *determinative measurement*. It is designed to analyze a given state or to verify a probability distribution predicted by the quantum-mechanical formalism.

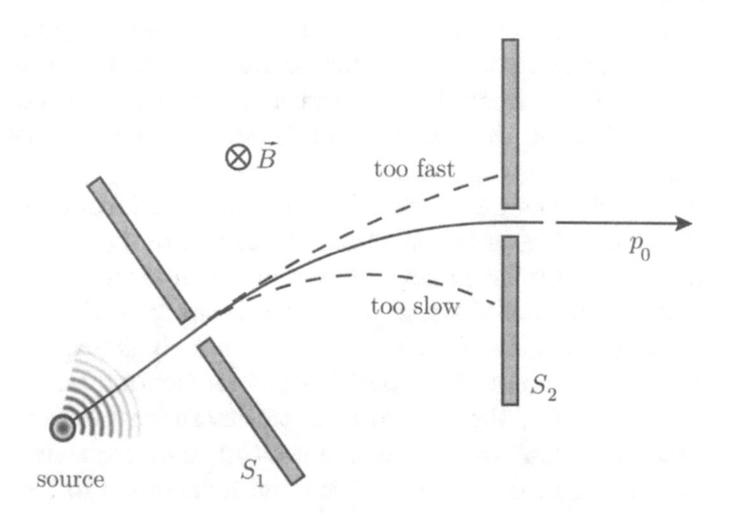


FIGURE 3.5. A monochromator prepares particles with momentum sharply concentrated around $\mathbf{p}_0$. A constant magnetic field between the screens $S_1$ and $S_2$ deflects the charged particles emitted by the source. The classical trajectories in a constant magnetic field are circles with a radius depending on the velocity. Therefore, only particles with a certain momentum $\mathbf{p}_0$ can emerge from the hole in the second screen $S_2$. This arrangement can be used for determinative experiments by replacing the hole with a detector.

If the measurement is not destructive, how do we describe the state of the particle after the measurement? The answer to this question certainly depends on the details of the interaction with the measurement device. Here we only consider a very simple model of a nondestructive measurement: a filtering procedure. For example, we can set up a momentum measurement in such a way that only particles with momentum $\mathbf{p}_0$ can pass the measurement device while all other particles are absorbed (see Fig. 3.5). The particles surviving this measurement all have the momentum $\mathbf{p}_0$ with an uncertainty depending on the quality of the apparatus. We can say that the apparatus prepares particles in a state with average momentum $\mathbf{p}_0$, and it is reasonable to assume that the ensemble of particles leaving the measurement device is described by a wave function which has a narrow peak at $\mathbf{p}_0$ in momentum space (whereas the particles entering the device might have had a wide momentum distribution or even might have been in a statistical mixture of states).

The effect of a measurement of this type is to prepare the system in such a way that it has certain properties. This is called a *preparatory measurement* or *preparation of the state*.

We note that the uncertainty relation refers to preparatory rather than determinative measurements. According to the uncertainty relation, it is not possible to perform simultaneous preparatory measurements of position and momentum. We say that the observables position and momentum are not *compatible*.

In order to verify the uncertainty relation, one conducts determinative measurements of position and momentum. According to the statistical interpretation, one has to perform many measurements of position and momentum on identically prepared systems. A priori it is not forbidden by quantum mechanics, to determine the values of $x$ and $p$ in a single experiment, but the simultaneous measurements of position and momentum tend to disturb each other. Fortunately, there is no need to determine position and momentum of the individual systems in simultaneous measurements. We may perform separate measurements of position and momentum in the course of many elementary experiments with identical systems. Quantum mechanics predicts that the values will be scattered around their mean values so that the uncertainty relation $\Delta x \, \Delta k \geq n/2$ is satisfied.

## 3.5. Continuity Equation

The wave function $\psi(\mathbf{x}, t)$ at a certain time $t$ contains the complete information about the state of motion of the particle. Hence it should be possible to predict the future time evolution just by looking at a picture of the wave function at time $t$. Here we show how this can be done at least for a few time-steps.

Assume that $\psi$ is a smooth solution of the Schrödinger equation. Let us calculate the time derivative of the position probability density $|\psi|^2$:

$$\begin{aligned}\frac{\partial}{\partial t}|\psi|^2 &= \frac{\partial \overline{\psi}}{\partial t}\psi + \overline{\psi}\frac{\partial \psi}{\partial t} = \text{(using the Schrödinger equation)}\\ &= -\frac{\mathrm{i}}{2}(\Delta\overline{\psi})\,\psi + \frac{\mathrm{i}}{2}\overline{\psi}\,(\Delta\psi)\\ &= -\frac{\mathrm{i}}{2}\nabla\cdot\big((\nabla\overline{\psi})\,\psi - \overline{\psi}\,(\nabla\psi)\big).\end{aligned}$$

If we define the *current vector*

$$\vec{\jmath} = \frac{\mathrm{i}}{2}\big((\nabla\overline{\psi})\,\psi - \overline{\psi}\,(\nabla\psi)\big), \tag{3.62}$$

we obtain the *continuity equation*

$$\frac{\partial}{\partial t}|\psi(\mathbf{x},t)|^2 = -\nabla\cdot\vec{\jmath}(\mathbf{x},t), \tag{3.63}$$

which states that a change in the position probability density is related to the divergence of a vector field $\vec{\jmath}(\mathbf{x},t)$. Integrating Eq. (3.63) over a volume $V \subset \mathbb{R}^3$ and using the divergence theorem of Gauss we find

$$\frac{d}{dt}\int_V |\psi(\mathbf{x},t)|^2\,d^3x = -\int_V \nabla\cdot\vec{\jmath}(\mathbf{x},t)\,d^3x = -\int_{\partial V}\vec{\jmath}\cdot\vec{n}\,df, \tag{3.64}$$

where $\vec{n}$ is the unit vector normal to the surface element $df$ and points in the positive (outside) direction. The change of the position probability in $V$ is equal to the flux of $\vec{\jmath}$ through the boundary $\partial V$ of $V$.

Using the polar form of complex numbers we write

$$\psi(\mathbf{x},t) = |\psi(\mathbf{x},t)|\,\exp(\mathrm{i}\,\varphi(\mathbf{x},t)). \tag{3.65}$$

The phase function $\varphi(\mathbf{x},t)$ is well defined up to multiples of $2\pi$ at all points $(\mathbf{x},t)$ where $|\psi(\mathbf{x},t)| \neq 0$. A little calculation shows that

$$\vec{\jmath} = |\psi|^2\,\nabla\varphi. \tag{3.66}$$

We learn from this that in a wave packet there is always a flow in the direction of increasing phase.

While the current vector $\vec{\jmath}$ is useful to describe a local change of the position probability density of a wave function, it does not correspond to the flow of a physical density (such as matter or charge density). The definition of $\vec{\jmath}$ is not at all unique. If $\vec{v}$ is the curl of any differentiable vector field, then $\nabla\cdot\vec{v} = 0$ and the current $\vec{\jmath}_1 = \vec{\jmath} + \vec{v}$ also satisfies the continuity equation (3.63). The "flow of probability" through a closed surface, see Eq. (3.64), is independent of the choice of the current vector. But the flow of $\vec{j}$ through a surface element $\vec{j}\cdot\vec{n}\,df$ depends on the choice of $\vec{j}$ and is hence not observable.

If a wave function is plotted using the standard color map to indicate its phase, then the direction of increasing phase is the counterclockwise direction in the color circle. For example, the wave packet always flows from red to yellow. Hence if a yellow region is surrounded by red, then the wave function will increase in the yellow region. If red is surrounded by yellow, then the wave packet will decrease in the red region. (Of course, a similar observation can be made with other colors that are neighbors in the color circle).

CD 3.11 illustrates the flow from red to yellow using special initial functions; see also Color Plate 12. The animations show how peaks are formed in the yellow regions of the initial wave function. As soon as the yellow peaks emerge, they start to decay again, thereby forming a more complicated interference pattern.

**Bohm's quantum mechanics**: In D. Bohm's pilot wave theory a particle is described by a wave function $\psi(\mathbf{x}, t)$ *and* by a position vector $\mathbf{r}(t)$. The wave function (a solution of Schrödinger's equation) is regarded as a sort of force field influencing the pointlike particle. By assuming that the motion of the position vector is described by the condition

$$\frac{d}{dt}\mathbf{r}(t) = \frac{1}{|\psi(\mathbf{r}(t), t)|}\vec{j}(\mathbf{r}(t), t) = \nabla\varphi(\mathbf{r}(t), t) \tag{3.67}$$

one arrives at a theory with a hidden variable $\mathbf{r}$ that is compatible with quantum mechanics. In Bohm's interpretation the position has a classical meaning and it makes sense to speak of a particle's trajectory. But the classical equation for the motion of the position variable is replaced by an ad hoc law with the quantum wave function serving as a pilot wave for the particle. There appear to be no experimental consequences of Bohm's interpretation. Therefore, many physicists prefer the probabilistic interpretation, which contains no unproven additional hypotheses. Moreover, the transition to quantum field theory is not straightforward within this framework. Nevertheless, Bohm's theory is important, because it has long been thought that a hidden variable theory was impossible.

## 3.6. Special Topic: Asymptotic Time Evolution

While the short-time behavior of wave packets can be guessed by analyzing the current as described in the previous section, the long-time behavior of a free particle can be predicted from the Fourier transform of the initial wave packet.

Using the explicit form (3.53) of the time evolution kernel $K$, we may rewrite Eq. (3.50) as

$$\begin{aligned}\psi(\mathbf{x},t) &= \frac{1}{(\mathrm{i}t)^{n/2}}\exp\Bigl(\mathrm{i}\frac{x^2}{2t}\Bigr)\frac{1}{(2\pi)^{n/2}}\int \exp\Bigl(-\mathrm{i}\mathbf{y}\cdot\frac{\mathbf{x}}{t}\Bigr)\exp\Bigl(\mathrm{i}\frac{y^2}{2t}\Bigr)\psi_0(\mathbf{y})\,d^n y \\ &= \frac{1}{(\mathrm{i}t)^{n/2}}\exp\Bigl(\mathrm{i}\frac{x^2}{2t}\Bigr)\bigl(\mathcal{F}\phi_t\bigr)\Bigl(\frac{\mathbf{x}}{t}\Bigr),\end{aligned}$$

where

$$\phi_t(\mathbf{y}) = \exp\Bigl(\mathrm{i}\frac{y^2}{2t}\Bigr)\psi_0(\mathbf{y}). \tag{3.68}$$

Now, given an initial state $\psi_0$, we define the function

$$\xi(\mathbf{x},t) = \frac{1}{(\mathrm{i}t)^{n/2}}\exp\Bigl(\mathrm{i}\frac{x^2}{2t}\Bigr)\hat{\psi}_0\Bigl(\frac{\mathbf{x}}{t}\Bigr) \tag{3.69}$$

and calculate

$$\begin{aligned}\|\psi(\cdot,t)-\xi(\cdot,t)\|^2 &= \int |\psi(\mathbf{x},t)-\xi(\mathbf{x},t)|^2 d^n x \\ &= \int \Bigl|\bigl(\mathcal{F}\phi_t\bigr)\Bigl(\frac{x}{t}\Bigr) - \bigl(\mathcal{F}\psi_0\bigr)\Bigl(\frac{x}{t}\Bigr)\Bigr|^2 \frac{d^n x}{t^n} \\ &= \int \bigl|(\mathcal{F}\phi_t)(y) - (\mathcal{F}\psi_0)(y)\bigr|^2 d^n y \\ &= \int |\phi_t(\mathbf{y}) - \psi_0(\mathbf{y})|\, d^n y \\ &= \int \Bigl|\exp\Bigl(\mathrm{i}\frac{y^2}{2t}\Bigr) - 1\Bigr|^2 |\psi_0(\mathbf{y})|^2\, d^n y.\end{aligned}$$

Here we used the variable substitution $\mathbf{y} = \mathbf{x}/t$, $d^n y = d^n x/t^n$, and the Fourier–Plancherel relation. But the last expression tends to zero, as $t \to \pm\infty$ (the integrand tends to zero, and because the integrand is bounded by $4|\psi_0(\mathbf{y})|^2$, the integral and the limit can be exchanged).

Hence, asymptotically in time, the wave function $\psi(\mathbf{x},t)$ can be replaced by the much simpler expression $\xi(\mathbf{x},t)$. Equation (3.69) shows that the modulus of $\psi$ asymptotically decays like $t^{-n/2}$; see also the discussion in Section 3.3.2.

**Wave functions for $|t| \to \infty$:**

Let $\psi(t)$ be any solution of the free Schrödinger equation. Let $\hat{\psi}_0$ be the Fourier transform of the initial function and define $\xi(t)$ as in Eq. (3.69). Then

$$\psi(t) - \xi(t) \longrightarrow 0, \quad \text{as } |t| \to \infty, \tag{3.70}$$

where the limit is approached in the quadratic mean (i.e., with respect to the norm in the Hilbert space). An equivalent way of writing (3.70) is

$$\lim_{t \to 0} \int |\psi(\mathbf{x}, t) - \xi(\mathbf{x}, t)|^2 d^n x = 0. \tag{3.71}$$

The *asymptotic position probability density* described by $\xi$ has the following property:

$$\int_{Bt} |\xi(x,t)|^2 d^n x = \int_{Bt} |\mathcal{F}\psi_0(x/t)|^2 \, d^n x / t^n = \int_B |\hat{\psi}_0(\mathbf{k})|^2 \, d^n k. \tag{3.72}$$

This result can be used to justify our interpretation of the Fourier transform as a momentum probability amplitude. In principle, the momentum of a particle can be determined by measuring the distance the particle moves within a certain time. Assuming that the particle is localized initially near $\mathbf{x} = 0$, we perform a position measurement after a time $t$. If we find the particle at $\mathbf{x}$ we would say that the particle had the momentum $\mathbf{p} = \mathbf{v} \approx \mathbf{x}/t$ (for particles with mass $m \neq 1$ the momentum is $\mathbf{p} = m\mathbf{v}$). This momentum measurement by the time of flight method will have an uncertainty caused by the uncertainty of the initial position. This uncertainty can be made small by waiting long enough because $(\mathbf{x} - \mathbf{x}_0)/t \approx \mathbf{x}/t$ for $|t|$ large. The probability for finding the position in a small volume $\Delta^n x = \Delta x_1 \, \Delta x_2 \cdots \Delta x_n$ around $\mathbf{x} \in \mathbb{R}^n$ is therefore equal asymptotically to the probability for finding the momentum in the volume $\Delta^n p = \Delta^n x / t^n$ around $\mathbf{p} = \mathbf{x}/t$. From the asymptotic form of the wave function we conclude that this probability is given by

$$|\psi(\mathbf{x}, t)|^2 \, \frac{\Delta^n x}{t^n} = |\hat{\psi}_0(\mathbf{x}/t)|^2 \, \frac{\Delta^n x}{t^n}. \tag{3.73}$$

The momentum probability amplitude is thus given by the Fourier transform of the wave function.

The result (3.72) simply describes the following fact:

**Asymptotic time evolution of free particles:**
The probability of finding the particle in a region $Bt = \{\mathbf{x}t \mid \mathbf{x} \in B\} \subset \mathbb{R}^n$ is asymptotically for $|t| \to \infty$ equal to the probability of finding the momentum of the particle in $B \subset \mathbb{R}^n$.

If $B$ is a cone with apex at the origin, then $Bt$ is the same cone. Hence the probability that a particle is finally found in a cone is equal to the probability that its momentum lies in that cone.

In this sense the momentum distribution of the initial function determines the asymptotic distribution of the wave packet in position space. The function $\xi$ describes a classical behavior of the wave function at large times because the wave packet is asymptotically where we expect it to be according to its momentum distribution.

At large times the distribution of the wave packet in position space is determined by the distribution of momenta in the initial function. CD 3.13 gives a particularly striking example; see also Color Plates 13 and 14.

## 3.7. Schrödinger Cat States

### 3.7.1. Superposition of two Gaussian functions

Here we apply the results of the previous section to a superposition of two Gaussian wave functions which are localized in different regions of space. We consider the one-dimensional case and assume that the two Gaussians have the same average momentum $p$. The wave function at the initial time $t = 0$ is thus given by

$$\psi(x) = N\,\mathrm{e}^{\mathrm{i}px}\left(\mathrm{e}^{-(x-x_1)^2/2} + \mathrm{e}^{-(x-x_2)^2/2}\right). \tag{3.74}$$

We assume that the distance $d = |x_1 - x_2|$ between the Gaussians is much larger than the spread of the Gaussians. Therefore, the position probability density has the shape shown in Fig. 3.6a.

EXERCISE 3.8. *Assuming that the distance between the two Gaussian wave packets in Eq. (3.74) is large, derive an approximate expression for the normalization constant $N$ such that $\|\psi\| = 1$. Why do you need the assumption on the distance between the Gaussians?*

Let us calculate the long-time behavior of the wave function. First we need the Fourier transform of $\psi$, which can be calculated easily. The two parts of the wave function in momentum space are both Gaussian functions centered at the average momentum $p$. The shift to $x_i$ in position space corresponds to the multiplication by a phase factor $\exp(ikx_i)$ in momentum space. Hence the Fourier transform of $\psi$ is obtained as

$$\begin{aligned}\hat{\psi}(k) &= N\,\mathrm{e}^{-ikx_1}\exp\Bigl(-\frac{(k-p)^2}{2}\Bigr) + N\,\mathrm{e}^{-ikx_2}\exp\Bigl(-\frac{(k-p)^2}{2}\Bigr)\\ &= 2N\exp\Bigl(-ik\,\frac{x_1+x_2}{2}\Bigr)\cos\Bigl(k\,\frac{x_2-x_1}{2}\Bigr)\exp\Bigl(-\frac{(k-p)^2}{2}\Bigr).\end{aligned}$$

The momentum distribution $|\hat{\psi}(k)|^2$ (which is constant in time) is shown in Fig. 3.6b. According to the results of the previous section, the wave function in the distant future can be approximated by

$$\psi(x,t) \approx \frac{1}{\sqrt{it}}\exp\Bigl(i\frac{x^2}{2t}\Bigr)\,\hat{\psi}\Bigl(\frac{x}{t}\Bigr). \tag{3.75}$$

Apart from a phase factor, this expression is just a scaled version of the wave function in momentum space. The corresponding position distribution is plotted in Fig 3.7.

It is tempting to interpret the two parts of the wave function at $t = 0$ as two pieces of a particle or even as two separate particles. But according to the statistical interpretation of Section 3.4, wave functions always describe

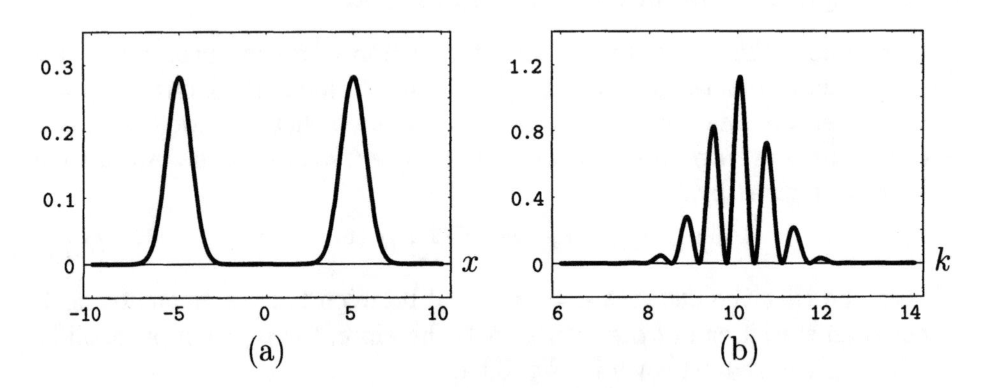


FIGURE 3.6. (a) Position probability density of a Schrödinger cat state consisting of two separated Gaussian peaks at $x_1 = -5$ and $x_2 = 5$; see Eq. (3.74). (b) Momentum distribution of the Schrödinger cat state. The two parts of the wave packet have the same average momentum $p = 10$, that is, they are located in the same region of momentum space. This causes the interference pattern in the momentum distribution.

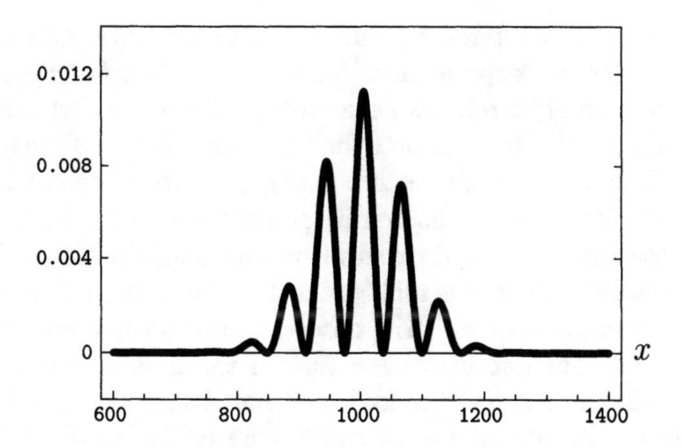


FIGURE 3.7. The position probability density of the Schrödinger cat state at some late time (here $t = 100$) is a scaled copy of the momentum distribution.

the states of one single particle.[1] The two peaks of the position probability density describe two distinct possibilities for the location of the particle. Quantum mechanics and the Schrödinger equation cannot predict which of these possibilities will actually be realized in an experiment. If we perform many measurements of the position in this state, then half of the results will show a position near $x_1$, while the other half locates the particle near $x_2$.

It is also tempting to assume that the particle is in fact either near $x_1$ or near $x_2$, and that we just don't know where it actually is until we perform a position measurement revealing the true position. In this interpretation the wave function describes primarily the state of knowledge of the observer and not so much a property of the system under observation. But this point of view does not take into account that the two possibilities described by the wave function can interfere at some later time.

In our case the interference is visible in Fig 3.7 and prevents that the particle can reach certain locations in the future. If the true initial position of the particle would be near $x_1$, an observer with a better knowledge would describe it by a wave function localized near $x_1$, and no interference pattern would show up in the asymptotic wave function. Hence in the future it would be possible to detect the particle in places where it cannot be found if the initial state also contains a possibility to find it near $x_2$. Thus, the "interference of possibilities" may lead to observable consequences that make it difficult to maintain the subjective interpretation of the wave function as a description of the knowledge of the observer.

[1] States of two or more particles are described in a more complicated way by wave functions depending on the position coordinates of all particles.

In this context it is interesting that the state Eq. (3.74) can be realized in neutron interference experiments. In a neutron interferometer a beam of neutrons can be split into two macroscopically (= a few centimeters) separated parts which are then brought together again to observe the self-interference. This interference pattern changes whenever one of the partial beams is manipulated. Here the density of neutrons in the beam is so low that this is actually a repeated experiment with single neutrons. (When a neutron is detected, the next is still confined in the nucleus of a radioactive atom in the radiation source). The observation of an interference pattern for beams of such low density proves that in the interferometer the wave function of a single neutron is indeed split into two parts. Now the wave packet in one of the partial beams can be shifted (delayed) so that after the reunion of the beams the state is described by two separated Gaussian peaks as above. The modulation of the momentum distribution exhibited in Fig. 3.6(b) has also been observed experimentally.

A state which is a superposition of two very distinct states is often called a *Schrödinger cat state.*

### 3.7.2. Schrödinger's cat

E. Schrödinger pointed out a possible paradox provoked by superpositions of states that are different on a macroscopic scale. He wanted to show that the wave function cannot be a complete description of nature. In order to illustrate his point by a thought experiment, Schrödinger suggested the following malicious arrangement: A poor little cat is locked into a dark cage with opaque walls. The cage also contains an apparatus with one atom of a radioactive material, a Geiger counter, and some poisonous substance. It is assumed that the atom decays with a probability of one half within an hour. If a decay occurs, it will be detected by the Geiger counter and the apparatus will set free the poison, killing the cat immediately.

In this arrangement the quantum uncertainty is transferred to a macroscopic system. The radioactive decay of an atom is a quantum-mechanical process. In the quantum-mechanical description the state of the atom after one hour will be a superposition of two possibilities: the intact atom and the atom after the decay. If quantum mechanics is a fundamental theory, then it should be possible to describe everything inside the cage in the language of quantum mechanics. The quantum-mechanical wave function of the whole system would describe the state of the atom, the apparatus, and the cat. The experimental setup lets the state of the atom interact with the state of the whole macroscopic system. Therefore, after an hour, the wave function of the whole system will be in a superposition of the two very distinct possibilities: cat dead or cat alive, that is,

$$\psi = \frac{1}{\sqrt{2}} (\psi_{\text{dead}} + \psi_{\text{alive}}). \tag{3.76}$$

We can make a measurement of the state of the cat simply by opening the cage and looking into it. Of course, one never experiences a superposition of states in the macroscopical world. So the cat will be either dead or alive. As soon as we notice the state of the cat, we cannot say that the state is a superposition any longer. So the wave function of the cat "collapses" to one of the two possibilities. After the measurement, we either have $\psi = \psi_{\text{dead}}$ or $\psi = \psi_{\text{alive}}$. We can carry the reasoning even further: The physicist $A$ who performs the experiment is a quantum-mechanical system himself. What is his state after an hour? As long as nobody opens the door of the lab, a fellow physicist $B$ outside the lab would have to describe the physicist $A$ by a superposition of two states: One state describes a sad physicist, the other shows him happy. In fact, if quantum mechanics would be a complete description of nature, every alternative of two quantum-mechanical events happening with comparable probability would put the whole universe in a superposition corresponding to these alternatives. The universe and everything in it would remain in that superposition until a hypothetical final observer looks to see which state is actually realized.

From the point of view of the statistical interpretation, the situation in itself is not paradoxical. Quantum mechanics predicts a probability distribution. Looking into the cage is an elementary experiment. One has to perform this experiment very often and will certainly find half of the cats dead, which is precisely the prediction of quantum mechanics. The problem is how to interpret the superposition state of the cat. One cannot say that the cat is either dead or alive at any time prior to the opening of the cage. The cat is in a superposition state and, in principle, the simultaneous presence of the two possibilities could produce some interference later (as it was the case with the superposition of the Gaussian wave packets). For example, it could be prohibited by interference that the surviving cat is found to be in a state where it wants to drink milk, while this state could well be reached without the poison in the cage.

Moreover, Schrödinger's paradox raises the question of what a measurement of a state really is. It has been proposed that a measurement consists in the interaction of a quantum-mechanical system with a classical system. A classical system is some device which can be described with reasonable accuracy within the framework of classical (i.e., nonquantum) physics. In particular, different states of a classical system cannot interfere with each other. In Schrödinger's example, the cat could be interpreted as a classical system that is incapable of getting in a superposition of states with the possibility of self-interference. The interaction of the radioactive atom with the cat is a measurement process that changes the state of the cat. The presence of a human observer is irrelevant in this context. The cat as a classical system is either dead or alive (not both) and the task of the observer is just to read off the result of the measurement. In this interpretation of the measurement process it is not clearly stated under which conditions a system can be regarded as classical. Also, it seems curious that a quantum-mechanical theory, which is assumed to generalize the classical theory, should need the classical theory for its justification.

In the case of the superposition (3.74) of Gaussian functions, the experimental results also indicate that the coherence of the two parts of the wave packet (i.e., the ability of self-interference) is very sensitive to perturbations and decreases with the separation of the parts. This could be a possible explanation of what is meant by a classical system: For superpositions of macroscopically separated states the perturbations and quantum fluctuations will dominate, and the possibility to observe any interference of the partial states is negligible. Such a system would appear as a classical system for which we can assume that one of the possibilities is actually realized.

A more complete investigation of the quantum measurement process would involve the interaction of a quantum system with a macroscopic (many-body) quantum system, the measurement device. The close correlation between the system and the measurement device after the measurement is called quantum entanglement. A careful analysis of the flow of information between an entangled system and the observer will be necessary to achieve a satisfactory description of the measurement process. These discussions are still going on and are beyond the scope of this book.

## 3.8. Special Topic: Energy Representation

This final section concludes our examination of the free motion with a more mathematical topic. You will need to know about the energy representation when you read Chapter 9 about scattering theory.

A wave packet in one space dimension can be written as an integral over the momentum space

$$\psi(x) = \frac{1}{\sqrt{2\pi}} \int_{-\infty}^{\infty} \mathrm{e}^{ikx}\, \hat{\psi}(k)\, dk, \tag{3.77}$$

where $\hat{\psi}$ is the Fourier transform of $\psi$. It is sometimes useful to write this as an integral over the energies of the system. This can be achieved by a variable substitution. We use the fact that energy and momentum of a free particle are related by $E = k^2/2$. In one dimension each value of the energy gives two possible values of the momentum,

$$k = \pm k(E), \quad k(E) = \sqrt{2E}, \quad 0 \leq E < \infty. \tag{3.78}$$

Hence we split the integral in Eq. (3.77) into two parts,

$$\psi(x) = \frac{1}{\sqrt{2\pi}} \int_{-\infty}^{0} e^{ikx}\, \hat{\psi}(k)\, dk + \frac{1}{\sqrt{2\pi}} \int_{0}^{\infty} e^{ikx}\, \hat{\psi}(k)\, dk.$$

With the substitution $k \to -k$, the first integral can be rewritten as

$$\frac{1}{\sqrt{2\pi}} \int_{0}^{\infty} e^{-ikx}\, \hat{\psi}(-k)\, dk.$$

Now we write $k = \sqrt{2E} = k(E)$, $dk = dE/k(E)$, and obtain

$$\psi(x) = \frac{1}{\sqrt{2\pi}} \int_{0}^{\infty} \frac{1}{\sqrt{k(E)}} \Big( e^{-ik(E)x}\, g_-(E) + e^{ik(E)x}\, g_+(E) \Big)\, dE, \tag{3.79}$$

where we defined the functions

$$g_\pm(E) = \frac{\hat{\psi}(\pm k(E))}{\sqrt{k(E)}} = \frac{1}{\sqrt{2\pi}} \int_{-\infty}^{\infty} \frac{1}{\sqrt{k(E)}}\, e^{\mp ik(E)x}\, \psi(x)\, dx. \tag{3.80}$$

We combine the functions $g_+$ and $g_-$ into a two-component wave function

$$g = \begin{pmatrix} g_+ \\ g_- \end{pmatrix}, \tag{3.81}$$

which is called the *energy representation* of the quantum-mechanical state, or simply the *wave function in energy space.* The transformation (3.80) mapping a wave function $\psi$ to $g$ is one-to-one. (The inverse transformation which expresses $\psi$ in terms of $g$ is given by Eq. (3.79)).

CD 3.9 shows the energy representation of a Gaussian wave packet. The first animation shows how the components $g_+$ and $g_-$ depend on the average momentum of the wave packet, the second animation visualizes the free time evolution in energy space.

The set of all two-component wave functions $g$ with square-integrable components $g_\pm$ forms a Hilbert space if we define the scalar product

$$\langle g^{(1)}, g^{(2)} \rangle = \int_{0}^{\infty} \Big( \overline{g_+^{(1)}(E)}\, g_+^{(2)}(E) + \overline{g_-^{(1)}(E)}\, g_-^{(2)}(E) \Big)\, dE. \tag{3.82}$$

If $g$ is the energy representation of a state $\psi$, then

$$\begin{aligned} \|\psi\|^2 &= \int_{-\infty}^{0} |\hat{\psi}(k)|^2\, dk + \int_{0}^{\infty} |\hat{\psi}(k)|^2\, dk \\ &= \int_{0}^{\infty} |\hat{\psi}(k(E))|^2\, \frac{1}{k(E)}\, dE + \int_{0}^{\infty} |\hat{\psi}(-k(E))|^2\, \frac{1}{k(E)}\, dE \\ &= \langle g, g \rangle = \|g\|^2. \end{aligned} \tag{3.83}$$

Indeed, the transition to the energy representation preserves the scalar product, that is,

$$\langle \psi^{(1)}, \psi^{(2)} \rangle = \langle g^{(1)}, g^{(2)} \rangle. \tag{3.84}$$

The time evolution in the energy representation is simply given by

$$g(E,t) = g(E)\,\mathrm{e}^{-\mathrm{i}Et} = \begin{pmatrix} g_+(E)\,\mathrm{e}^{-\mathrm{i}Et} \\ g_-(E)\,\mathrm{e}^{-\mathrm{i}Et} \end{pmatrix}. \tag{3.85}$$

Here $g(E)$ is the initial wave function in the energy space.

The components $g_+$ and $g_-$ of a state in the energy representation describe a decomposition according to the direction of motion. In position space,

$$\psi_+(x,t) = \frac{1}{\sqrt{2\pi}} \int_0^\infty \frac{1}{\sqrt{k(E)}}\, \mathrm{e}^{\mathrm{i}k(E)x - \mathrm{i}Et}\, g_+(E)\, dE \tag{3.86}$$

is a wave packet which moves to the right because only positive momenta are used in the superposition. Similarly, the part

$$\psi_-(x,t) = \frac{1}{\sqrt{2\pi}} \int_0^\infty \frac{1}{\sqrt{k(E)}}\, \mathrm{e}^{-\mathrm{i}k(E)x - \mathrm{i}Et}\, g_-(E)\, dE \tag{3.87}$$

describes a wave packet moving to the left. Obviously,

$$\psi(x,t) = \psi_+(x,t) + \psi_-(x,t) \tag{3.88}$$

is a decomposition of a wave packet $\psi$ into two orthogonal parts,

$$\langle \psi_+, \psi_- \rangle = \left\langle \begin{pmatrix} g_+ \\ 0 \end{pmatrix}, \begin{pmatrix} 0 \\ g_- \end{pmatrix} \right\rangle = 0. \tag{3.89}$$

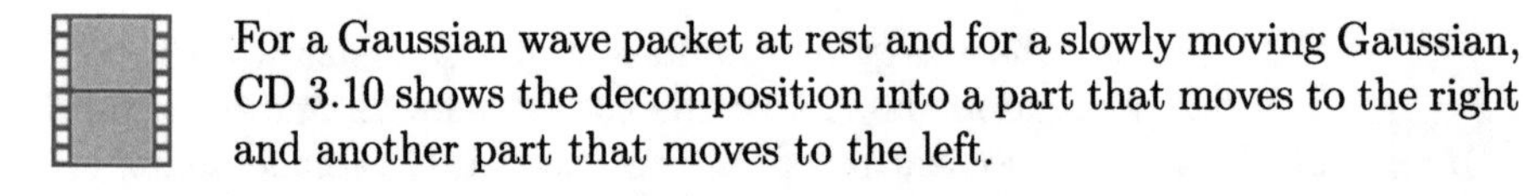
For a Gaussian wave packet at rest and for a slowly moving Gaussian, CD 3.10 shows the decomposition into a part that moves to the right and another part that moves to the left.

EXERCISE 3.9. *What happens in the energy representation when the wave function in position space is differentiated twice?*

*Chapter 4*

# States and Observables

**Chapter summary**: This chapter describes the basic structure of any quantum-mechanical theory. The first step in the mathematical formulation is to associate a suitable Hilbert space with a physical system. The vectors in that Hilbert space (or rather the one-dimensional subspaces spanned by these vectors) describe the possible states of the system. Certain linear operators correspond to the physically observable quantities. The expectation value of an observable defines the link between the mathematical quantities and physical experiments. It is interpreted as the mean value of many measurements of that observable.

For the single-particle systems described in this book, the Hilbert space is always represented as a set of square-integrable functions. You will learn how to find the operators corresponding to the classical observables, and you will learn how to predict the probability for measuring certain values of an observable in a given state.

The transition from classical to quantum mechanics can be formally achieved by replacing the classical observables with linear operators according to the following substitution rule. Always replace the position variable $x$ with the operator of multiplication by $x$ and the momentum $p$ by the differential operator $-i\hbar\nabla$. Unlike the classical quantities, the quantum-mechanical operators of position and momentum do not commute. Instead, they satisfy canonical commutation relations, where the commutator replaces the classical Poisson brackets. A pair of noncommuting operators leads to a generalized uncertainty relation and limits the accuracy with which the values of incompatible observables can be predicted simultaneously.

The quantum-mechanical observable corresponding to the kinetic energy of a particle is the Laplace operator in the Schrödinger equation. We can obtain the Schrödinger equation in an electromagnetic field by replacing the Laplace operator with the operator describing the energy of a particle in an electromagnetic field. The classical energy is described by the Hamiltonian function, which expresses the energy in terms of the position and momentum variables. An application of the substitution rule leads to the Hamiltonian operator of quantum mechanics. We describe this process for electric and magnetic fields and discuss the gauge freedom related to the nonuniqueness of the electromagnetic potentials. Finally, projection operators are introduced to describe properties of a physical system. They are used to determine the probability that the measured value of an observable is found within a given set of possible values.

# 4.1. The Hilbert Space of Wave Functions

## 4.1.1. State vectors

In classical mechanics, the state of a particle at time $t$ is described by its position $x(t)$ and momentum $p(t)$, that is, by a point in phase space. A quantum-mechanical particle occupies a certain region in phase space whose minimal size is determined by the uncertainty relation (see CD 3.9). What is the state according to quantum mechanics?

Quite generally, the *state* of a physical system is meant to be a collection of properties giving a complete description of the system. The set of information that constitutes a complete description depends on which aspects of the system one is interested in. In any case, the information should be complete in the sense that the state at any future time can be obtained from the description of the state at the initial time. The time development of the state is usually given by a dynamical law—the *evolution equation* or *equation of motion*. In classical mechanics, the dynamical law is given by the Hamiltonian equations.

In quantum mechanics, the wave function contains all information about the position and momentum distributions of the particle. The information provided by the wave function is also complete in the sense that if we know the wave function at time $t = 0$, then the wave function at any future time is completely determined by the Schrödinger equation. Hence the quantum-mechanical state of a particle is usually associated with its wave function, and the Schrödinger equation is the evolution equation.

There are physical quantities such as mass and electric charge which affect the behavior of the particle but which are usually not considered part of the state. These quantities remain unchanged in a whole set of experiments, hence they are characteristic for the system itself rather than for its state. In the formalism, these quantities appear as parameters whose numerical value is usually kept constant.

Sometimes it may happen that refined experiments reveal important properties that have been forgotten in the description of the state. For example, the *spin* is a property of electrons that cannot be described by complex-valued wave functions. In order to take into account the spin, our description of states in terms of wave functions will have to be modified. These problems will be discussed in Book Two.

It may also turn out that the evolution equation has to be modified to describe the temporal behavior of states in extreme situations. For example, if the velocity of particles is close to the velocity of light, the Schrödinger equation becomes inaccurate and has to be replaced by some relativistically invariant equation of motion—which might also require a change in the concept of a state. For example, the Dirac equation of relativistic quantum mechanics requires $C^4$-valued functions to describe the state of an electron.

In nonrelativistic quantum mechanics, the state of a (spinless) particle at time $t$ is described by a nonzero, square-integrable wave function $\psi(\cdot, t)$.

The correspondence between the states and the wave functions is not one-to-one. For example, if we multiply a wave function by a real number, it would nevertheless describe the same state. The real number would go away when normalizing the wave function in order to apply the probabilistic interpretation (Section 3.4.1). Moreover, the equations (3.55) and (3.56) describing the physical content of the wave function are insensitive to a multiplication of the wave function by a phase factor. (A *phase factor* is a complex number with absolute value 1). Hence we may conclude: After normalization, a wave function $\psi$ and its scalar multiple $c\psi$ (with arbitrary $c \in \mathbb{C}$, $c \neq 0$) only differ by a phase factor and therefore lead to the same physical predictions, for example, about probabilities of finding the particle somewhere in position or momentum space. In this sense, $\psi$ and $c\psi$ both describe the same state.

In principle, any square-integrable nonzero function could be used to define a quantum-mechanical state. The set of all square-integrable wave functions forms the Hilbert space $L^2(\mathbb{R}^n)$ (see Section 2.2). Obviously, this Hilbert space is associated with an elementary particle moving in an $n$-dimensional configuration space.

It is one of the basic assumptions of the quantum-mechanical formalism that the states of every physical system are given by vectors in a suitable Hilbert space, and that two vectors describe the same state if one is a scalar multiple of the other. We formulate this assumption in the following box.

**States of a physical system**:

The states of a quantum-mechanical system can be described by vectors in a suitable Hilbert space. Two vectors $\psi$ and $\phi$ describe the same state if and only if $\phi = c\psi$. Therefore, the physical states correspond precisely to the one-dimensional subspaces

$$[\psi] = \{c\psi \mid c \in \mathbb{C}\}, \qquad \psi \neq 0, \tag{4.1}$$

of the Hilbert space. Any nonzero element $\psi$ of this subspace can be used to represent the state $[\psi]$, physical predictions do not depend on the choice of $\psi$.

The state of a system is best represented by a normalized wave function $\psi$, $\|\psi\| = 1$. This allows a direct application of the interpretation rules. Fortunately, the time evolution generated by the Schrödinger equation has the property that the norm of the solution is left invariant, $\|\psi(\cdot,t)\| = \|\psi_0\|$; thus, it is sufficient to normalize the initial state.

### 4.1.2. Superposition principle

Given two wave functions $\psi_1$ and $\psi_2$, any linear combination $\phi = c_1\psi_1 + c_2\psi_2$ defines a new possible state of the system. This *superposition principle* just expresses the linearity of the Hilbert space. Here we would like to add the following remarks:

**1**: Although the wave functions $\psi$ and $c\psi$ describe the same state, this is not true for the wave functions $\psi_1 + \psi_2$ and $c_1\psi_1 + c_2\psi_2$, unless $c_1 = c_2$,

$$[\psi_1 + \psi_2] \neq [c_1\psi_1 + c_2\psi_2]. \tag{4.2}$$

**2**: For two wave functions $\psi_1$ and $\psi_2$ the position probability density of the sum $\psi_1 + \psi_2$ is different from the sum of the individual densities (*interference*),

$$|\psi_1(x) + \psi_2(x)|^2 \neq |\psi_1(x)|^2 + |\psi_2(x)|^2. \tag{4.3}$$

## 4.2. Observables and Linear Operators

Observables are physical quantities such as position, momentum, and energy that one might want to measure in order to learn about the state of the system. In quantum mechanics, observables are described as linear operators in the Hilbert space of the physical system. Linear operators are rather abstract mathematical objects (see Section 2.5). We will have to specify how to extract the experimentally verifiable information from the operators. We start by considering some examples.

### 4.2.1. The position operator

A quantity that is related to the position and that (at least in principle) can be determined experimentally is the expectation value of the position. For a particle moving in one space dimension, it can be written as

$$\langle x \rangle_\psi = \int_{-\infty}^{\infty} x\,|\psi(x)|^2\,dx = \langle \psi, x\psi \rangle; \tag{4.4}$$

see Sections 2.8.1 and 3.4.3. Here and in the following it is always assumed that the wave function is normalized, $\|\psi\| = 1$. From a mathematical point of view this expression is the scalar product between $\psi$ and the wave function $\xi$ given by

$$\xi(x) = x\,\psi(x). \tag{4.5}$$

The mapping between $\psi$ and $\xi$ is a linear operator in the Hilbert space of square-integrable functions. Obviously, we just need this linear operator to extract the information about the position from the state of the particle. Hence we choose this linear operator to represent the observable quantity position in the formalism of quantum mechanics.

**Position operator:**

The position observable for a particle in one dimension is represented by the *position operator* $x$, which is defined as the linear operator that multiplies the wave function $\psi$ with the variable $x$.

We will follow the dangerous but usual convention of denoting both the linear operator and the position variable by the letter $x$.

$\Psi$ **Domain of definition**: The definition of a linear operator would not be complete without specifying its domain. The domain of the multiplication operator $x$ is the linear subspace of all $\psi \in \mathfrak{H}$ with $\int x^2 |\psi(x)|^2\, dx < \infty$. The operator $x$ can only be applied to the functions in this subspace because otherwise the image $x\psi$ is not in the Hilbert space.

How does a wave function look, if it is not in the domain of the position operator $x$? Such a wave function vanishes rather slowly, as $|x| \to \infty$, because the function $x\psi(x)$ is not square-integrable. It may even happen that the position has no finite expectation value (if $\sqrt{x}\psi(x)$ is not square-integrable). Still, as long as $\psi$ belongs to the Hilbert space of square-integrable functions, we can maintain the interpretation of $|\psi(x)|^2$ as a position probability density.

**Position operator in $n$ dimensions**: For particles in an $n$-dimensional configuration space, the position observables are represented by an $n$-tuple of linear operators which are written as a vector,

$$\mathbf{x} = (x_1, \ldots, x_n). \tag{4.6}$$

When we speak of the position operator for a particle in $\mathbb{R}^3$ we mean in fact the triple formed by the three operators of multiplication by $x_i$, $i = 1, 2, 3$.

### 4.2.2. Momentum operator

For a particle in one dimension the expectation value of the momentum is given by

$$\langle p\rangle_\psi = \int_{-\infty}^{\infty} k\,|\hat{\psi}(k)|^2\,dk. \tag{4.7}$$

Again, it is possible to write this as a scalar product in the form

$$\langle p\rangle_\psi = \langle \psi, p\,\psi\rangle \tag{4.8}$$

with a linear operator $p$. Just recall the following property of the Fourier transformation $\mathcal{F}$:

$$\mathcal{F}(-\mathrm{i}\psi') = k\,\mathcal{F}\psi = k\,\hat{\psi}, \tag{4.9}$$

with $\psi' = \frac{d}{dx}\psi$. Using this and the Fourier–Plancherel relation (2.64) we obtain

$$\langle p\rangle_\psi = \int_{-\infty}^{\infty} \overline{\psi(x)}\,(-\mathrm{i})\,\psi'(x)\,dx = \langle \psi, -\mathrm{i}\psi'\rangle. \tag{4.10}$$

Hence it is meaningful to define the *momentum operator* as the linear operator $p$ with

$$p\,\psi = -\mathrm{i}\frac{d}{dx}\psi. \tag{4.11}$$

The derivative is understood as a generalized derivative in the sense of Eq. (2.90). The momentum operator is thus defined on the domain of square-integrable functions $\psi$ for which $k\hat{\psi}(k)$ is also square-integrable.

**Notation**: For particles in an $n$-dimensional configuration space, we form a vector $\mathbf{p}$ from the components of the momentum in the coordinate directions and write the momentum operator as

$$\mathbf{p} = -\mathrm{i}(\partial_1, \ldots, \partial_n), \quad \text{with} \quad \partial_k = \frac{\partial}{\partial x_k}. \tag{4.12}$$

### 4.2.3. Kinetic energy

The Laplace operator $-\Delta$ which appears in the Schrödinger equation can be written as

$$-\Delta = \mathbf{p}\cdot\mathbf{p} = -\sum_{i=1}^{n} \partial_i^2. \tag{4.13}$$

Because $\mathbf{p}\cdot\mathbf{p}/2 = p^2/2$ is just the classical expression for the kinetic energy in terms of the momentum of a particle (with mass $m = 1$), we regard the linear operator

$$H_0 = -\frac{1}{2}\Delta \tag{4.14}$$

as the operator representing the observable kinetic energy.

$\boxed{\Psi}$ **The domain of $H_0$**: The operator $H_0$ has to be defined in the Hilbert space $L^2(\mathbb{R}^n)$. We could choose a domain for $H_0$ that consists of twice differentiable functions $\psi$ for which $\Delta\psi$ is again square-integrable, but it is more appropriate to understand differentiability in the general sense discussed in the chapter on the Fourier transform (Section 2.6.2). Thus, a square-integrable function $\psi$ belongs to $\mathfrak{D}(H_0) = \mathfrak{D}(\Delta)$ if and only if the function $\mathbf{k} \to k^2\hat{\psi}(\mathbf{k})$ is square-integrable. For $\psi \in \mathfrak{D}(\Delta)$, the action of $H_0$ is given by

$$H_0\psi = \mathcal{F}^{-1}\frac{k^2}{2}\mathcal{F}\psi. \tag{4.15}$$

The action of $H_0$ on a wave function $\psi$ takes its simplest form in the energy representation (Section 3.8). The transition from momentum space to energy space is performed with the help of the variable substitution $k \to \sqrt{2E}$. In momentum space, the action of $H_0$ amounts to the multiplication by $k^2/2$, which becomes just multiplication by $E$ in the energy space. Hence, if $g(E) = (g_-(E), g_+(E))$ is the energy representation of a wave function $\psi(x)$, then $E\,g(E) = (E\,g_-(E), E\,g_+(E))$ is the energy representation of the wave function $H_0\,\psi$.

## 4.3. Expectation Value of an Observable

In quantum mechanics, any observable of a physical system is represented by a suitable operator in the Hilbert space of the system. The expectation value of an arbitrary observable is defined in complete analogy to the expectation values of position and momentum:

**Expectation value**:

For any linear operator $A$ that represents a physical observable, the *expectation value* of $A$ in the state $\psi$ (with $\|\psi\| = 1$) is

$$\langle A\rangle_\psi = \langle\psi, A\psi\rangle. \tag{4.16}$$

This is interpreted as the mean value of many measurements performed on identically prepared copies of the physical system.

Only those operators can represent observables for which the expectation value is a real number because physical measurements should always produce real numbers. These operators are called *symmetric*.

DEFINITION 4.1. A densely defined linear operator is called *symmetric* or *Hermitian* if all expectation values are real:

$$A \text{ is symmetric if and only if } \quad \langle\psi, A\psi\rangle \in \mathbb{R} \quad \text{for all } \psi \in \mathfrak{D}(A).$$

There are more restrictions on the class of operators describing observables. An operator that is useful as a candidate for an observable must be self-adjoint. The self-adjoint operators are a subset of the symmetric operators. We will return to this question in Chapter 6.

EXAMPLE 4.3.1. In the energy representation the operator of kinetic energy $H_0$ is just multiplication by $E$. Using Eq. (3.84) we find for the expectation value of the energy the expression

$$\begin{aligned}\langle H_0\rangle_\psi &= \langle\psi, H_0\psi\rangle = \langle g, Eg\rangle \\ &= \int_0^\infty E(|g_+(E)|^2 + |g_-(E)|^2)\, dE.\end{aligned}$$

The function $|g_+(E)|^2 + |g_-(E)|^2$ is therefore interpreted as an energy probability density: Let $G$ be a subset of the positive real numbers. The integral

$$\int_G (|g_+(E)|^2 + |g_-(E)|^2)\, dE$$

is the probability that the kinetic energy is found in $G$ if a measurement is performed in the state $\psi$.

**Uncertainty:**

For an observable $A$ the quantity

$$\Delta_\psi A \equiv \|(A - \langle A\rangle_\psi)\psi\| = \sqrt{\langle (A - \langle A\rangle_\psi)^2\rangle_\psi} \tag{4.17}$$

is called the *uncertainty* of $A$ in the state $\psi$ (which is assumed to be normalized). The uncertainty describes the dispersion of the actually measured values of the observable $A$ around the mean value $\langle A\rangle_\psi$.

We note that in general the expectation value and the uncertainty of an observable $A$ are not defined for every $\psi$ in the Hilbert space, but only for those $\psi$ that are in the domain of the operator $A$.

EXERCISE 4.1. *Show the formula*

$$(\Delta_\psi A)^2 = \langle A^2 \rangle_\psi - \langle A \rangle_\psi^2. \tag{4.18}$$

## 4.4. Other Observables

### 4.4.1. The substitution rule

Up to now, we have only introduced linear operators for the observables position, momentum, and kinetic energy. We want to find the operators corresponding to other observables (e.g., angular momentum, potential energy, etc.) In many cases the following procedure has proved to be a successful way to guess the right operators.

Classically, an observable is a function of the position and the momentum, that is, a function on the phase space. Very often (but not always), it is possible to obtain a suitable quantum-mechanical operator by simply replacing in the classical expression each component of $p$ by the corresponding differential operator

$$p_k \to -\mathrm{i}\,\frac{\partial}{\partial x_k}, \qquad k = 1, \ldots, n, \tag{4.19}$$

and each component of $\mathbf{x}$ by the corresponding multiplication operator,

$$x_k \to \text{multiplication by } x_k, \qquad k = 1, \ldots, n. \tag{4.20}$$

(Here $n$ is the dimension of the configuration space). If we use units where $\hbar$ is not equal to 1, then the substitution rule for the momentum observable has to be changed into

$$p_k \to \mathrm{i}\hbar\frac{\partial}{\partial x_k}, \quad \text{or} \quad \mathbf{p} \to -\mathrm{i}\hbar\nabla. \tag{4.21}$$

### 4.4.2. Functions of x

If $V(\mathbf{x})$ is a real-valued function of the position $\mathbf{x}$, then the corresponding quantum-mechanical observable is the operator of multiplication by $V(\mathbf{x})$,

$$V : \psi(\mathbf{x}) \to \phi(\mathbf{x}) = V(\mathbf{x})\,\psi(\mathbf{x}). \tag{4.22}$$

Very often, the operator $V$ represents the potential energy of a particle.

**The domain of $V$:** If $V(\mathbf{x})$ is an unbounded function (e.g., $V(\mathbf{x}) = |\mathbf{x}|^2$), then the domain of the multiplication operator $V(\mathbf{x})$ consists only of those square-integrable functions, for which the integral

$$\int |V(\mathbf{x})\,\psi(\mathbf{x})|^2\, d^n x$$

is finite. If the function $V$ is bounded (i.e., there exists a constant $M$ such that $V(\mathbf{x}) \leq M$ for all $\mathbf{x}$), then the operator $V$ is defined everywhere. In this case,

$$\int |V(\mathbf{x})\,\psi(\mathbf{x})|^2\, d^n x \leq M^2 \|\psi\|^2 < \infty, \quad \text{for all } \psi,$$

and the domain of the multiplication operator $V(\mathbf{x})$ is the entire Hilbert space $L^2(\mathbb{R}^n)$. The reader may check that the expectation value of this operator is always a real number.

### 4.4.3. Functions of p

If we apply the substitution rule $\mathbf{p} \to -\mathrm{i}\nabla$ to a function $f(\mathbf{p})$, we have to give a meaning to the expression $f(-\mathrm{i}\nabla)$. At first sight it might appear difficult to define a function of a differential operator, but as we learned in Section 2.6.2, this is actually quite straightforward. A function of the momentum operator can easily be defined with the help of the Fourier transform:

$$f(-\mathrm{i}\nabla)\psi(\mathbf{x}) = \frac{1}{(2\pi)^{n/2}} \int \mathrm{e}^{\mathrm{i}\mathbf{k}\cdot\mathbf{x}}\, f(\mathbf{k})\, \hat{\psi}(\mathbf{k})\, d^n k. \tag{4.23}$$

EXERCISE 4.2. *Try to describe the domain of definition of the linear operator* $f(-\mathrm{i}\nabla)$.

### 4.4.4. Angular momentum

For a classical particle in three dimensions the angular momentum in $z$-direction is a function of the position and momentum coordinates,

$$L_3 = x_1\, p_2 - x_2\, p_1. \tag{4.24}$$

The other components of the angular momentum are defined in a similar way,

$$L_1 = x_2\, p_3 - x_3\, p_2, \qquad L_2 = x_3\, p_1 - x_1\, p_3. \tag{4.25}$$

This is usually compactly written in vector form as

$$\mathbf{L} = \mathbf{x} \times \mathbf{p}. \tag{4.26}$$

($\times$ is the vector product or cross product of the two vectors).

Now, let us apply the substitution rule to the angular momentum. In that way we obtain the *angular momentum operator* $\mathbf{L} = (L_1, L_2, L_3)$ of a particle in three dimensions,

$$L_j = -\mathrm{i}\Big(x_k\, \frac{\partial}{\partial x_l} - x_l\, \frac{\partial}{\partial x_k}\Big), \tag{4.27}$$

where $(j, k, l)$ is a cyclic permutation of $(1, 2, 3)$.

EXERCISE 4.3. *Calculate the commutators of the components of the angular momentum operator:* $[L_1, L_2]$, $[L_2, L_3]$, *and* $[L_3, L_1]$.

It was easy to translate the angular momentum operator to quantum mechanics, but how do we proceed for more general functions of the position and momentum observables?

## 4.5. The Commutator of $x$ and $p$

The application of the substitution rule for the transition to quantum mechanics becomes ambiguous for functions involving products of the position and momentum coordinates. The origin of this problem is that the momentum and position operators do not commute.

Let us calculate, in one dimension, the composition of the linear operators $p = -\mathrm{i}d/dx$ and $x$. Using the product rule for differentiation we find

$$p\,x\,\psi(x) = -\mathrm{i}\frac{d}{dx}\,x\,\psi(x) = -\mathrm{i}\,\psi(x) - \mathrm{i}\,x\,\frac{d}{dx}\,\psi(x) \tag{4.28}$$

$$= -\mathrm{i}\,\psi(x) + x\,p\,\psi(x). \tag{4.29}$$

Hence we find for the commutator of $x$ and $p$,

$$[x,p]\,\psi(x) = x\,p\,\psi(x) - p\,x\,\psi(x) = \mathrm{i}\,\psi(x). \tag{4.30}$$

In higher dimensions, the corresponding relations are as follows.

**Commutation relations for position and momentum:**

The components of the position and the momentum operators in $L^2(\mathbb{R}^n)$ satisfy the relations

$$[x_j, x_k] = [p_j, p_k] = 0, \qquad [x_j, p_k] = \mathrm{i}\,\delta_{jk}\,\mathbf{1}. \tag{4.31}$$

These relations are very similar to the Poisson bracket relations for the positions and momenta of a classical mechanical system. This has led to the conjecture that the transition from classical to quantum mechanics can be achieved by replacing any classical observable (i.e., any function on the classical phase space) by an operator in such a way that the Poisson bracket of two classical observables is i times the commutator of the corresponding operators. Unfortunately, this program cannot be carried through in that generality because it leads to inconsistencies in the algebra of observables. Moreover, a quantum analog to an observable $f(x,p)$ need not exist due to the lack of a suitable domain of definition.

The noncommutativity of the position and momentum operators means that the classically identical observables $xp$ or $px$ would correspond to different operators in quantum mechanics.

EXERCISE 4.4. *Show that the operator $xp = -\mathrm{i}x\, d/dx$ is not suitable as a quantum-mechanical observable because for certain states its expectation value is not a real number.*

The exercise above suggests that we replace the classical expression $xp$ by the symmetrized expression $(xp + px)/2$. Performing the transition to quantum mechanics on the symmetric expression gives the operator

$$D = -\frac{\mathrm{i}}{2}\,(\mathbf{x} \cdot \nabla + \nabla \cdot \mathbf{x}) \tag{4.32}$$

which has only real expectation values.

For the definition of the angular momentum operator $\mathbf{L}$ the noncommutativity of $x$ and $p$ is not a problem because $L_i$ contains only products of the commuting operators $x_j$ and $p_k$ $(j \neq k)$.

## 4.6. Electromagnetic Fields

The transition from classical to quantum mechanics by substitution of operators for classical functions on phase space can be used to motivate the form of the Schrödinger equation in the presence of electromagnetic fields. The first step is to consider the classical Hamiltonian function $H(\mathbf{x}, \mathbf{p})$, that is, the total energy of a classical particle in an electromagnetic field as a function of position and momentum. For the energy operator of the corresponding quantum-mechanical system one usually tries the expression $H = H(\mathbf{x}, -\mathrm{i}\hbar\nabla)$ according to the substitution rule described in Section 4.4. One postulates that the operator of total energy $H$ determines the time evolution of particles in an electric field in the same way as the free kinetic energy operator determines the time evolution of free particles. Thus, the Schrödinger equation for an electromagnetic field has the general form of an initial-value problem

$$\mathrm{i}\hbar \frac{d}{dt}\,\psi = H\,\psi, \qquad \psi(t=0) = \psi_0, \tag{4.33}$$

where $\psi_0$ is some initial state in the Hilbert space of the system. Because the energy operator $H$ corresponds to the classical Hamiltonian function, it is usually called the *Hamiltonian operator* or simply the *Hamiltonian* of the system.

### 4.6.1. Electric potentials

In the presence of an electric field $\vec{E}(\mathbf{x})$, a charged particle at the point $\mathbf{x}$ feels a force

$$\vec{F}(\mathbf{x}) = -q\,\vec{E}(\mathbf{x}), \tag{4.34}$$

where $q$ is the charge of the particle. The electric field $\vec{E}(\mathbf{x})$ can be described in terms of a scalar potential field $V(\mathbf{x})$,

$$\vec{E}(\mathbf{x}) = -\nabla V(\mathbf{x}). \tag{4.35}$$

The quantity $q\,V(\mathbf{x})$ is just the potential energy of the charged particle at the point $\mathbf{x}$. If we add the potential energy to the kinetic energy $\mathbf{p}^2/2m$ we obtain the Hamiltonian function which describes the total energy as a function of the position and momentum variables,

$$H(\mathbf{x},\mathbf{p}) = \frac{\mathbf{p}^2}{2m} + q\,V(\mathbf{x}). \tag{4.36}$$

The Hamiltonian function is a function on the classical phase space. We may now apply the substitution rule

$$\mathbf{x} \longrightarrow \text{multiplication by } \mathbf{x},$$
$$\mathbf{p} \longrightarrow -\mathrm{i}\hbar\nabla,$$

which also works for functions of $\mathbf{x}$ and $\mathbf{p}$ as described in Section 4.4, as long as these functions do not contain products of the position and momentum variables. In that way the potential energy is replaced by the *operator* $V(\mathbf{x})$ of multiplication by the *function* $V(\mathbf{x})$, and the kinetic energy $\mathbf{p}^2/2m$ is replaced by the Laplace operator $-(\hbar^2/2m)\,\Delta$. Hence the Hamiltonian function $H(\mathbf{x},\mathbf{p})$ can be translated into the *Hamiltonian operator*

$$H = -\frac{\hbar^2}{2m}\,\Delta + q\,V(\mathbf{x}). \tag{4.37}$$

The operator $H$ is the observable of total energy for a quantum-mechanical particle in an electric field. It replaces the kinetic energy operator $H_0$ in the Schrödinger equation for free particles.

**Schrödinger equation in an electric field:**

The Schrödinger equation for a particle with charge $q$ in an electric field $\vec{E}(\mathbf{x}) = -\nabla V(\mathbf{x})$ is given by

$$\mathrm{i}\hbar\,\frac{\partial}{\partial t}\,\psi(\mathbf{x},t) = -\frac{\hbar^2}{2m}\,\Delta\,\psi(\mathbf{x},t) + q\,V(\mathbf{x})\,\psi(\mathbf{x},t). \tag{4.38}$$

Differential operators of the form $-\Delta + V$ are usually called *Schrödinger operators* by mathematicians. In the second half of the twentieth century a great deal of effort has been spent on investigating the properties of these operators for general classes of potentials. The interested reader will find more information on this subject in the many excellent books on Schrödinger operators listed in the bibliography.

### 4.6.2. Magnetic fields

For particles in magnetic fields, the transition to quantum mechanics is achieved by the same heuristic procedure.

A magnetic field $\vec{B}(\mathbf{x})$ can be described by a magnetic vector potential $\vec{A}(\mathbf{x})$. In three dimensions we have

$$\vec{B}(\mathbf{x}) = \operatorname{curl} \vec{A}(\mathbf{x}), \quad \mathbf{x} = (x_1, x_2, x_3) \in \mathbb{R}^3. \tag{4.39}$$

Hence any magnetic field automatically satisfies the condition

$$\operatorname{div} \vec{B}(\mathbf{x}) = 0. \tag{4.40}$$

The vector potential is needed in order to define the classical Hamiltonian function for a particle in a magnetic field,

$$H(\mathbf{p}, \mathbf{x}) = \frac{1}{2m} \left(\mathbf{p} - \frac{q}{c}\vec{A}(\mathbf{x})\right)^2. \tag{4.41}$$

With the help of the formal analogy described in Section 4.4, we define the Hamiltonian operator as

$$H = \frac{1}{2}\left(-\mathrm{i}\nabla - \frac{q}{c}\vec{A}(\mathbf{x})\right)^2, \tag{4.42}$$

thereby returning to the habit of setting $\hbar/m = 1$.

The formal substitution rule $\mathbf{p} \to -\mathrm{i}\nabla$, $\mathbf{x} \to$ (multiplication by) $\mathbf{x}$ is ambiguous for this situation. This becomes obvious if we expand the square in Eq (4.41). We find that the classical Hamiltonian function contains a product of $\vec{A}(\mathbf{x})$ and $\mathbf{p}$:

$$\left(\mathbf{p} - \frac{q}{c}\vec{A}(\mathbf{x})\right)^2 = \mathbf{p}^2 - 2\frac{q}{c}\mathbf{p} \cdot \vec{A}(\mathbf{x}) + \vec{A}(\mathbf{x})^2. \tag{4.43}$$

The expressions

$$2\mathbf{p} \cdot \vec{A}(\mathbf{x}), \quad 2\vec{A}(\mathbf{x}) \cdot \mathbf{p}, \quad \text{and} \quad \mathbf{p} \cdot \vec{A}(\mathbf{x}) + \vec{A}(\mathbf{x}) \cdot \mathbf{p} \tag{4.44}$$

all represent the same function of the classical variables $\mathbf{x}$ and $\mathbf{p}$. After applying the formal substitution rule they correspond to different operators because

$$\begin{aligned}\mathbf{p}\cdot\vec{A}(\mathbf{x})\,\psi(\mathbf{x}) &= -\mathrm{i}\nabla\cdot\vec{A}(\mathbf{x})\,\psi(\mathbf{x})\\ &= -\mathrm{i}(\nabla\cdot\vec{A}(\mathbf{x}))\,\psi(\mathbf{x}) - \mathrm{i}\vec{A}(\mathbf{x})\cdot\nabla\psi(\mathbf{x})\\ &= (-\mathrm{i}\,\mathrm{div}\,\vec{A}(\mathbf{x}) + \vec{A}(\mathbf{x})\cdot\mathbf{p})\,\psi(\mathbf{x})\\ &\neq \vec{A}(\mathbf{x})\cdot\mathbf{p}\,\psi(\mathbf{x}) \qquad (\text{unless } \mathrm{div}\,\vec{A}=0).\end{aligned} \tag{4.45}$$

If we expand the square in the quantum-mechanical Hamiltonian (4.42) we have to take into account that the product of the operators $\mathbf{p} = -\mathrm{i}\nabla$ and $\vec{A}(\mathbf{x})$ is not commutative. The expression (4.42) is thus equivalent with

$$H = -\frac{1}{2}\Delta + \mathrm{i}\,\frac{q}{c}\,(\nabla\cdot\vec{A}(\mathbf{x}) + \vec{A}(\mathbf{x})\cdot\nabla) + \frac{q^2}{c^2}\vec{A}(\mathbf{x})^2. \tag{4.46}$$

The Hamiltonian operator (4.42) thus corresponds to the choice of the symmetric classical expression $\mathbf{p}\cdot\vec{A}(\mathbf{x}) + \vec{A}(\mathbf{x})\cdot\mathbf{p}$. Unsymmetric operators $f(x)g(p)$ do not represent observables because they may lead to complex expectation values (see Exercise 4.4).

In principle, there is no problem with defining the Hamiltonian for a charged particle in a time-dependent electromagnetic field. By analogy, we obtain the time-dependent Hamiltonian

$$H(t) = \frac{1}{2}\Big(-\mathrm{i}\nabla - \frac{q}{c}\vec{A}(\mathbf{x},t)\Big)^2 + q\,V(\mathbf{x},t). \tag{4.47}$$

# 4.7. Gauge Fields

## 4.7.1. Nonuniqueness of the wave function

The description of quantum-mechanical states in terms of wave functions is not unique. This nonuniqueness is partly due to our interpretation.

Recall the discussion in Section 4.1.1: Wave functions have to be normalized before applying the interpretation rules, but a normalized wave function $\psi$ is still not unique. It can be multiplied with a phase factor,

$$\psi \longrightarrow \mathrm{e}^{\mathrm{i}\lambda}\psi \quad (\text{with some real number } \lambda),$$

without changing the interpretation. The phase factor $\mathrm{e}^{\mathrm{i}\lambda}$ drops out of all formulas related to the physical interpretation of the wave function (e.g., the position probability density, expectation values, etc.).

In our method of visualization the presence of a phase factor $\mathrm{e}^{\mathrm{i}\lambda}$ changes the color of the wave function. All colors of the colored plane (Color Plate 3) are rotated through an angle $\lambda$.

### 4.7.2. Nonuniqueness of the Schrödinger equation

Another ambiguity in the quantum-mechanical formalism stems from the nonuniqueness of the electromagnetic potentials. A *gauge transformation* of the potentials changes the classical Hamiltonian function and thus the Schrödinger equation without changing the physics of the system.

The simplest example of the so-called *gauge freedom* is given by the Schrödinger equation in a constant electric potential. A constant electric potential, say $qV(\mathbf{x}) = K$, describes a zero electric field $\vec{E}(\mathbf{x}) = \nabla V(\mathbf{x}) = 0$ (for all $\mathbf{x}$). Hence the motion of a particle in a constant electric potential should be physically indistinguishable from the free motion. If $\psi$ is a solution of the free Schrödinger equation

$$\mathrm{i}\,\frac{\partial}{\partial t}\,\psi(\mathbf{x},t) = -\frac{1}{2}\,\Delta\psi(\mathbf{x},t), \tag{4.48}$$

then the function

$$\phi(\mathbf{x},t) = \mathrm{e}^{-\mathrm{i}Kt}\psi(\mathbf{x},t) \tag{4.49}$$

is a solution of

$$\mathrm{i}\,\frac{\partial}{\partial t}\,\phi(\mathbf{x},t) = -\frac{1}{2}\,\Delta\phi(\mathbf{x},t) + K\,\phi(\mathbf{x},t). \tag{4.50}$$

The function $\phi$ obviously describes the same physical state as the original function $\psi$. From a physical point of view the two descriptions are completely equivalent.

The presence of a phase factor $\mathrm{e}^{-\mathrm{i}Kt}$ changes the phase velocity of the wave packet. Hence the phase velocity of a wave function is not gauge invariant. It cannot represent an observable quantity because it depends on the chosen description.

CD 3.19 shows the motion of a Gaussian wave packet in a constant potential $V(x) = K$ and compares the phase velocities for different values of $K$.

### 4.7.3. Gauge transformations of magnetic fields

A more interesting example of the gauge freedom occurs for a particle in a magnetic field. The choice of the vector potential for a magnetic field is by no means unique. For any differentiable function $g$ we have $\operatorname{curl}\nabla g = 0$. Hence, changing the vector potential $\vec{A}$ to $\vec{A}' = \vec{A} + \nabla g$ does not change the magnetic field. The two vector potentials $\vec{A}$ and $\vec{A}'$ describe the same physical situation. One says that the two vector potentials are related by a gauge transformation.

The freedom of choosing the vector potential can sometimes be used to simplify the mathematical description. It is always possible to choose the vector potential

$$\vec{A}(\mathbf{x}) = \int_0^1 s\vec{B}(\mathbf{x}s) \times \mathbf{x}\, ds \qquad \text{(in three dimensions).} \tag{4.51}$$

(Here "$\times$" denotes the vector product.) This vector potential $\vec{A}$ is said to be in the *Poincaré gauge*. The Poincaré gauge is characterized by the property

$$\vec{A}(\mathbf{x}) \cdot \mathbf{x} = 0. \tag{4.52}$$

Another possible choice of gauge is the *Coulomb gauge* characterized by

$$\operatorname{div} \vec{A}(x) = 0. \tag{4.53}$$

If the vector potential $\vec{A}$ has this property, then the calculation (4.45) shows that the order of the operators $\mathbf{p}$ and $\vec{A}$ obviously does not matter:

$$\mathbf{p} \cdot \vec{A}(\mathbf{x}) = \vec{A}(\mathbf{x}) \cdot \mathbf{p} \quad \text{if } \operatorname{div} \vec{A} = 0. \tag{4.54}$$

### 4.7.4. Gauge transformation of the Schrödinger equation

The quantum-mechanical description of particles in a magnetic field involves the vector potential and has to be changed after a gauge transformation. For example, the Hamiltonian

$$H = \frac{1}{2}\big(-\mathrm{i}\nabla - (\nabla g)\big)^2, \tag{4.55}$$

also describes the free motion because the vector potential $(\nabla g)$ corresponds to a field strength zero. Let us assume that $\psi$ is a solution of the free Schrödinger equation (4.48). Consider the function

$$\phi(\mathbf{x}, t) = \mathrm{e}^{\mathrm{i}g(\mathbf{x})}\psi(\mathbf{x}, t). \tag{4.56}$$

It is easy to see that

$$\big(-\mathrm{i}\nabla - (\nabla g)\big)\phi = \mathrm{e}^{\mathrm{i}g}(-\mathrm{i}\nabla\psi), \qquad \big(-\mathrm{i}\nabla - (\nabla g)\big)^2\phi = -\mathrm{e}^{\mathrm{i}g}(\Delta\psi), \tag{4.57}$$

and therefore,

$$\mathrm{i}\,\frac{\partial}{\partial t}\,\phi = \mathrm{i}\,\mathrm{e}^{\mathrm{i}g}\,\frac{\partial}{\partial t}\,\psi = -\frac{1}{2}\,\mathrm{e}^{\mathrm{i}g}\,\Delta\psi = \frac{1}{2}\big(-\mathrm{i}\nabla - (\nabla g)\big)^2\phi. \tag{4.58}$$

Hence $\phi$ is a solution of the Schrödinger equation with the pure gauge field $\nabla g$. Both $\psi$ and $\phi$ describe the same physical state of the free particle, but they are solutions of different equations.

**Gauge freedom:**

If a vector potential $\vec{A}(\mathbf{x})$ is replaced by

$$\vec{A}'(\mathbf{x}) = \vec{A}(\mathbf{x}) + \nabla g(\mathbf{x}), \tag{4.59}$$

then every wave function $\psi$ has to be multiplied by a phase factor,

$$\psi(\mathbf{x}, t) \longrightarrow \phi(\mathbf{x}, t) = \mathrm{e}^{\mathrm{i}g(\mathbf{x})}\psi(\mathbf{x}, t). \tag{4.60}$$

If $\psi$ is a solution of the Schrödinger equation with vector potential $\vec{A}$, then $\phi$ is the corresponding solution (describing the same physical state) of the Schrödinger equation with vector potential $\vec{A}'$.

We see that the complex-valued wave function does not describe the quantum-mechanical process in a unique way. In particular the effects visible in the complex phase can only be interpreted correctly if a particular gauge has been fixed. For example, we will always assume that in the force-free case the electromagnetic potentials are given by $V(\mathbf{x}) = 0$ and $A(\mathbf{x}) = 0$.

CD 3.20 shows the motion of Gaussian wave packets in pure gauge fields in one and two dimensions. The effect of the gauge potential is only visible in the phase of the wave function.

## 4.8. Projection Operators

### 4.8.1. An example

Let $I = (a, b)$ be an interval of real numbers. Consider the linear operator $P_I$ which multiplies wave functions by the *characteristic function* of I,

$$\chi_I(x) = \begin{cases} 1, & x \in I, \\ 0, & x \notin I. \end{cases} \tag{4.61}$$

It can be defined everywhere in the Hilbert space of square-integrable functions,

$$P_I : \psi \to \chi_I \psi \qquad \text{all } \psi \in L^2(\mathbb{R}^3), \tag{4.62}$$

and since $P_I$ is multiplication by a real-valued function, it is a symmetric operator (in the sense of Definition 4.1). It is easy to see that the operator $P_I$ satisfies

$$P_I^2 = P_I, \tag{4.63}$$

that is, $P_I$ is *idempotent*. The expectation value

$$\langle P_I \rangle_\psi = \int_{-\infty}^{\infty} \overline{\psi(x)}\, \chi_I(x)\, \psi(x)\, dx = \int_a^b |\psi(x)|^2\, dx$$

gives just the probability of finding the particle in the interval $I$.

What physical observable is represented by $P_I$? Let us have a look at the expectation value. In order to determine this quantity experimentally, one has to perform many elementary experiments (Section 3.4.4). $\langle P_I \rangle_\psi$ is (approximately) given by the fraction of events where the particle is found in $I$. The point is that in every elementary experiment one only has to determine whether the particle is in $I$ and to record the answer "yes" (particle is in $I$) or "no" (particle is elsewhere). Obviously, the observable quantity that is measured is the property of being localized in $I$.

DEFINITION 4.2. An everywhere defined, symmetric, and idempotent operator is called a *projection operator*. Projection operators are observables describing properties. A *property* is a physical quantity whose measurement gives either the result "yes" or "no" (thus, respectively, "true" or "false", "1" or "0".)

In Section 3.8 we considered the decomposition of a one-dimensional wave packet according to the direction of motion. The mapping from $\psi$ onto the part $\psi_+$ is a projection operator (likewise $\psi \to \psi_-$). Hence the movie CD 3.10 visualizes the action of a projection operator on the wave packet.

### 4.8.2. Measurements

A projection operator can be used to describe the effect of a (preparatory) measurement on a physical state. Let us discuss this for position measurements.

As before, we denote by $P_I$ the operator of multiplication by the characteristic function of the region $I \subset \mathbb{R}$. This projection operator corresponds to the observable property of being localized in $I$. A measurement apparatus for this property is a device that gives a signal whenever it detects a particle within $I \subset \mathbb{R}$. It is possible to detect the particle in $I$ whenever the wave function $\psi$ has the property that $\langle P_I \rangle_\psi \neq 0$.

If the measurement is nondestructive, it seems reasonable to say that right after the measurement the particle is in a state that is in harmony with the outcome of the measurement. Hence the measurement device performs a state preparation in the sense discussed in Sect 3.4.4. The particles that leave the measuring device with result "yes" have the property of being localized within $I$. Hence their state is described by a wave function $\psi_1$, for which we must have

$$\langle P_I \rangle_{\psi_1} = 1. \tag{4.64}$$

This implies (since $\psi_1$ is normalized)

$$\int_{\mathbb{R}} |\psi_1(x)|^2 \, dx = \int_I |\psi_1(x)|^2 \, dx, \tag{4.65}$$

and hence $\psi_1$ vanishes outside $I$. The wave function $\psi_1$ is thus in the range of the projection $P_I$, i.e, $\psi_1 = P_I \psi_1$.

Usually one goes even further by assuming the following projection postulate, which describes the wave function $\psi_1$ more precisely. This postulate assumes that a measurement detecting the particle in $I$ does not change the part of the wave function inside $I$.

**Projection postulate for position measurements:**

An ideal measurement device for the property "the particle is in $I$" acts as a black box with one input and two outputs: If the result is "yes", the measurement changes the state from $\psi$ to $cP_I\psi$, if the result is "no", the state after the measurement is described by $c'(1 - P_I)\psi$. (Here $c$ and $c'$ are appropriate normalization constants.)

EXERCISE 4.5. *Show that for any projection operator $P$, the operator $1 - P$ is again a projection operator. For the projection $P_I$ defined above, show that*

$$1 - P_I = P_{\mathbb{R}\setminus I}. \tag{4.66}$$

*Hence $1 - P_I$ describes the property "the particle is not in $I$."*

**Collapse of the wave packet:** While the Schrödinger equation describes a continuous time-evolution, the projection postulate introduces a radically different method to change the state of a physical system. The observation of a property prepares individual systems into a state that is in harmony with the outcome ("yes" or "no") of the measurement. This state preparation procedure is probabilistic (one cannot say when it will happen) and discontinuous—it influences the wave function of the particle instantaneously (collapse). This is not only typical for position measurements, but for all observations in quantum mechanics, as you can see from our discussion of Schrödinger's cat in Section 3.7.

You should be aware that during the observation the particle is by no means an isolated system. In fact, it undergoes a very complex interaction with a (ususally macroscopic) measurement device. The description of the state preparation by projection is thus a simple model for a complicated process that takes place in a large combined quantum system. The projection postulate just describes the total change of a partial system after its separation from the large system. We obtain just one bit of information about the change of the state of the measurement device. A very simple model indeed. At this stage we cannot go into further details because we don't yet have the background to describe composite quantum systems.

### 4.8.3. The general projection postulate

Let $A$ be a physical observable. Because a physical observable can only have real values, we can ask whether in a given state the observable has a value in an interval $I \subset \mathbb{R}$, that is, whether the property "$A$ has a value in $I$" is true. The operator representing this property is a projection operator denoted by $P_I(A)$. A measuring apparatus for $P_I(A)$ is a detector which gives a signal whenever it finds the value of $A$ in $I$, and another signal if it finds that the value of $A$ is not in $I$. The general projection postulate is often formulated as follows.

**The general projection postulate**:

For any observable $A$ we can define a projection operator $P_I(A)$, which measures the property "the value of $A$ is in a subset $I$ of the real numbers." If the state of the particle entering the measurement device is $\psi$ (with $\|\psi\| = 1$), then the probability for the result "yes" is

$$\|P_I(A)\,\psi\|^2 = \langle \psi, P_I(A)\,\psi \rangle \tag{4.67}$$

and the probability for the result "no" is $\|(1 - P_I(A))\,\psi\|^2$. Right after the measurement, the particle is in the state $cP_I(A)\,\psi$ if the result has been "yes," and in the state $c'(1 - P_I(A))\,\psi$ if the result has been "no" (with suitable normalization constants $c$ and $c'$).

EXERCISE 4.6. *Define the projection operator which measures the property: "The particle has a momentum $p$ in the interval $I \subset \mathbb{R}$." Do the same for the kinetic energy.*

You have learned that measurements can serve to prepare particles with well defined properties for further experiments. The measurement of a property (that is, the application of a projection operator $P$) filters the particles in two output channels. If we block the output channel for which the result is "no," all particles emerging from the measurement device are in the subspace $\operatorname{Ran} P$ (the range of the projection operator).

Usually, one does not know much about the state of the particles prior to this preparatory measurement. After the measurement the particles still don't have a well defined state, because all we can say is that their wave function is in the range of $P$. But eventually, it is possible to find a property for that $\operatorname{Ran} P$ is one-dimensional. In this case, the state preparation procedure leads to a *pure state* that is described by a wavefunction $\psi$ (unique up to a multiplicative constant). I am going to discuss the corresponding projection operator in the next section.

## 4.9. Transition Probability

For any normalized vector $\psi$ in a Hilbert space you can verify that

$$P_\psi = \langle \psi, \cdot \rangle \, \psi \tag{4.68}$$

is a projection operator. In this notation, the dot is a placeholder for the vector to which $P_\psi$ is applied. The action of $P_\psi$ on a vector $\phi$ is thus obtained by calculating the scalar product of $\psi$ with $\phi$ and by forming the vector $\langle \psi, \phi \rangle \, \psi$. If the states $\psi$ and $\phi$ are normalized, the projection operator $P_\psi$ just calculates the component of $\phi$ in the direction of $\psi$. This is done in the same way you would calculate the component of a vector $\mathbf{x}$ in the direction of a given vector $\mathbf{y}$ in the three-dimensional space $\mathbb{R}^3$. Obviously, the range of $P_\psi$ is the one-dimensional subspace generated by (scalar multiples of) $\psi$.

Quantum-mechanically, the projection operator $P_\psi$ describes the property of being in the state $\psi$. A preparatory measurement of this property in a state $\phi$ is described by the application of the projection operator. This changes the wave function from $\phi$ to $c\psi$, where $c$ is the scalar product between $\psi$ and $\phi$.

We may calculate the expectation value of $P_\psi$ in the state $\phi$,

$$\begin{aligned} \langle P_\psi \rangle_\phi &= \langle \phi, P_\psi \, \phi \rangle \\ &= \langle \phi, \langle \psi, \phi \rangle \psi \rangle \\ &= \langle \psi, \phi \rangle \, \langle \phi, \psi \rangle = |\langle \psi, \phi \rangle|^2. \end{aligned} \tag{4.69}$$

This quantity describes the probability that a state $\phi$ is in the state $\psi$. It is called the *transition probability* from $\phi$ to $\psi$. According to the statistical interpretation, the transition probability describes the fraction of elementary experiments in which a system is prepared in the state $\phi$ and detected in the state $\psi$. The transition probability from $\psi$ to $\psi$ is $\|\psi\|^2 = 1$.

**Transition probability**:

The transition probability between two states $\phi$ and $\psi$ is given by

$$p_{\phi \to \psi} = |\langle \psi, \phi \rangle|^2. \tag{4.70}$$

Here it is assumed that $\psi$ and $\phi$ are normalized to 1. The transition probability from $\phi$ to $\psi$ is the same as the transition probability from $\psi$ to $\phi$.

If the vectors representing the physical states are not normalized, the expression for the transition probability between the two states has to be replaced by

$$p_{\phi \to \psi} = \frac{|\langle \psi, \phi \rangle|^2}{\|\psi\|^2 \, \|\phi\|^2}. \tag{4.71}$$

At first sight it seems paradoxical that a state $\psi$ should have a certain probability to be in another state $\phi$. Indeed, this can only happen in quantum mechanics. Classically, a particle which is in one state cannot be in any other state. But in quantum mechanics, only orthogonal states are different enough to exclude each other.

If two states are orthogonal, $\langle \psi, \phi \rangle = 0$, then the transition probability between $\phi$ and $\psi$ is zero.

*Chapter 5*

# Boundary Conditions

**Chapter summary**: In this chapter we describe the elastic reflection of particles in the presence of impenetrable obstacles. Instead of describing walls and obstacles by electrostatic forces (which would have to be infinitely strong and concentrated on the surface of the obstacle), it is more appropriate to interpret an impenetrable barrier as a boundary condition. Starting with the simplest example—a solid wall in one dimension—we discuss Dirichlet boundary conditions, which exert a strongly repulsive influence, and Neumann boundary conditions, which are more neutral toward the particle.

A very interesting problem is the description of particles in a box. The surrounding walls confine the particle for all times to a finite region. Thus, the behavior of a particle in a box is quite different from the free motion. Instead of propagating wave packets we find an orthonormal basis of stationary states, which can be described as eigenvectors of the Hamiltonian operator. As a consequence, the quantum-mechanical energy of a particle in a box cannot have arbitrary values. The only possible energies are given by a discrete set of eigenvalues of the Hamiltonian operator—a fact that cannot be understood by classical mechanics. In particular, the lowest possible energy (the energy of the ground state) is greater than zero, that is, a confined particle is never really at rest. By forming superpositions of eigenstates, we can describe the motion of arbitrary initial states. The motion is always periodic in time and can be very complicated, as illustrated by the mathematically interesting example showing the unit function in a Dirichlet box.

The accompanying CD-ROM contains many movies of wave packets hitting walls and obstacles in various geometric configurations. Of particular interest is the double slit—a wall with two holes through which the particle can reach the other side. Behind the wall, the wave function shows a nice interference pattern which vanishes as soon as one of the slits is closed. More generally, one can say that the interference vanishes as soon as one attempts to determine through which of the holes the particle actually goes. We use this behavior to illustrate once more how quantum mechanics contradicts the classical picture of localized particles.

## 5.1. Impenetrable Barrier

### 5.1.1. Dirichlet boundary conditions

A very simple way to influence the motion of free particles in a way that cannot be described by electromagnetic potentials (Section 4.6) is to impose boundary conditions on the wave function. As a first example we describe in this section the behavior of a quantum-mechanical particle hitting a solid wall.

We can assume that a solid wall is an impenetrable barrier to the particle. In classical terms this corresponds to an infinitely strong repulsive force acting at a single point. In quantum mechanics this can be described by requiring that the wave function of the particle is identically zero behind the wall (and, assuming continuity, on the wall). Hence an impenetrable barrier can be interpreted as a boundary condition.

Consider the following example in one dimension: Let there be a solid wall at $x = 0$ and assume that the particle is on the left side ($x < 0$). We use the free Schrödinger equation to describe the motion of the particle in the domain $x < 0$ and impose the following boundary condition (*Dirichlet boundary condition*) at $x = 0$,

$$\psi(0,t) = 0, \qquad \text{for all } t. \tag{5.1}$$

The Hilbert space of this system is $L^2((-\infty, 0])$, that is, the set of complex-valued functions that are square-integrable on the interval $-\infty < x \leq 0$. The Hamiltonian of the system is a linear operator that acts like the free Schrödinger operator

$$H = -\frac{1}{2}\frac{d^2}{dx^2}, \tag{5.2}$$

but which is defined on a domain restricted by the boundary condition. The domain of $H$ consists of all square-integrable functions, for which $k^2\hat{\psi}(k)$ is also square-integrable, and which satisfy the boundary condition $\psi(0) = 0$.

CD 4.1 shows the time evolution of various Gaussian wave packets near a Dirichlet boundary condition. CD 4.2 shows the behavior of these wave packets in momentum space. The motion is best described as an elastic reflection at a wall. This process turns the average momentum of the wave packet into its negative. During the collision with the wall the shape of the wave packet is distorted by complex oscillations, but soon after the impact the Gaussian shape is restored.

### 5.1.2. Plane waves

Any solution of the free equation can be described as a superposition of plane waves. But none of the plane waves satisfies the boundary condition at $x = 0$! However, it is easy to find a linear combination of two plane waves that does satisfy the boundary condition:

$$v_k(x,t) = \begin{cases} \mathrm{e}^{ikx-\mathrm{i}k^2t/2} - \mathrm{e}^{-ikx-\mathrm{i}k^2t/2} = 2\mathrm{i}\,\sin(kx)\mathrm{e}^{-\mathrm{i}k^2t/2}, & x < 0, \\ 0, & x \geq 0. \end{cases} \tag{5.3}$$

Obviously, $v_k$ is (for all $k \in \mathbb{R}$) a solution of the free Schrödinger equation that satisfies the boundary condition $v_k(0,t) = 0$.

The solution $v_k$ is a linear combination of the free plane waves $u_k(x,t)$ and $-u_{-k}(x,t)$; see Eq. (3.21). If $k > 0$, then $u_k(x,t)$ moves to the right (because it has positive momentum = velocity), and $-u_{-k}(x,t)$ moves to the left. We may also think of an incoming wave which moves toward the wall and a reflected wave which moves away from it.

### 5.1.3. Wave packets

The function $v_k$ is not square-integrable and hence has no probability interpretation in quantum mechanics. In the same way, as we used the solution $u_k(x,t)$ of the free equation to form wave packets in the Hilbert space of the system, we can now form superpositions of the $v_k$'s in order to obtain square-integrable solutions of the equation with boundary condition. We write this superposition in the form

$$\psi(x,t) = \frac{1}{\sqrt{2\pi}} \int_{-\infty}^{\infty} v_k(x,t)\,\tilde{\psi}(k)\,dk, \tag{5.4}$$

where $\tilde{\psi}$ is a suitable complex-valued function (we assume that $\tilde{\psi}$ is integrable and square-integrable). Defining

$$\phi(x,t) = \frac{1}{\sqrt{2\pi}} \int_{-\infty}^{\infty} \mathrm{e}^{ikx}\,\mathrm{e}^{-\mathrm{i}k^2t/2}\,\tilde{\psi}(k)\,dk, \tag{5.5}$$

we find that

$$\psi(x,t) = \begin{cases} \phi(x,t) - \phi(-x,t), & x < 0, \\ 0, & x \geq 0. \end{cases} \tag{5.6}$$

This method of constructing a solution $\psi$ of the Schrödinger equation with boundary condition from a solution $\phi$ of the free Schrödinger equation will be referred to as *the method of mirrors*.

CD 4.3 explains the method of mirrors by showing the wave packet $\phi(x,t)$ together with the mirror function $-\phi(-x,t)$. The superposition fulfills the Dirichlet boundary condition at $x = 0$.

### 5.1.4. Reflection of a Gaussian wave packet

As an example, we consider the amplitude function

$$\tilde{\psi}(k) = N\, e^{ika}\, e^{-(k-p_0)^2/2}. \tag{5.7}$$

Let us assume that $a > 0$ is sufficiently large. Then

$$\phi(x,0) = N\, e^{-(x+a)^2/2}\, e^{ixp_0} \tag{5.8}$$

is a Gaussian wave packet that is well localized around $x = -a$ in the physical region (in the half-interval $(-\infty, 0]$). The contribution of the summand $-\phi(-x,0)$ to the wave function $\psi(x,0)$ in that region is very small and can be neglected.

Let us choose the average momentum $p_0$ to be positive and large enough, so that mostly positive momenta are used to build the wave packet. Hence the initial function $\psi(x,0) \approx \phi(x+a)$ will start moving to the right (toward the wall) with average momentum $p_0$. The function

$$\phi(-x,0) = N\, e^{-(x-a)^2/2}\, e^{-ixp_0} \tag{5.9}$$

is a Gaussian with center at $x = a$ behind the wall. It will move to the left with average momentum $-p_0$ and will finally enter the physical region. The rather complicated shape of the wave function that emerges during the collision with the wall thus has a very simple explanation. It is just the interference pattern of a Gaussian function and a mirror Gaussian. Far in the future, the part of the function $\phi(x,t)$ in the interval $(\infty, 0]$ becomes more and more negligible and only the summand $-\phi(-x,t)$ that moves to the left will contribute to $\psi(x,t)$. The motion of $\psi(x,t)$ in the physical region $(-\infty, 0]$ obviously describes the behavior of a particle that hits the wall at $x = 0$ and gets elastically reflected.

## 5.2. Other Boundary Conditions

Another type of boundary condition is the *Neumann boundary condition*

$$\psi'(0) = \frac{d}{dx}\psi(x)\Big|_{x=0} = 0. \tag{5.10}$$

Here the wave function $\psi$ itself is not forced to vanish at the wall. Instead, it is required that the derivative $\psi'$ of the wave function be zero at $x = 0$.

For any solution $\phi(x,t)$ of the free Schrödinger equation we can obtain a solution that satisfies a Neumann boundary condition at $x = 0$,

$$\psi(x,t) = \begin{cases} \phi(x,t) + \phi(-x,t), & \text{for } x \le 0, \\ 0, & \text{for } x > 0. \end{cases} \tag{5.11}$$

CD 4.4 shows the behavior of Gaussian wave packets near a barrier represented by a Neumann boundary condition. The Neumann wall is more neutral in comparison to the strongly repulsive Dirichlet wall. The probability of finding the particle very close to the wall is much larger than in the case of a Dirichlet boundary condition. Nevertheless, the wall is impenetrable in both cases and causes a total reflection of the particle at $x = 0$.

Dirichlet and Neumann boundary conditions have natural generalizations for higher dimensions. Let $S$ be a sufficiently smooth surface in three dimensions (or curve in two dimensions). A Dirichlet boundary condition consists in the specification of certain boundary values (usually zero) of the wave function $\psi(x,t)$ for $x \in S$. A Neumann boundary condition prescribes the values of the normal derivative $\nabla\psi(x) \cdot \vec{n}(x)$ at $S$. Here $\vec{n}(x)$ is the unit vector normal to the surface $S$ at the point $x \in S$. Unfortunately, the theory of higher-dimensional boundary value problems is beyond the scope of this text.

Beginning with CD 4.14 we present several movies showing the reflection of wave packets at walls and obstacles in two dimensions. The walls and obstacles are realized by Dirichlet boundary conditions on certain curves. The scattering at obstacles is described by the free Schrödinger equation in the region outside a closed curve with a boundary condition on the curve.

## 5.3. Particle in a Box

### 5.3.1. Gaussian wave packet between two walls

Let us put walls on both sides of a localized initial state and see what happens.

CD 4.6 shows the one-dimensional motion of a Gaussian wave packet between two walls (Dirichlet boundary conditions). After a few reflections the wave function occupies the whole available space and shows complicated oscillations. This behavior can be investigated with several methods. CD 4.5 illustrates the method of mirrors for this situation. This will be described next.

We assume that the walls are situated at $x = 0$ and $x = L$ (with $L > 0$). In the Schrödinger equation the presence of the walls is described by the Dirichlet boundary conditions

$$\psi(0,t) = \psi(L,t) = 0, \quad \text{for all } t. \tag{5.12}$$

We can try to explain the motion of a Gaussian wave packet between two walls using the method of mirrors as described previously (Section 5.1.3). Let us start with a Gaussian function that is well localized between the walls at an average initial position $x_0$ near $L/2$. In order to satisfy the boundary condition at $x = L$, we have to assume a mirror Gaussian centered at $2L - x_0$. Another mirror Gaussian at $-x_0$ will take care of the boundary condition at 0. The superposition of the original Gaussian with its mirror at $2L - x_0$ would now destroy the boundary condition at 0, unless we introduce a new mirror Gaussian at $-2L + x_0$. Likewise, we need a mirror Gaussian at $2L + x_0$ that cancels the value at $L$ of the Gaussian located at $-x_0$. By the same argument every new mirror Gaussian again needs a mirror Gaussian with respect to the opposite wall. Hence we end up with an infinite number of mirror wave packets (imagine the situation of a person standing between two parallel optical mirrors). The physical state is described as the infinite sum of the original Gaussian and all its mirror images. Of course, only the part of the wave function within the interval $[0, L]$ has physical relevance. Because of the exponential decay of each Gaussian function, the contribution of the infinite sum of mirror Gaussians in the original interval is very small.

The time evolution is now described easily by the time evolution of the infinite sum of Gaussians. The first reflections perhaps can be described with sufficient accuracy as the interference pattern arising from the superposition of a Gaussian with its respective mirror image. But as soon as the wave packet has spread over a region larger than the interval $[0, L]$, more and more Gaussians will contribute significantly to the interference pattern, which will hence become more and more complicated.

### 5.3.2. Method of mirrors

In a more formal way the motion of a particle in a box can be described as follows. Take an initial function $\phi(x)$, $0 \le x \le L$, which satisfies the Dirichlet boundary conditions at 0 and $L$. Let $\varphi(x,t)$ be the solution of the free Schrödinger equation on the whole line with initial value

$$\varphi(x,0) = \begin{cases} \phi(x), & 0, \le x \le L, \\ 0, & \text{elsewhere}. \end{cases} \tag{5.13}$$

**Step 1**: Define an odd function $\psi$ on the interval $[-L, L]$ by setting

$$\psi(x) = \begin{cases} -\phi(-x), & \text{for } -L \le x \le 0, \\ \phi(x), & \text{for } 0 \le x \le L. \end{cases} \tag{5.14}$$

**Step 2**: Extend this definition to the whole real axis by periodic continuation:

$$\psi(2jL + x) = \psi(x), \qquad j = \pm 1, \pm 2, \pm 3, \ldots . \tag{5.15}$$

In terms of the function $\varphi$ we can write $\psi$ as

$$\psi(x) = \sum_{j=\pm1,\pm2,\dots} \bigl(\varphi(2jL + x, 0) - \varphi(2jL - x, 0)\bigr). \tag{5.16}$$

**Step 3**: The time evolution of the initial state $\phi$ according to the Schrödinger equation with boundary conditions is now given by

$$\psi(x,t) = \sum_{j=\pm1,\pm2,\dots} \bigl(\varphi(2jL + x, t) - \varphi(2jL - x, t)\bigr), \quad \text{for } x \in [0, L]. \tag{5.17}$$

Formally, $\psi(x,t)$ is a solution of the free Schrödinger equation that satisfies the boundary conditions at 0 and $L$ for all $t$.

An example of a solution obtained by the method of mirrors is shown in Color Plate 5. This method is useful in particular for small times, as long as only a few mirror waves contribute to the solution inside the physical region $0 \le x \le L$.

### 5.3.3. A special set of solutions

Now we try something different. We want to satisfy the boundary conditions (5.12) by superpositions of plane waves. As in Eq. (5.3), the following superposition of two plane waves (with momenta $\pm k$),

$$\sin(kx)\exp\Bigl(-\mathrm{i}\frac{k^2}{2}\,t\Bigr), \tag{5.18}$$

automatically satisfies the boundary condition at 0, but it will not satisfy the boundary condition at $L$ for all $k$. At least there are *some* $k$ for which the boundary conditions at 0 *and* at $L$ can be fulfilled. This is the case for the wave numbers

$$k_n^{(L)} = n\,\frac{\pi}{L}, \quad n = 1, 2, \dots, \tag{5.19}$$

for which $\sin(k_n^{(L)} x) = 0$ not only at $x = 0$ but also at $x = L$. Using the function (5.18) with $k = k_n^{(L)}$ we find an exact solution of the Schrödinger equation for a particle in a box:

$$\psi_n^{(L)}(x,t) = \begin{cases} \sqrt{\frac{2}{L}}\,\sin(k_n^{(L)}x)\,\exp(-\mathrm{i}E_n^{(L)}t), & 0 \le x \le L, \\ 0, & x \le 0, \quad \text{or} \quad x \ge L, \end{cases} \tag{5.20}$$

where

$$E_n^{(L)} = \frac{1}{2}\,(k_n^{(L)})^2 = \frac{\pi^2}{2L^2}\,n^2, \quad n = 1, 2, \dots. \tag{5.21}$$

The functions $\psi_n^{(L)}$, $n = 1, 2, \dots$, form an infinite set of solutions. Moreover, any linear combination of these functions is again a solution because the Schrödinger equation (together with the boundary conditions) is linear.

In our investigation of the free Schrödinger equation we had to form wave packets by integrating the plane waves over a continuum of $k$-values. Only by this continuous superposition were we able to obtain square-integrable wave functions, which could be interpreted as states of a single quantum particle. For the particle in a box, only a discrete number of $k$-values is at our disposal and it is impossible to form continuous superpositions. Fortunately, we don't have to, because all the functions $\psi_n^{(L)}$ are already square-integrable.

Because the particle can only exist between the walls at 0 and $L$, the behavior of the wave function in the region outside the box is not of interest. Hence we will simply ignore this part of the solution (which is zero anyway) and consider only functions defined on the interval $[0, L]$ and square-integrable. This amounts to choosing the Hilbert space $L^2([0, L])$ as the state space for this system. The norm of a wave function in this Hilbert space is

$$\|\psi\| = \Bigl(\int_0^L |\psi(x)|^2 \, dx\Bigr)^{1/2}. \tag{5.22}$$

The factor $(2/L)^{1/2}$ in the definition of $\psi_n^{(L)}$ makes this norm equal to 1,

$$\|\psi_n^{(L)}(\cdot, t)\| = 1, \quad \text{for all } n \text{ and all } t. \tag{5.23}$$

This means that the particle is inside the box at any time with certainty (with probability 1).

The initial functions

$$\psi_n^{(L)}(x) \equiv \psi_n^{(L)}(x, 0) = \sqrt{\frac{2}{L}} \sin(k_n^{(L)} x), \quad x \in [0, L], \tag{5.24}$$

are a set of so-called *eigenfunctions* of the Hamiltonian $H$ for the particle in the box. Eigenfunctions are of central importance to quantum mechanics. So it is worthwhile explaining this concept in some detail. This will be done in the next section.

## 5.4. Eigenvalues and Eigenfunctions

### 5.4.1. Eigenvectors of linear operators

The following definition is perhaps familiar from linear algebra, but there is a difference. The linear operators in finite-dimensional vector spaces are usually defined everywhere. In function spaces, however, the action of linear operators often cannot be defined for every vector in the vector space. One has to restrict its domain of definition to a suitable subspace.

DEFINITION 5.1. Let $A$ be a linear operator, defined on a domain $\mathfrak{D}(A)$ which is a linear subspace of a vector space $\mathfrak{H}$. A vector $\phi \in \mathfrak{D}(A)$, $\phi \neq 0$, is called an *eigenvector* if there exists a (possibly complex) number $\lambda$ such that

$$A\phi = \lambda\phi. \tag{5.25}$$

The number $\lambda$ is called *eigenvalue* of $A$. If the underlying vector space is a function space, then the eigenvectors are also called *eigenfunctions* of $A$.

Assume that the linear operator $A$ represents an observable of a physical system. If the state of the system is given by an eigenvector $\phi$ of $A$ belonging to an eigenvalue $\lambda \in \mathbb{R}$, then the expectation value of $A$ is equal to $\lambda$ and the uncertainty is zero.

$$\langle A \rangle_\phi = \lambda, \qquad \Delta_\phi A = 0, \qquad \text{if and only if } A\phi = \lambda\phi. \tag{5.26}$$

We leave the proof of this result as an exercise to the reader. According to the statistical interpretation, this means that whenever the state of the system is an eigenvector of $A$, a measurement of $A$ yields the eigenvalue $\lambda$ with certainty. We say, the observable $A$ has the value $\lambda$.

> **Value of an observable in an eigenstate:**
>
> If the state $\phi$ of a physical system is described by an eigenvector of an observable $A$,
>
> $$A\phi = \lambda\phi, \tag{5.27}$$
>
> then any measurement of $A$ gives the value $\lambda$.

$\boxed{\Psi}$ There are linear operators which have no eigenvalue at all. Indeed, this is the case for the position and momentum operators (and hence also for the kinetic energy of free particles). An eigenvector of the position must have uncertainty zero, which is not possible in the Hilbert space of square integrable wave functions. The Dirac delta function $\delta(x)$ is sort of a "generalized eigenfunction" of the operator $x$, but $\delta(x)$ is not in the domain of $x$, not even in the Hilbert space of the system. Similarly, the plane waves $u_k(x) = \exp(ikx)$ formally satisfy the eigenvalue equation $p\,u_k(x) = k\,u_k(x)$ with the momentum operator $p = -i d/dx$. But the plane waves are not square-integrable and do not belong to the Hilbert space either. We note that the generalized eigenvectors can be given a rigorous meaning in the framework of distribution theory.

It is possible that several eigenvectors belong to the same eigenvalue.

DEFINITION 5.2. The linear subspace spanned by all linear combinations of eigenvectors belonging to the same eigenvalue $\lambda$ is called the *eigenspace* of $\lambda$. The dimension of the eigenspace is called the *multiplicity* or *degeneracy* of that eigenvalue. An eigenvalue is called *degenerate* if its multiplicity is greater than one, otherwise it is called *nondegenerate*.

The multiplicity counts how many mutually orthogonal eigenvectors belong to some eigenvalue.

EXERCISE 5.1. *Let $P$ be a projection operator. What are its eigenvalues? What is their multiplicity?*

EXERCISE 5.2. *Show that the uncertainty $\Delta_\phi A$ of an observable $A$ is zero, whenever $\phi$ is an eigenvector of $A$.*

### 5.4.2. Eigenfunctions in a box

For the particle in a box we have to deal with a differential operator in the Hilbert space $L^2([0, L])$. The Hamiltonian $H_\mathrm{D}$ of this system is the operator of kinetic energy

$$H_\mathrm{D}\psi(x) = -\frac{1}{2}\frac{d^2}{dx^2}\psi(x). \tag{5.28}$$

Here we added the subscript D in order to indicate that $H_\mathrm{D}$ is defined on a subspace of functions that satisfy the Dirichlet boundary conditions (5.12). The functions in the domain of $H_\mathrm{D}$ also must be twice differentiable (in a generalized sense) such that the derivatives are again functions in $L^2([0, L])$,

$$\mathfrak{D}(H_\mathrm{D}) = \{\psi \in L^2 \mid \psi', \psi'' \in L^2, \psi(0) = \psi(L) = 0\}. \tag{5.29}$$

Thus, the functions $\psi_n^{(L)}$ defined in Eq. (5.24) are obviously in the domain of $H$. Moreover, for all $n$,

$$H_\mathrm{D}\psi_n^{(L)} = E_n^{(L)}\psi_n^{(L)}, \qquad E_n^{(L)} = \frac{\pi^2}{2L^2}n^2, \quad n = 1, 2, \ldots. \tag{5.30}$$

Hence the functions $\psi_n^{(L)}$ form a set of eigenfunctions of $H_\mathrm{D}$.

I would like to stress once again that the boundary conditions describe the domain of the operator $H_\mathrm{D}$. The eigenvalues of $H_\mathrm{D}$ depend very much on these domain properties. The choice of other boundary conditions would lead to a different set of eigenvalues (see, e.g., Section 5.6 below).

EXERCISE 5.3. *Find the eigenfunctions and eigenvalues of the operator $H_\mathrm{N}$ which is the operator of kinetic energy on a domain with Neumann boundary conditions in the Hilbert space $L^2([0, L])$.*

### 5.4.3. Time dependence of eigenfunctions

In quantum mechanics, the eigenfunctions of the Hamiltonian are of particular importance. If $H$ is the Hamiltonian of a quantum-mechanical system, then the eigenvalue equation $H\phi = E\phi$ is called the *stationary Schrödinger equation*. When you know a solution of the stationary Schrödinger equation (an eigenvector of $H$), you can obtain a corresponding solution of the time-dependent Schrödinger equation quite easily. The time-dependent equation is an initial-value problem of the form

$$\mathrm{i}\,\frac{d}{dt}\,\psi(t) = H\,\psi(t), \qquad \psi(0) = \phi. \tag{5.31}$$

If $H\phi = E\phi$, then you can verify by differentiation that

$$\psi(t) = \phi\,\mathrm{e}^{-\mathrm{i}Et} \tag{5.32}$$

is a solution of Eq. (5.31). Note that if $\psi(0) = \phi$ is an eigenstate of $H$, then $\psi(t)$ is an eigenstate of $H$ with the same eigenvalue for all times.

CD 4.9 shows the first eigenfunctions for a Dirichlet box. The time dependence of these states is only visible in the phase of the wave function. Also shown are oscillating states formed by superpositions of two eigenstates.

The time dependence of an eigenstate is rather trivial. It is completely described by the phase factor $\exp(-\mathrm{i}Et)$. The wave function at a time $t_1$ differs only by a phase factor from the wave function at another time $t_2$. Hence, according to our quantum-mechanical interpretation, the wave functions at different times represent the same physical state. When in an eigenstate, the system remains in that eigenstate forever. Therefore, the eigenfunctions of the Hamiltonian are also called *stationary states* of the physical system.

**Stationary states:**

If the state of a quantum-mechanical system is initially described by an eigenvector of the Hamiltonian $H$, then the system remains in that state forever.

All physically measurable quantities that can be calculated from the stationary states do not depend on time. For example, the position and momentum distributions remain stationary:

$$|\psi_n(x,t)|^2 = |\phi(x)|^2, \qquad |\hat{\psi}_n(p,t)|^2 = |\hat{\phi}_n(p)|^2. \tag{5.33}$$

An eigenstate of $H$ that sits in some region of space has to stay there forever. This behavior is very different from the propagation of free wave packets.

Therefore the eigenstates of the Hamiltonian (and their linear combinations) are often called *bound states.*

### 5.4.4. Eigenfunction expansion

Because of the linearity of the time-dependent Schrödinger equation (5.31) it is easy to obtain the solution for any initial state that is a linear combination of the eigenvectors.

**Solution of the time-dependent Schrödinger equation:**

Let $\psi_n$ be a set of eigenvectors belonging to eigenvalues $E_n$ of the Hamiltonian $H$. Let $\phi$ be an initial state that can be written as a linear combination of the $\psi_n$, that is,

$$\phi = \sum_n c_n \psi_n, \tag{5.34}$$

with suitable (complex) constants $c_n$. Then the unique solution of the initial-value problem

$$\mathrm{i}\frac{d}{dt}\psi(t) = H\,\psi(t), \qquad \psi(0) = \phi, \tag{5.35}$$

is given by

$$\psi(t) = \sum_n c_n \psi_n \mathrm{e}^{-\mathrm{i}E_n t}. \tag{5.36}$$

How large is the set of initial functions that are linear combinations of eigenfunctions? For the particle in a box we can give a very convenient answer: It turns out that the eigenfunctions $\psi_n^{(L)}$ form an orthonormal basis in the Hilbert space $L^2([0, L])$.

This can be seen as follows. Let us extend each function $f$ in $L^2([0, L])$ to an odd function on the larger interval $[-L, L]$ just by setting $f(-x) = -f(x)$. In that way the subspace of odd functions of $L^2([-L, L])$ can be identified with $L^2([0, L])$. According to our considerations in the chapter about Fourier analysis this subspace is spanned by the functions $\sin(k_n^{(L)} x)$.

Using the fact that the eigenfunctions $\psi_n^{(L)}$ form a basis of the Hilbert space, we can obtain a complete solution of the initial-value problem for the particle in a box. Let $\phi$ be an arbitrary initial state. First we expand the wave function $\phi$ in the orthonormal basis. This is just a Fourier expansion.

$$\phi = \sum_{n=1}^{\infty} c_n \psi_n^{(L)} \tag{5.37}$$

Here the coefficients of the linear combination are given by

$$c_n = \langle \psi_n^{(L)}, \phi \rangle = \int_0^L \psi_n^{(L)}(x)\, \phi(x)\, dx, \tag{5.38}$$

and the convergence of the sum is meant with respect to the topology of the Hilbert space $L^2([0, L])$ (*convergence in the mean*). The reader should remember that the formula for the coefficients $c_n$ is just a consequence of the fact that the $\psi_n^{(L)}$ form an orthonormal basis. Another consequence is

$$\|\phi\|^2 = \sum_{n=1}^{\infty} |c_n|^2. \tag{5.39}$$

The time evolution of the initial state $\phi$ is described by the formula

$$\phi = \sum_{n=1}^{\infty} c_n \exp(-\mathrm{i} E_n^{(L)} t)\, \psi_n^{(L)}. \tag{5.40}$$

$\boxed{\Psi}$ Note that we can describe the time evolution of any square-integrable initial state in this way. We can even consider initial states $\phi$ that do not satisfy the boundary conditions or that are not differentiable (i.e., wave functions that are not in the domain of the Hamiltonian $H$). For these initial states the solution $\psi(t)$ will not be differentiable at $t = 0$ and the differential equation is not satisfied in a pointwise sense. The next section presents an example.

## 5.5. Special Topic: Unit Function in a Dirichlet Box

As an example of the eigenfunction expansion we consider again the particle in a box, which is described by the Hamiltonian $H_\mathrm{D}$ with Dirichlet boundary conditions at $x = 0$ and $x = 1$. As we have seen we can define a time evolution for any initial function that is square-integrable. For $L = 1$ the eigenvalues of $H_\mathrm{D}$ are

$$E_n = \frac{n^2 \pi^2}{2}, \tag{5.41}$$

and the functions $\psi_n(x) = \sqrt{2}\, \sin(n\pi x)$ form an orthonormal basis of eigenfunctions. The Schrödinger equation gives the time evolution

$$\psi(x,t) = \sqrt{2} \sum_{n=0}^{\infty} c_n \sin(n\pi x) \exp\Bigl(-\mathrm{i}\, \frac{n^2\pi^2}{2}\, t\Bigr), \tag{5.42}$$

where the constants $c_n$ are the coefficients of the Fourier expansion of the initial function.

It is evident from Eq. (5.42) that $\psi(x,t)$ is periodic in time. For $t = 4/\pi$ we find

$$\exp\Bigl(-\mathrm{i}\,\frac{n^2\pi^2}{2}\,\frac{4}{\pi}\Bigr) = \exp(-\mathrm{i}\,n^2 2\pi) = 1, \quad \text{for } n = 1,2,3,\ldots \tag{5.43}$$

and hence $\psi(x, t + 4/\pi) = \psi(x,t)$.

**Periodicity in time**:

Every state of a particle in a Dirichlet box depends periodically on time with period $T = 4/\pi$.

EXERCISE 5.4. *What is the period for the motion of a particle in a box of length $L$?*

Here we want to stress that Eq. (5.42) can be used to define a time evolution for any square-integrable initial function—even for a function that does not fulfill the boundary conditions. We illustrate this observation by determining the time evolution of the initial function

$$\psi(x, t\!=\!0) = 1, \quad \text{for all } x \in [0,1]. \tag{5.44}$$

This initial state describes a particle with a uniform position probability density in the interval $[0,1]$. We are going to prove the following astonishing fact about $\psi(x,t)$ (see also Color Plate 16):

**Time evolution of the unit function**:

If $\psi(x,0) = 1$, then $\psi(x,t)$ is a step function at every time $t$ for which $t/T$ is a rational number.

CD 4.11 shows the spectacular behavior of the unit function subjected to Dirichlet boundary conditions. The wave function is not differentiable, neither with respect to $x$ nor with respect to $t$.

The first step toward a solution of the initial-value problem is to determine the Fourier expansion of the initial function $\psi = 1$. We first note that in Eq. (5.42) only the summands with $n$ odd can be nonzero because of the symmetry properties of the initial function. Indeed, according to Eq. (5.38) the Fourier coefficients are

$$c_n = \sqrt{2}\int_0^1 \sin(n\pi x)\,dx = \begin{cases} 2\sqrt{2}/(n\pi), & \text{for } n = 1,3,5,\ldots, \\ 0, & \text{for } n = 2,4,6,\ldots. \end{cases} \tag{5.45}$$

Hence we have the Fourier representation

$$\psi(x,0) = 1 = \frac{4}{\pi} \sum_{k=1}^{\infty} \frac{1}{2k-1} \sin((2k-1)\pi x). \tag{5.46}$$

Of course, the Fourier series does not converge to 1 at $x = 0$ and $x = 1$ (where every summand is zero). But all that is required is convergence in the topology of the Hilbert space.

Inserting the time dependence of the eigenfunctions in the Fourier expansion of the initial state we obtain

$$\psi(x,t) = \frac{4}{\pi} \sum_{k=1}^{\infty} \frac{1}{2k-1} \exp\Bigl(-\mathrm{i}\,\frac{(2k-1)^2\pi^2}{2}\,t\Bigr) \sin((2k-1)\pi x). \tag{5.47}$$

EXERCISE 5.5. *We can use this result to show that it is sufficient to know the time evolution of $\psi$ for $t$ in the interval $[0, 1/(4\pi)]$. Replace $t$ by $t + 1/(2\pi)$ in Eq. (5.47) in order to prove*

$$\psi\Bigl(x, t + \frac{1}{2\pi}\Bigr) = \exp\Bigl(-\mathrm{i}\,\frac{\pi}{4}\Bigr)\,\psi(x,t). \tag{5.48}$$

*Replace $t$ by $1/(2\pi) - t$ and deduce*

$$\psi\Bigl(x, \frac{1}{2\pi} - t\Bigr) = \exp\Bigl(-\mathrm{i}\,\frac{\pi}{4}\Bigr)\,\overline{\psi(x,t)}. \tag{5.49}$$

What is special about $t/T$ (or equivalently $t\pi$) being a rational number? Let us write

$$t = \frac{q}{p\pi} \tag{5.50}$$

in Eq. (5.47), where we assume that $q$ and $p$ are integers. We obtain

$$\psi(x,t) = \frac{4}{\pi} \sum_{k=1}^{\infty} \frac{a_k}{2k-1} \sin((2k-1)\pi x). \tag{5.51}$$

with

$$a_k = \exp\Bigl(-2\pi\mathrm{i}\,\frac{(2k-1)^2}{4}\,\frac{q}{p}\Bigr), \quad k = 1, 2, \ldots. \tag{5.52}$$

We can check that the sequence $(a_k)$ is periodic in $k$. If

$$\frac{(2(k+r)-1)^2}{4}\,\frac{q}{p} = \frac{(2k-1)^2}{4}\,\frac{q}{p} + n \tag{5.53}$$

holds for some integer $n$, then (using the $2\pi$-periodicity of the exponential function) the numbers $a_k$ and $a_{k+r}$ are equal. Equation (5.53) is fulfilled for all integers $q$ if we choose $r = p$. Hence we obtain

$$a_k = a_{k+p}, \quad \text{for all } k = 1, 2, \ldots. \tag{5.54}$$

Moreover, we have the relation

$$a_r = a_{p-r+1}, \quad \text{for } r = 1, 2, \dots, p. \tag{5.55}$$

Hence the solution at time $t = q/(p\pi)$ is uniquely determined by the phase factors

$$\left\{ a_r \;\middle|\; r = 1, 2, \dots, \left[\frac{p+1}{2}\right] \right\}. \tag{5.56}$$

Here $[x]$ is the greatest integer less than or equal to $x$.

We are going to prove the following fact:

THEOREM 5.1. *An expression of the form* (5.51) *with coefficients satisfying* (5.54) *and* (5.55) *is the Fourier series of a step function $t$. The step function is symmetric in the interval* $[0, 1]$ *and has at most $p$ steps of equal length, that is, $t$ is of the form*

$$t(x) = \sum_{r=1}^{\left[\frac{p+1}{2}\right]} c_r \left( \chi_{\left[\frac{r-1}{p}, \frac{r}{p}\right]}(x) + \chi_{\left[1-\frac{r}{p}, 1-\frac{r-1}{p}\right]}(x) \right), \tag{5.57}$$

*with suitable constants $c_r \in \mathbb{C}$. (Here $\chi_I(x)$ is the characteristic function of the interval $I$ ).*

PROOF. We first investigate the Fourier series of especially simple step functions. Given the integer $p$, we denote

$$p_m = \left[\frac{p+1}{2}\right]. \tag{5.58}$$

For $r = 1, 2, \dots, p_m$ we define

$$t_r(x) = \chi_{\left[\frac{r-1}{p}, 1-\frac{r-1}{p}\right]}(x) = \begin{cases} 1, & \text{for } \frac{r-1}{p} \le x \le 1 - \frac{r-1}{p}, \\ 0, & \text{elsewhere.} \end{cases} \tag{5.59}$$

Obviously, $t_1(x) = 1$ and every $t_r$ is a simple example of a step function of the type (5.57). Moreover, the $p_m$ functions $t_r$ are linearly independent and every symmetric step function of the form (5.57) can be written as a linear combination of the $t_r$. It is easy to calculate explicitly the Fourier series of $t_r$. Because each $t_r$ is symmetric in the interval $[0, 1]$, the Fourier series is of the form

$$t_r(x) = \sqrt{2} \sum_{k=1}^{\infty} c_{r,k} \sin((2k-1)\pi x). \tag{5.60}$$

The Fourier coefficients are given by

$$c_{r,k} = \sqrt{2} \int_{\frac{r-1}{p}}^{1-\frac{r-1}{p}} \sin((2k-1)\pi x)\, dx \tag{5.61}$$

$$= \frac{2\sqrt{2}}{\pi} \frac{1}{2k-1} \cos\Big((2k-1)\pi \frac{r-1}{p}\Big). \tag{5.62}$$

We finally obtain

$$t_r(x) = \frac{4}{\pi} \sum_{k=1}^{\infty} \frac{d_{r,k}}{2k-1} \sin((2k-1)\pi x), \tag{5.63}$$

where we introduced the abbreviation

$$d_{r,k} = \cos\Big((2k-1)\pi \frac{r-1}{p}\Big). \tag{5.64}$$

It is easy to see that the constants $d_{r,k}$ have the property

$$d_{r,k+p} = d_{r,k}, \qquad \text{for all } k = 1, 2, \ldots, \tag{5.65}$$

$$d_{r,j} = d_{r,p-j+1}, \qquad \text{for all } j = 1, 2, \ldots, p_m.. \tag{5.66}$$

Hence the step function $t_r$ can be represented in a unique way by the vector

$$d_r = (d_{r,1}, d_{r,2}, \ldots, d_{r,p_m}) \in \mathbb{R}^{p_m}. \tag{5.67}$$

There are $p_m$ different vectors $d_r$ corresponding to the $p_m$ different step functions $t_r$. By the uniqueness of the Fourier series, the linear independence of the $t_r$ carries over to the coefficient vectors $d_r$. Hence every vector in $a \in \mathbb{C}^{p_m}$ can be written as a (complex) linear combination of the vectors $d_r$,

$$a_k = \sum_{r=1}^{p_m} b_r d_{r,k}. \tag{5.68}$$

The coefficients $(b_1, \ldots, b_{p_m})$ can be determined by multiplying the row vector $a$ with the inverse of the matrix $D = (d_{r,k})$, that is, $b = aD^{-1}$. The matrix $D$ is invertible because its rows are the linearly independent vectors $d_r$. But the $p_m$-dimensional vector $a$ can be used to represent any sequence $(a_k)$ of coefficients with the properties (5.54) and (5.55). This shows that the Fourier series (5.51) is a linear combination of the functions $t_r$,

$$\frac{4}{\pi} \sum_{k=1}^{\infty} \frac{a_k}{2k-1} \sin((2k-1)\pi x) = \sum_{r=1}^{p_m} b_r t_r(x), \tag{5.69}$$

and hence it is a step function with the required properties. □

## 5.6. Particle on a Circle

We consider a free particle on an interval $[0, L]$ described by the Hamiltonian

$$H_{\rm p} = -\frac{1}{2}\,\frac{d^2}{dx^2}. \tag{5.70}$$

The domain of $H_{\rm p}$ consists of differentiable functions and we assume that they satisfy the boundary conditions

$$\psi(0) = \psi(L), \qquad \psi'(0) = \psi'(L). \tag{5.71}$$

A function $\psi$ that is defined on an interval and that satisfies these boundary conditions can easily be continued to the whole real axis by setting $\psi(x + nL) = \psi(x)$. In this way we obtain a periodic function with period $L$. Therefore, boundary conditions of this type are called *periodic boundary conditions.*

The periodic boundary conditions lead to an identification of the end points 0 and $L$ of the interval. This identification allows us to treat the variable $\varphi = 2\pi x/L$ as an angle coordinate. Hence it appears to be natural to visualize the configuration space as a circle. (Topologically, a circle is indeed defined as an interval $[0, L]$ with end points glued together).

The operator $H_{\rm p}$ on $[0, L]$ with periodic boundary conditions has the eigenfunctions

$$\psi_n^{(L)}(x) = \frac{1}{\sqrt{L}}\exp\Bigl(\mathrm{i}\,n\,\frac{2\pi}{L}\,x\Bigr), \quad n = 0, \pm 1, \pm 2, \dots . \tag{5.72}$$

These eigenfunctions belong to the energy eigenvalues

$$E_n^{(L)} = \frac{1}{2}\,n^2\Bigl(\frac{2\pi}{L}\Bigr)^2 = 2(k_n^{(L)})^2, \quad \text{where} \quad k_n^{(L)} = \frac{n\pi}{L}. \tag{5.73}$$

Because the energy $E_n^{(L)}$ does not depend on the sign of the quantum number $n$, there are two orthogonal eigenvectors $\psi_n$ and $\psi_{-n}$ for each energy eigenvalue, except for $E_0^{(L)}$. In quantum mechanics, the lowest possible energy is always nondegenerate, that is, there is (up to multiplication by a constant) only one eigenfunction. The unique state with the lowest possible energy is called the *ground state.* For the particle on a circle, all other energies (with quantum numbers $n \neq 0$) are eigenvalues with multiplicity 2 (the eigenspace is two-dimensional). In each eigenspace we can form the linear combinations

$$\frac{1}{\sqrt{2}}\,(\psi_n^{(L)}(x) + \psi_{-n}^{(L)}(x)) = \sqrt{\frac{2}{L}}\,\cos k_n^{(L)}\,x, \quad n = 0, 1, 2, \dots , \tag{5.74}$$

which also satisfy the Neumann boundary conditions, and

$$\frac{-\mathrm{i}}{\sqrt{2}}\,(\psi_n^{(L)}(x) - \psi_{-n}^{(L)}(x)) = \sqrt{\frac{2}{L}}\,\sin k_n^{(L)}\,x, \quad n = 1, 2, 3, \dots , \tag{5.75}$$

which are the eigenfunctions satisfying Dirichlet boundary conditions (see Section 5.4.2).

EXERCISE 5.6. *Show that the* $\psi_n^{(L)}$ *form an orthonormal basis.*

EXERCISE 5.7. *Prove that every state on a circle is periodic in time with period* $T = L^2/\pi$.

Because the eigenfunctions form a basis in the Hilbert space of the system, we can form arbitrary wave packets from the eigenfunctions. CD 4.12 shows Gaussian wave packets moving on a circle. The periodic boundary condition has the effect that a wave packet that leaves the interval on one side reenters the interval from the other side. The natural spreading of wave packets according to the free time evolution soon leads to self-interference on the circle. From the interference pattern localized states reemerge after certain fractions of the period $T$. You can watch the formation of a Schrödinger cat state at $T/2$.

EXERCISE 5.8. *Consider the angular momentum operator for a particle moving on a circle. Assume that the circle is in the* $x_1x_2$*-plane with center at the origin. Thus, it is sufficient to consider only the third component of the angular momentum,* $L_3 = x_1p_2 - x_2p_1$. *Using the angle* $\varphi$ *with the* $x_1$*-axis as the position coordinate, describe the action of* $L_3$ *on functions of* $\varphi$.

## 5.7. The Double Slit Experiment

The movies CD 4.16–4.19 show wave packets partially penetrating a screen with one or several slits. The screen is an obstacle realized by a Dirichlet boundary condition in the Schrödinger equation. The interference pattern that is visible behind the screen is often used to explain the mysteries of quantum mechanics.

Very often, people feel dissatisfied with quantum mechanics because in general it makes only probabilistic statements and one obtains no predictions about individual experiments. Moreover, some results of quantum mechanics seem to contradict our conception of the nature of reality. Usually, the double slit experiment is considered as a simple example where all these difficulties can be explained in a most transparent way. In his book [19], Richard Feynman introduces the chapter about the double slit experiment with the following words: "In this chapter we shall tackle immediately the basic element of the mysterious behavior in its most strange form. We choose to examine a phenomenon which is impossible, *absolutely* impossible, to explain in any classical way, and which has in it the heart of quantum mechanics."

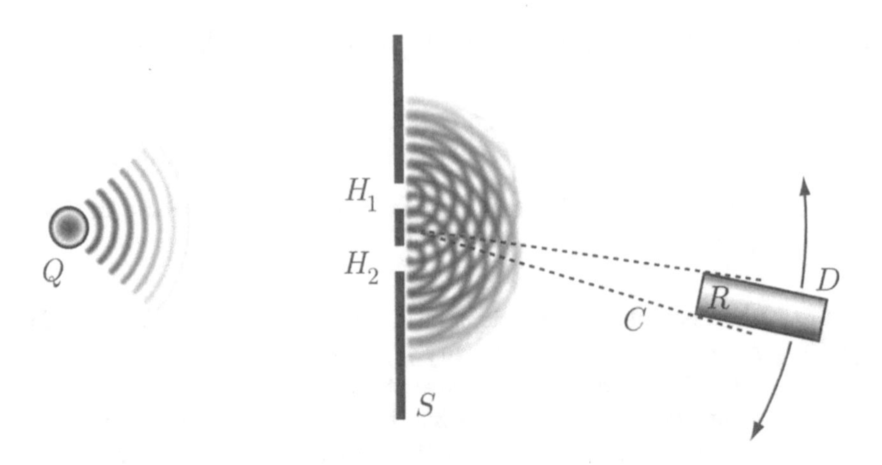


FIGURE 5.1. The double-slit experiment

### 5.7.1. The experimental setup

A particle source $Q$, a screen $S$, and a detector $D$ are arranged as in Fig. 5.1. The screen is assumed to be an impenetrable barrier except that it has two narrow slits or holes that are denoted by $H_1$ and $H_2$. The detector is positioned far behind the screen. It has a circular opening $R$ and gives a signal whenever a particle enters the detector through $R$.

In an elementary experiment, a particle is emitted from the source at time $t = 0$ and the physicist waits to see whether the particle is finally detected at $R$. The goal is to determine the probability for finding the particle at $R$. Thus, we have to perform the elementary experiment many times. By moving the detector around we can determine how the probability of finding the particle depends on the position of the detector.

Until very recently, the experiment described here has been considered an idealized thought experiment. Meanwhile very similar experiments have been actually performed and the predictions of quantum mechanics have been confirmed. We also note that the special realization of the experiment is not so crucial for what we want to point out. The double slit experiment is just a concrete example of a type of experiments where the particle goes through an intermediate state which is a superposition of two (or more) orthogonal states.

Now we proceed to make a quantum-mechanical model of this experiment.

### 5.7.2. Quantum-mechanical description

In order to simplify the analysis, it is assumed that the source $Q$ always generates a particle in the same quantum-mechanical initial state $\psi_0$. In Color Plate 18 we show a typical situation. The initial state has an uncertainty in position that is larger than the distance of the two holes. The average momentum points toward the screen.

For the Hamiltonian we choose the operator of kinetic energy with Dirichlet boundary conditions on the surface of the screen. Thus, the wave function cannot enter the material of the screen, but it may enter the region behind the screen through the two holes. Let $\psi(t)$ be the unique solution of the Schrödinger equation with $\psi(0) = \psi_0$. What quantity should represent the probability that the particle enters the detector through the surface $R$?

Let us assume that the detector is sufficiently far away and points toward the origin (which is situated somewhere near the two holes). Consider the cone $C$ formed by all straight lines from the origin through the disk $R$. Essentially all the particles scattered into the cone $C$ will go through $R$ and will thus be collected by the detector. The probability $p(C)$ that the particle is finally found in the cone $C$ is given by

$$p(C) = \lim_{t\to\infty} \int_C |\psi(\mathbf{x},t)|^2 \, d^3x. \tag{5.76}$$

Notice that we do not speak about the probability $p(R)$ that the particle goes through the opening $R$. Instead, we consider the probability $p(C)$ that the particle can be found asymptotically in a cone. The reason is that the quantity $p(R)$ is somewhat ill defined in quantum mechanics. In quantum mechanics, the motion of a particle from one region in space to another is described by a change of the position probability for these regions. Nothing is said about the actual motion between the regions. Therefore, it is not clear what it means if we say that a particle goes through a surface $R$. For smooth wave functions one might nevertheless try to define $p(R)$ as the expression

$$\int_0^\infty \Big( \int_R \vec{j}(\mathbf{x},t) \cdot \vec{n}(\mathbf{x}) \, df \Big) \, dt. \tag{5.77}$$

Here $\vec{j}$ is the probability current defined in Section 3.5, $\vec{n}$ is a unit normal vector to the surface $R$, pointing into the detector, and $df$ is a surface element of $R$. Equation (5.77) describes the time-integrated flux of the position probability density through the surface $R$. But in general this quantity does not have the properties of a probability. For an arbitrary surface it might be larger than one if some interaction causes the wave packet to go through the surface $R$ several times. Or it might be negative if the wave packet goes through the surface from the wrong direction.

The movie CD 4.17 shows the time evolution of the initial state $\psi_0$ shown in Color Plate 18. Two frames from this movie are shown in Color Plate 19. We see how a part of the wave function gets reflected at the screen while another part penetrates the holes. The parts of the wave function recombine behind the screen to form an interference pattern In certain directions, the probability of finding the particle is very small.

### 5.7.3. Comparing theory with experiment

Experimentally, the probability $p(C)$ is determined by counting the number $n(C)$ of experiments where the particle is scattered into the cone $C$ and dividing this number by the total number $n$ of experiments. This quotient is the relative frequency

$$\frac{n(C)}{n} = \frac{\text{number of particles finally entering } C}{\text{number of particles emitted from } Q}. \tag{5.78}$$

It is believed that for very large $n$ the relative frequency approximates the probability $p(C)$.

According to the best of our knowledge, quantum mechanics seems to make correct predictions about the outcome of this or similar experiments. But as discussed earlier, quantum mechanics gives no information about the path of the particle between source and detector. There is no mathematical quantity representing the probability that the quantum particle goes through one of the holes in the screen.

But how did the particle get from the source $Q$ to the region $C$? Did the particle go through $H_1$ or through $H_2$? These questions are of course motivated by our classical understanding of particles and their propagation through space. Can we understand the results of quantum mechanics in terms of these classical ideas?

There is one situation where it makes sense to say that the particle goes through one of the holes, say through $H_1$. Namely, if the other hole is closed and the only way to go to the region behind the screen is by passing the hole $H_1$. In this case we may identify the probability of going through the hole with the probability of finding the particle asymptotically (for $t \to \infty$) behind the screen.

Therefore, when we want to determine quantum-mechanically the probabilities of going through one of the holes, we must modify our experiment. We assume that the holes in the screen can be closed or opened individually and consider the following variants of our experimental setup:

**Situation 1:** $H_1$ is open; $H_2$ is closed.
**Situation 2:** $H_2$ is open; $H_1$ is closed.
**Situation 3:** Both holes open.

### 5.7.4. The predictions of quantum mechanics

Let $\psi_1$, $\psi_2$, and $\psi$ denote the solutions of the Schrödinger equation for the situations 1, 2, and 3, respectively, each with the same initial condition (see Color Plate 20). Because the propagation of waves according to the Schrödinger equation is a linear and local phenomenon, we are led to the conjecture that in the region behind the screen (and sufficiently far away from the holes) the wave function $\psi$ is very well approximated by the sum of the solutions $\psi_1$ and $\psi_2$:

$$\psi(\mathbf{x}, t) \approx \psi_1(\mathbf{x}, t) + \psi_2(\mathbf{x}, t) \quad \text{behind the screen.} \tag{5.79}$$

Thus, we regard the wave function behind the screen simply as the sum of the two approximately spherical waves emerging from the individual holes. (This is essentially an application of Huygens' principle). Indeed, the numerical solution of the corresponding Schrödinger equations confirms this assumption. Hence

$$|\psi(\mathbf{x}, t)|^2 \approx |\psi_1(\mathbf{x}, t) + \psi_2(\mathbf{x}, t)|^2. \tag{5.80}$$

We want to compare the quantity

$$p(C) = \lim_{t \to \infty} \int_C |\psi(\mathbf{x}, t)|^2 \, d^3x \tag{5.81}$$

with the expression

$$p_1(C) + p_2(C) = \lim_{t \to \infty} \int_C (|\psi_1(\mathbf{x}, t)|^2 + |\psi_2(\mathbf{x}, t)|^2) \, d^3x. \tag{5.82}$$

But the interference of the two wave functions will cause the sum $\psi_1 + \psi_2$ to vanish in certain directions where the individual summands are equal in size but have opposite phases. In these regions, $|\psi_1|^2 + |\psi_2|^2$ will be large, while the summands in $|\psi_1 + \psi_2|^2$ cancel out. Hence in general,

$$p(C) \neq p_1(C) + p_1(C) \tag{5.83}$$

and the difference between the expressions $p(C)$ and $p_1(C) + p_2(C)$ is not small or negligible.

Because of the interference of the two wave functions in (5.80) the outcome of the theory (which is in agreement with the experiment) is very strange. In some directions the probability of finding the particle *increases* if one of the holes is closed. With both holes open, there are more possibilities to go from the source to the detector, but in certain directions, we find no particles at all!

CD 4.17.2 shows the solution with one hole closed. It is approximately a spherical wave behind the screen; see also Color Plate 20. CD 4.17.3 shows the superposition of two spherical waves. The comparison with the interference pattern in the double-slit experiment justifies the assumption (5.79).

## 5.8. Special Topic: Analysis of the Double Slit Experiment

### 5.8.1. Events and probability

In the present context, an elementary experiment (see Section 3.4.2) is just the emission of a particle by the source $Q$. Eventually, the particle will be detected somewhere behind the screen. In a single experiment it cannot be predicted where the particle goes, no matter how much care we invest in preparing or knowing the initial state. According to the statistical interpretation, elementary experiments are not reproducible and therefore not the subject of quantum mechanics. What can be predicted are the probability distributions of certain observables, which have to be determined by repeating the elementary experiment many times.

Perhaps it is possible to refine the theory, so that it gives more information about the outcome of a single experiment? Can we assume that at least the particle itself knows where it goes?

Whenever probability theory is applied, it is essential to describe the events very precisely. First we define a *probability space* $\mathcal{X}$, the set of all elementary experiments. The subsets of the probability space are usually simply called *events* in probability theory. Here we are interested mainly in the following events (set of experiments):

- $\mathcal{X}$—the set of all experiments where the particle is emitted by $Q$. This is our probability space.
- $\mathcal{C}$—the set of all experiments where the particle is emitted by $Q$ and is finally detected in the cone $C$.

The main result of quantum mechanics is to associate a probability to the event $\mathcal{C}$. (The probability of the event $\mathcal{X}$ is 1.)

We are conducting experiments in different setups or situations, as described in Section 5.7.3. A priori, the probability spaces of different situations have nothing to do with each other. We append an index—$\mathcal{X}_j$, $\mathcal{C}_j$, $(j = 1, 2, 3)$—in order to distinguish between the situations. The reader should always remember that these are collections of experiments in different situations and are therefore different sets.

In probability theory, a probability is associated to each event. The probability of being scattered into the cone $C$ in Situation $j$ (i.e., the probability of the event $\mathcal{C}_j$) will be simply called $p_j(C)$. According to our earlier notation, we write $p(C)$ for $p_3(C)$.

One might also wish to consider the following events:

- $\mathcal{H}_1$—the set of all experiments where the particle goes through $H_1$.
- $\mathcal{H}_2$—the set of all experiments where the particle goes through $H_2$.
- $\mathcal{X}_>$—the set of all experiments where the particle finally gets behind the screen.

According to the conditions of the experimental setup, it is reasonable to identify

$$(\mathcal{X}_>)_j = (\mathcal{H}_j)_j \quad (j = 1, 2). \tag{5.84}$$

As discussed earlier, in quantum mechanics it is difficult to define the probability of going through a surface, and hence this identification may be considered a definition. Hence we *define* the probability of going through the screen as the probability of finding the particle asymptotically somewhere behind the screen. The latter can be expressed quantum-mechanically as

$$p_j(\mathcal{H}_j) = \lim_{t\to\infty} \int_{\mathbb{R}_>} |\psi_j(\mathbf{x}, t)|^2 \, d^3x, \quad j = 1, 2, \tag{5.85}$$

where $\psi_j$ is the solution of the Schrödinger equation in Situation $j$ and $\mathbb{R}_>$ is the half space behind the screen. In principle, we can determine the probabilities $p_j(\mathcal{H}_j)$ by positioning the detector $R$ directly behind (or even inside) the corresponding hole.

It makes sense to define the event $\mathcal{H}_2$ even for Situation 1 where the hole $H_2$ is closed (there, it is just the empty set $\emptyset$, that is, an impossible event). Hence we write

$$(\mathcal{H}_1)_2 = (\mathcal{H}_2)_1 = \emptyset \tag{5.86}$$

and the corresponding probabilities are zero.

A priori it is not allowed to speak of $(\mathcal{H}_i)_3$ because in Situation 3 no attempt is made to determine whether the particle goes through one of the holes. Any assumption about the event $\mathcal{H}_j$ in Situation 3 makes an identification of events in different experimental setups. All we can say a priori (as a consequence of the geometric setup of the experiment) is that a particle can only reach the cone $C$ if it reaches the half-space behind the screen. As a formula this statement reads

$$\mathcal{C}_3 = (\mathcal{X}_>)_3 \cap \mathcal{C}_3, \tag{5.87}$$

which just states that the event $\mathcal{C}_3$ is a subset of $(\mathcal{X}_>)_3$.

The Situations 1 and 2 are designed to make sure through which hole the particle has passed when it is finally detected. In these situations every particle that is detected in $C$ must have passed the open hole. This is expressed in

$$\mathcal{C}_j = (\mathcal{H}_j)_j \cap \mathcal{C}_j \quad \text{or equivalently,} \quad \mathcal{C}_j \subset (\mathcal{H}_j)_j, \quad j = 1, 2. \tag{5.88}$$

One has tried to design other situations in order to determine where the particle goes without closing one of the holes. For example, one might think of a measuring apparatus that detects flying particles through their interaction with photons (i.e., just by looking at the particle). But it has been shown that (as a consequence of the disturbing influence of the measurement) any such attempt to obtain a "which-way" information has essentially the same effect as closing one of the holes altogether: It destroys the interference pattern of the wave function behind the screen.

### 5.8.2. Classical consideration

The spreading of the wave function over a large region of space reflects the uncertainty of position. But the particle itself is never seen to be smeared out. An electron is always detected as a mass point containing the whole of its mass and charge. No matter how small the opening $R$ or how narrow the cone $C$, a particle is always detected as a whole. Thus, it seems reasonable to assume that the particle remains pointlike in the time between emission and detection. If the particle is indeed pointlike, then it makes sense to assume that it actually has passed through one and only one of the holes before it arrives in the region behind the screen. Even if the physicist does not (or cannot) know through which of the holes the particle goes, let us assume that at least the particle itself knows. (This amounts to saying that the spreading of the wave function just reflects the incomplete knowledge of the observer and is not a fundamental property of the particle itself).

**Addition of probabilities**:

The assumption

(A) "*Whenever a particle is detected behind the screen, then it has passed through one and only one of the two holes*"

and the elementary rules of logic, set theory, and probability theory imply

$$p(C) = p_1(C) + p_2(C). \tag{5.89}$$

The assumption (A) stated above is about the behavior of an individual particle during an elementary experiment. Quantum mechanics makes no statement about the validity of this assumption. The location of the particle when it penetrates the screen may be regarded as a hidden variable.

In quantum mechanics, the relation (5.89) is replaced by Eq. (5.83). For those who believe in the classical rules of logic and probability theory, the experimentally confirmed violation of (5.89) just shows that this type of hidden variables is prohibited in a quantum theory.

Historically, there have been various attempts to change the rules of logic or probability theory in order to be able to keep the assumption (A). In order to see which of these rules are involved, we try to give a proof of (5.89) based on assumption (A). The derivation given below might seem difficult—in fact, it is very elementary. What makes it difficult is to list all subconscious assumptions about reality and logic involved in this derivation.

We split the derivation of Eq. (5.89) into several steps. Because this relation is violated by quantum mechanics, see Eq. (5.83), and because the quantum-mechanical result is confirmed experimentally, either the assumption or one of the following steps is wrong.

1. We make the following assumption: When the particle is detected behind the screen in Situation 3, then it has passed through one or both of the holes $H_1$ and $H_2$. Written as a formula, this assumption means
$$(\mathcal{H}_1)_3 \cup (\mathcal{H}_2)_3 = (\mathcal{X}_>)_3. \tag{5.90}$$
This contains the implicit assumption that it makes sense to speak of the events $\mathcal{H}_1$ and $\mathcal{H}_2$ even in Situation 3.
2. We use elementary set algebra and (5.87) to deduce from step 1:
$$((\mathcal{H}_1)_3 \cap \mathcal{C}_3) \cup ((\mathcal{H}_2)_3 \cap \mathcal{C}_3) = \mathcal{C}_3. \tag{5.91}$$
This, of course, just says that the particle can reach $C$ only through one or both of the holes in the screen.
3. The next assumption is: When the particle goes through the hole $H_1$, it does not matter, whether the hole $H_2$ is open or closed (and similarly with $H_2$ and $H_1$ exchanged). This assumption makes an identification of events in Situation 1 or 2 with events in Situation 3. More explicitly,
$$(\mathcal{H}_1)_3 = (\mathcal{H}_1)_1, \qquad (\mathcal{H}_2)_3 = (\mathcal{H}_2)_2 \tag{5.92}$$
and
$$(\mathcal{H}_1)_3 \cap \mathcal{C}_3 = (\mathcal{H}_1)_1 \cap \mathcal{C}_1, \quad (\mathcal{H}_2)_3 \cap \mathcal{C}_3 = (\mathcal{H}_2)_2 \cap \mathcal{C}_2. \tag{5.93}$$

4. This is crucial: We assume that in Situation 3 a particle can go through only one of the holes at a time. (The particle does not split or divide itself). Hence
$$(\mathcal{H}_1)_3 \cap (\mathcal{H}_2)_3 = \emptyset. \tag{5.94}$$
5. Using elementary set algebra, the previous step implies
$$((\mathcal{H}_1)_3 \cap \mathcal{C}_3) \cap ((\mathcal{H}_2)_3 \cap \mathcal{C}_3) = \emptyset, \tag{5.95}$$
6. From (5.91), (5.93), and (5.88) we obtain
$$\mathcal{C}_1 \cup \mathcal{C}_2 = \mathcal{C}_3, \tag{5.96}$$
and with (5.95)
$$\mathcal{C}_1 \cap \mathcal{C}_2 = \emptyset. \tag{5.97}$$
Thus, $\mathcal{C}_3$ is the union of two disjoint subsets $\mathcal{C}_1$ and $\mathcal{C}_2$. Disjoint events are called *independent*.
7. The event $\mathcal{C}_1$ in Situation 1 has the same probability as in Situation 3 (where it has been identified with $(\mathcal{H}_1)_3 \cap \mathcal{C}_3$), that is, we assume
$$p_1(\mathcal{C}_1) = p_3(\mathcal{C}_1), \quad p_2(\mathcal{C}_2) = p_3(\mathcal{C}_2). \tag{5.98}$$
8. In probability theory, one associates probabilities to events according to certain rules. One of these rules states that the joint probability of independent events (which are disjoint sets of experiments) is the sum of the individual probabilities. Thus, step 6 implies that
$$p_3(\mathcal{C}_3) = p_3(\mathcal{C}_1) + p_3(\mathcal{C}_2). \tag{5.99}$$
Using the previous step and the notation $p_i(\mathcal{C}_i) = p_i(C)$, $p_3(C) = p(C)$, we obtain the result (5.89).

Because this result is not in agreement with experiments, one of the above steps must be false. The quantum logic point of view rejects the use of the *tertium non datur* from classical logic when thinking about the quantum-physical reality.

In Bohm's quantum mechanics (Section 3.5) one would reject step 3, because the particle's pilot wave is changed by opening or closing the second slit.

If one thinks that the wave function is a property of an individual particle, then—considering the movies CD 4.17 and 4.18—one would reject both steps 3 and 4.

In view of our discussion in Section 5.8.1 we can already reject the first step, which puts events from the probability space of Situations 1 or 2 into the probability space of Situation 3.

Which of the steps above would you like to reject?

*Chapter 6*

# Linear Operators in Hilbert Spaces

**Chapter summary**: Having had some experience working with quantum-mechanical formalism in various situations, it is now time to investigate more deeply its mathematical properties. In particular, we are going to describe the time evolution and, more generally, symmetry transformations and their generators.

Time evolution according to the Schrödinger equation has the property that it conserves the norm of wave packets. In mathematical terms, the relation between the initial state (at time $t = 0$) and the state at time $t$ can be described as the action of a unitary operator $U(t)$. The set of operators $U(t)$ is called the unitary group generated by the Hamiltonian operator $H$. In the same way all self-adjoint operators generate unitary groups. For example, the momentum operator generates the unitary group of translations and the angular momentum operator generates rotations. For self-adjoint operators satisfying the canonical commutation relations, the corresponding unitary groups satisfy the Weyl relations. We shall have occasion to use the Weyl relations for calculating the time evolution of harmonic oscillator states in the next chapter.

A system is said to be symmetric with respect to a certain group of transformations, if these transformations commute with the time evolution. Equivalently, the generator of the symmetry transformation commutes with the Hamiltonian of the system. An immediate consequence is Noether's theorem. If the Schrödinger equation has a symmetry, then any physical observable corresponding to the generator of the symmetry is a conserved quantity (a constant of motion). For example, momentum is conserved in a system with translational symmetry.

## 6.1. Hamiltonian and Time Evolution

The Schrödinger equation has the general form

$$\mathrm{i}\,\frac{d}{dt}\,\psi(t) = H\,\psi(t) \tag{6.1}$$

where $H$ is a linear operator in a Hilbert space $\mathfrak{H}$ (usually called the *Hamiltonian operator* or simply the *Hamiltonian* of the system).

If $H$ were just a number, then a solution would be a function $t \to \psi(t)$ with complex values. Given an initial value $\psi_0$, the solution can be determined explicitly as $\exp(-\mathrm{i}Ht)\psi_0$.

But in our case $H$ is a linear operator in $\mathfrak{H}$ and hence the solution must be a function $t \to \psi(t)$ with values in $\mathfrak{H}$!

$\boxed{\Psi}$ **Strict solutions.** What other properties must a solution have? In order to give a meaning to the time derivative in Eq. (6.1) we have to assume that the limit

$$\lim_{h\to 0} \frac{\psi(t+h)-\psi(t)}{h} \quad \text{exists for all } t. \tag{6.2}$$

Of course, the limit is meant with respect to the topology of the Hilbert space. If it exists, the limit is called $d\psi(t)/dt$. It should be stressed that $d/dt$ is *not* a linear operator in the Hilbert space of the system, instead it acts on functions $t \to \psi(t)$ with values in the Hilbert space.

If you want to apply $H$ to $\psi(t)$ on the right-hand side of (6.1) you have to make sure that the solution at time $t$ is in the domain of definition of $H$. Hence a solution must also have the property

$$\psi(t) \in \mathfrak{D}(H), \quad \text{for all } t. \tag{6.3}$$

What we have described here is called a *strict solution*, that is, a solution in the literal sense. It is sometimes useful to consider more general solutions of the initial-value problem. But before we proceed with the abstract nonsense, let us consider the free time evolution.

**Free time evolution**: As you learned in Section 3.3.1, the solution of the free Schödinger equation with initial function $\psi_0$ can be expressed as

$$\psi(\cdot,t) = \mathcal{F}^{-1} \exp\Bigl(-\mathrm{i}\frac{k^2}{2}t\Bigr)\mathcal{F}\psi_0. \tag{6.4}$$

The free time evolution can therefore be described as the action of a linear operator on the initial function $\psi_0$. This operator is just the exponential function of the free Schrödinger operator $H_0 = p^2/2$, that is,

$$\mathrm{e}^{-\mathrm{i}H_0 t} = \mathcal{F}^{-1} \exp\Bigl(-\mathrm{i}\frac{k^2}{2}t\Bigr)\mathcal{F}. \tag{6.5}$$

Hence the free time evolution is the inverse Fourier transform of a bounded multiplication operator in momentum space.

We want to stress that it is not at all obvious how to define $\exp(-\mathrm{i}H_0 t)$ in terms of the familiar power series of the exponential function. The power series would be an infinite sum of unbounded operators. Moreover, for all positive integers $n$ the domain of $H_0^{n+1}$ is strictly smaller than the domain of $H_0^n$.

You know from Section 3.3.4 that the free time evolution is an integral operator,

$$(e^{-iH_0t}\phi)(\mathbf{x}) = \int_{\mathbb{R}^n} K(\mathbf{x}-\mathbf{y},t)\,\phi(\mathbf{y})\,d^n y, \tag{6.6}$$

with an integral kernel $K$ given by Eq. (3.53).

The free time evolution has the following properties:

**Some properties of** $\exp(-iH_0t)$:

1. The domain of $\exp(-iH_0t)$ consists of the entire Hilbert space.
2. $\|\exp(-iH_0t)\psi\| = \|\psi\|$ for all $\psi$.
3. If $\psi \in \mathfrak{D}(H_0)$, then $\psi(t) = \exp(-iH_0t)\psi$ is a strict solution of the Schrödinger equation.

These results can be proved with the help of Eq. (6.5).

The first property states that we can define a time evolution for any initial state in the Hilbert space.

CD 3.12 shows a solution $\psi(x,t) = \exp(-iH_0t)\psi_0(x)$ of the free Schrödinger equation where $\psi_0$ is the characteristic function of an interval. Hence the initial function is not differentiable with respect to $x$, not even in the weak sense described in Section 2.6.2. (Obviously, $\psi_0$ is not in the domain of the operator $d^2/dx^2$). Hence $\psi(x,t)$ is not a strict solution of the Schrödinger equation. This becomes apparent in the movie. The function $\psi(x,t)$ does not evolve continuously (in a pointwise sense) from $\psi(x,0)$. Instead, rapid oscillations are present after an arbitrary short time. Nevertheless, $\psi(\cdot,t)$ tends to $\psi_0$, as $t \to 0$, in the norm of $L^2$.

The second property states that the time evolution operator is for all $t$ a bounded (hence continuous) operator with norm 1. Hence the norm of $\psi(t)$ is independent of $t$, which is essential for our interpretation of wave functions. This property also implies that the solution depends on the initial condition in a continuous way. If we compare two time evolutions $\psi(t)$ and $\phi(t)$ corresponding to two different initial states $\psi_0$ and $\phi_0$, then

$$\|\psi(t) - \phi(t)\| = \|\exp(-iH_0t)\,(\psi_0 - \phi_0)\| = \|\psi_0 - \phi_0\|. \tag{6.7}$$

If the initial states are close together, then the states at time $t$ are also close together. The separation of the initial states remains constant in time.

The third property means that the domain of $H_0$ is invariant under the time evolution. If the initial state belongs to $\mathfrak{D}(H_0)$, so does the state at time $t$. Moreover, the mapping $t \to \psi(t)$ is differentiable with respect to $t$ in the sense of Eq. (6.2).

$\boxed{\Psi}$ There are wave functions $\psi$ in $L^2(\mathbb{R}^n)$ that do not belong to the domain of $H_0$ (any discontinuous function). The operator $\exp(-\mathrm{i}H_0t)$ nevertheless defines a time evolution for these initial states, and in a strict sense this does not give a solution of the Schrödinger equation (see CD 3.12). But it does give a solution that for all times is arbitrarily close to a strict solution. This follows from the fact that the domain of $H_0$ is dense in the Hilbert space. In functional analysis, a solution that is arbitrarily close to a strict solution (uniformly in time on compact time intervals) is called a *mild solution.*

## 6.2. Unitary Operators

The following definition is motivated by the properties 1 and 2 of the time evolution operator, which are stated in the previous section.

DEFINITION 6.1. A linear operator $U$ defined everywhere in a Hilbert space $\mathfrak{H}$ is called *isometric* if

$$\|U\psi\| = \|\psi\|, \qquad \text{for all } \psi \in \mathfrak{H}. \tag{6.8}$$

$U$ is called *unitary* if it maps onto all of $\mathfrak{H}$ (i.e., if the range of $U$ equals $\mathfrak{H}$, or $U\mathfrak{H} = \mathfrak{H}$).

The most important property of isometric and unitary operators is the following:

An isometric operator $U$ leaves the scalar product invariant,

$$\langle U\psi, U\phi\rangle = \langle \psi, \phi\rangle. \tag{6.9}$$

PROOF. Because $\|\psi\|^2 = \langle\psi,\psi\rangle$ for all $\psi$, you can see that

$$\langle U(\psi+\phi), U(\psi+\phi)\rangle = \langle \psi+\phi, \psi+\phi\rangle,$$

hence

$$\|U\psi\|^2 + \langle U\psi, U\phi\rangle + \langle U\phi, U\psi\rangle + \|U\phi\|^2 = \|\psi\|^2 + \langle\psi,\phi\rangle + \langle\phi,\psi\rangle + \|\phi\|^2,$$

and therefore, using $\langle\psi,\phi\rangle = \overline{\langle\phi,\psi\rangle}$ and Eq. (6.8),

$$\operatorname{Re}\langle U\psi, U\phi\rangle = \operatorname{Re}\langle\psi,\phi\rangle.$$

If you repeat the same calculation with $\psi + \mathrm{i}\psi$ you will obtain

$$\operatorname{Im}\langle U\psi, U\phi\rangle = \operatorname{Im}\langle\psi,\phi\rangle,$$

and the result (6.9) follows immediately. □

EXAMPLE 6.2.1. The Fourier transform $\mathcal{F}$ is a unitary operator on the Hilbert space $L^2(\mathbb{R}^n)$. This follows from the Fourier–Plancherel relation (2.64), and the fact that $\mathcal{F}$ and $\mathcal{F}^{-1}$ can be defined on all of $L^2(\mathbb{R}^n)$ (Section 2.5.4).

EXAMPLE 6.2.2. The time evolution operator $\exp(-\mathrm{i}H_0t)$ defined in (6.5) is unitary for all $t$. This can be seen from the unitarity of the Fourier transform, and the unitarity of the multiplication by the phase factor $\exp(-\mathrm{i}k^2t/2)$ in momentum space.

EXAMPLE 6.2.3. The relation (3.84) shows that the transition from $\psi(x)$ to the energy representation $g(E)$ is a unitary transformation.

EXERCISE 6.1. *Show that the gauge transformation defined in Eq.* (4.60) *is a unitary transformation.*

EXERCISE 6.2. *Show that for arbitrary $\lambda > 0$ the scaling transformation*

$$U_\lambda : \psi \longrightarrow U_\lambda \psi, \qquad \text{where} \quad (U_\lambda \psi)(x) = \sqrt{\lambda}\,\psi(\lambda \mathbf{x}). \tag{6.10}$$

*is a unitary operator in $L^2(\mathbb{R})$.*

## 6.3. Unitary Time Evolution and Unitary Groups

The quantum-mechanical time evolution leads to the definition of a linear time evolution operator $U(t)$.

DEFINITION 6.2. A *one-parameter unitary group* is a function $t \to U(t)$ from the real numbers $t \in \mathbb{R}$ into the set of bounded operators on a Hilbert space $\mathfrak{H}$ with the following properties:

1. $U(t)$ is a unitary operator (for all $t$). *(unitarity)*
2. $U(0) = \mathbf{1}$, $U(t)\,U(s) = U(t+s)$ (for all $s, t$). *(group property)*
3. $\lim_{t\to 0} U(t)\psi = \psi$, (for all $\psi \in \mathfrak{H}$). *(strong continuity)*

The time evolution has a generator, which appears in the evolution equation. The generator can be obtained from $U(t)$ by taking the derivative at $t = 0$. This is the content of the following definition:

DEFINITION 6.3. Let $U(t)$ be a one-parameter unitary group in a Hilbert space $\mathfrak{H}$. The *infinitesimal generator* $H$ of the unitary group is the linear operator given by

$$H\,\psi = \mathrm{i} \lim_{t\to 0} \frac{U(t) - \mathbf{1}}{t}\,\psi \tag{6.11}$$

on a domain consisting of all vectors $\psi$ for which this limit exists.

It can be shown that the domain $\mathfrak{D}(H)$ of the infinitesimal generator is dense in $\mathfrak{H}$. (See Definition 2.4).

We can draw some immediate conclusions from these definitions. The most important one states that the unitary group gives a solution of an appropriate Schrödinger equation.

**Time evolution equation**:

Let $U(t)$ be a unitary group with generator $H$ and write

$$\psi(t) = U(t)\,\psi. \tag{6.12}$$

Then for all $\psi$ in the domain of the generator, $\psi(t)$ is a solution of the initial-value problem

$$\mathrm{i}\,\frac{d}{dt}\,\psi(t) = H\,\psi(t), \qquad \psi(0) = \psi. \tag{6.13}$$

This statement includes the invariance of the domain $\mathfrak{D}(H)$ under the time evolution. More precisely, we have

$$U(t)\,\mathfrak{D}(H) \subset \mathfrak{D}(H), \qquad H\,U(t) - U(t)\,H = 0 \quad \text{on } \mathfrak{D}(H). \tag{6.14}$$

PROOF. If $\psi$ is in the domain $\mathfrak{D}(H)$, then the limit (6.11) exists. Because $U(t)$ is bounded and defined everywhere in $\mathfrak{H}$, you can interchange the action of the operator $U(t)$ with the limit,

$$U(t)\,\lim_{h\to 0}\frac{U(h)-\mathbf{1}}{t}\,\psi = \lim_{h\to 0} U(t)\,\frac{U(h)-\mathbf{1}}{t}\,\psi \tag{6.15}$$

$$= \lim_{h\to 0}\frac{U(t+h)-U(t)}{t}\,\psi \tag{6.16}$$

$$= \lim_{h\to 0}\frac{U(h)-\mathbf{1}}{t}\,U(t)\psi. \tag{6.17}$$

Hence $U(t)\psi$ is also in the domain of $H$ and $U(t)H\psi = HU(t)\psi$. For the second line in this calculation, Eq. (6.16), the group property from the definition of a unitary group is used. With $U(t)\psi = \psi(t)$ this expression is just the derivative of $\psi(t)$,

$$\lim_{h\to 0}\frac{U(t+h)-U(t)}{t}\,\psi = \frac{d}{dt}\,\psi(t), \tag{6.18}$$

and hence the result (6.13) follows immediately from (6.16) and (6.17). □

The unitarity of the time evolution operator $U(t)$ states that the scalar product of two states $\psi$ and $\phi$ (and hence the transition probability) is independent of time,

$$\langle \psi(t), \phi(t)\rangle = \langle \psi, \phi\rangle, \qquad \text{for all } t. \tag{6.19}$$

A little calculation shows that for $\psi$ and $\phi \in \mathfrak{D}(H)$

$$\mathrm{i}\,\frac{d}{dt}\,\langle\psi(t),\phi(t)\rangle = \langle\psi(t), H\,\phi(t)\rangle - \langle H\psi(t),\phi(t)\rangle. \tag{6.20}$$

Hence for $U(t)$ to be unitary it is necessary that

$$\langle\psi, H\,\phi\rangle = \langle H\psi,\phi\rangle, \qquad \text{for all } \phi,\psi \in \mathfrak{D}(H). \tag{6.21}$$

This property of $H$ is equivalent to the symmetry (see Definition 4.1).

## 6.4. Symmetric Operators

In Section 4.3, a symmetric operator was defined as an operator with only real expectation values. Let me now give another definition which is equivalent to the earlier definition.

DEFINITION 6.4. A linear operator $H$ in a Hilbert space $\mathfrak{H}$ is said to be *symmetric* if $H$ has a dense domain and

$$\langle\psi, H\,\phi\rangle = \langle H\,\psi,\phi\rangle \tag{6.22}$$

holds for all vectors $\phi$ and $\psi$ in the domain of $H$.

As an example, we will show that the operator $H_D = -\frac{1}{2}\frac{d^2}{dx^2}$ in the Hilbert space $L^2([0,L])$ is symmetric on the domain $\mathfrak{D}(H_D)$ consisting of twice differentiable functions satisfying Dirichlet boundary conditions. Using partial integrations we find

$$\langle\phi, H_D\,\psi\rangle = -\int_0^L \overline{\phi(x)}\,\psi''(x)\,dx \tag{6.23}$$

$$= -\overline{\phi(x)}\,\psi'(x)\Big|_0^L + \int_0^L \overline{\phi'(x)}\,\psi'(x)\,dx \tag{6.24}$$

$$= -\overline{\phi(x)}\,\psi'(x)\Big|_0^L + \overline{\phi'(x)}\,\psi(x)\Big|_0^L - \int_0^L \overline{\phi''(x)}\,\psi(x)\,dx \tag{6.25}$$

$$= -\overline{\phi(x)}\,\psi'(x)\Big|_0^L + \overline{\phi'(x)}\,\psi(x)\Big|_0^L + \langle H_D\,\phi,\psi\rangle. \tag{6.26}$$

$$\tag{6.27}$$

This calculation holds for all twice differentiable functions $\phi$ and $\psi$. If, moreover, both $\phi$ and $\psi$ are in the domain $\mathfrak{D}(H_D)$, they satisfy Dirichlet boundary conditions and therefore the boundary terms in the calculation above vanish. This proves that $H_D$ is symmetric on $\mathfrak{D}(H_D)$.

EXERCISE 6.3. *Define a suitable domain such that the momentum operator* $p = -\mathrm{i}d/dx$ *is symmetric in the Hilbert space* $L^2(\mathbb{R})$. *Do the same for the position operator* $x$ *and the kinetic energy* $p^2/2$

The following result is one of the main reasons for our talking about symmetric operators.

THEOREM 6.1. *A symmetric operator has only real eigenvalues. If $\psi_1$ and $\psi_2$ are eigenvectors of a symmetric operator $H$ belonging to* different *eigenvalues $E_1$ and $E_2$, then $\psi_1$ is orthogonal to $\psi_2$,*

$$\langle \psi_1, \psi_2 \rangle = 0. \tag{6.28}$$

PROOF. The proof is very simple. A scalar product in a Hilbert space has the property that exchanging the factors amounts to a complex conjugation

$$\langle \psi, \phi \rangle = \overline{\langle \phi, \psi \rangle}. \tag{6.29}$$

Therefore, using the symmetry of $H$, we find that

$$\langle H \rangle_\psi = \langle \psi, H\psi \rangle = \overline{\langle \psi, H\psi \rangle}. \tag{6.30}$$

This means that the expectation value of a symmetric operator is always a real number. But if $\psi$ is an eigenvector of $H$ with eigenvalue $E$, then the expectation value of $H$ is given by (assuming $\|\psi\| = 1$)

$$\langle H \rangle_\psi = \langle \psi, H\psi \rangle = E\langle \psi, \psi \rangle = E\|\psi\|^2 = E. \tag{6.31}$$

Now consider two eigenvectors of $H$, that is, $H\psi_i = E_i\psi_i$, $i = 1, 2$. The symmetry of $H$,

$$\langle \psi_1, H\psi_2 \rangle = \langle H\psi_1, \psi_2 \rangle, \tag{6.32}$$

implies immediately that

$$(E_2 - E_1)\langle \psi_1, \psi_2 \rangle = 0. \tag{6.33}$$

By assumption, $E_2 - E_1 \neq 0$, and we finally obtain $\langle \psi_1, \psi_2 \rangle = 0$. □

EXERCISE 6.4. *Assume that $\psi$ is a superposition of eigenfunctions $\phi_i$ of $H$, that is,*

$$\psi = \sum c_i \, \phi_i, \qquad H\phi_i = E_i\phi_i. \tag{6.34}$$

*Find the expectation value of $H$ in the state $\psi$.*

# 6.5. The Adjoint Operator

## 6.5.1. Adjoint of a bounded operator

If $T$ is a bounded linear operator defined everywhere in a Hilbert space $\mathfrak{H}$, then the adjoint of $T$, denoted[1] by $T^\dagger$, is the operator for which

$$\langle \phi, T\psi \rangle = \langle T^\dagger \phi, \psi \rangle, \qquad \text{all } \phi, \psi \in \mathfrak{H}. \tag{6.35}$$

By taking the adjoint, we can move an operator to the other side in a scalar product.

It can be shown that $T^\dagger$ is uniquely defined, bounded, and defined everywhere on $\mathfrak{H}$. The norm of $T^\dagger$ equals the norm of $T$,

$$\|T^\dagger\| = \|T\|. \tag{6.36}$$

The double-adjoint $(T^\dagger)^\dagger$, which is usually written as $T^{\dagger\dagger}$, is equal to $T$. If $S$ and $T$ are two bounded operators, then

$$(ST)^\dagger = T^\dagger S^\dagger. \tag{6.37}$$

By taking the complex conjugate in (6.35), we find that

$$\langle T\phi, \psi \rangle = \langle \phi, T^\dagger \psi \rangle, \qquad \text{all } \phi, \psi \in \mathfrak{H}. \tag{6.38}$$

## 6.5.2. Adjoint of a unitary operator

For a unitary operator, we have

$$\langle \phi, \psi \rangle = \langle U\phi, U\psi \rangle = \langle U^\dagger U \phi, \psi \rangle, \qquad \text{all } \phi, \psi \in \mathfrak{H}, \tag{6.39}$$

hence $U^\dagger U \phi = \phi$ for all $\phi$, which means $U^\dagger U = 1$. Because the range of $U$ is all of $\mathfrak{H}$, and $U^\dagger$ is defined everywhere, we have $U^\dagger = U^{-1}$. This leads to an equivalent definition of unitarity

**Unitarity (equivalent definition):**

A linear operator that is bounded and defined everywhere on $\mathfrak{H}$ is unitary if and only if

$$U^\dagger = U^{-1} \qquad (\text{i.e., } U^\dagger U = \mathbf{1} \text{ and } UU^\dagger = \mathbf{1}). \tag{6.40}$$

[1]In the mathematical literature, the adjoint of $T$ is usually denoted by $T^*$.

### 6.5.3. Special Topic: Adjoint of an unbounded operator

Typically, the operators that represent physical observables are unbounded. This requires a slightly modified definition of the adjoint operator.

DEFINITION 6.5. Let $T$ be a linear operator with domain $\mathfrak{D}(T)$ dense in $\mathfrak{H}$. The adjoint operator is a linear operator $T^\dagger$ with the property

$$\langle \phi, T\psi \rangle = \langle T^\dagger \phi, \psi \rangle \qquad \text{for all } \psi \in \mathfrak{D}(T) \text{ and all } \phi \in \mathfrak{D}(T^\dagger). \tag{6.41}$$

Here the domain of $T^\dagger$ is defined as follows: $\phi$ is in $\mathfrak{D}(T^\dagger)$ if and only if there is a vector $\xi \in \mathfrak{H}$ such that

$$\langle \xi, \psi \rangle = \langle \phi, T\psi \rangle \qquad \text{for all } \psi \in \mathfrak{D}(T). \tag{6.42}$$

The condition that $\mathfrak{D}(T)$ be dense guarantees that the vector $\xi$ in (6.42) is determined uniquely by $\phi$. Hence $T^\dagger$ can be defined by setting $T^\dagger \phi = \xi$.

The double adjoint exists if $\mathfrak{D}(T^\dagger)$ is dense. In this case, $T^{\dagger\dagger}$ is an extension of $T$ (i.e., an operator defined on a larger domain, which on the domain of $T$ coincides with $T$).

## 6.6. Self-Adjointness and Stone's Theorem

Self-adjoint operators are symmetric operators with a domain that is maximal in a certain sense. The distinction between self-adjoint and symmetric operators is subtle, but nevertheless important. The self-adjointness of the Hamiltonian operator is equivalent to the existence of a quantum-mechanical (unitary) time evolution.

DEFINITION 6.6. A linear operator $T$ is *self-adjoint* if

$$T^\dagger = T. \tag{6.43}$$

The definition of self-adjointness requires in particular the equality of the domains of $T$ and $T^\dagger$. If $T$ is only symmetric, then (by definition)

$$\langle \phi, T\psi \rangle = \langle T\phi, \psi \rangle, \qquad \text{for all } \phi, \psi \in \mathfrak{D}(T). \tag{6.44}$$

Recalling the definition of the domain of the adjoint operator, we conclude that any $\phi$ in $\mathfrak{D}(T)$ belongs to the domain of $T^\dagger$, and that

$$T^\dagger \phi = T\phi, \qquad \text{for all } \phi \in \mathfrak{D}(T). \tag{6.45}$$

This still does not imply that the operators $T$ and $T^\dagger$ are the same because the domain of $T^\dagger$ could be larger than $\mathfrak{D}(T)$. Hence the symmetry property just means that $T^\dagger$ is an extension of $T$ to a larger domain. The self-adjointness requires much more:

$$T \text{ symmetric}: \qquad T = T^\dagger \qquad \text{on } \mathfrak{D}(T) \subset \mathfrak{D}(T^\dagger), \tag{6.46}$$

$$T \text{ self-adjoint}: \qquad T = T^\dagger \qquad \text{on } \mathfrak{D}(T) = \mathfrak{D}(T^\dagger). \tag{6.47}$$

EXAMPLE 6.6.1. Consider the momentum operator on an interval $[0,1]$ with Dirichlet boundary conditions, that is,

$$T\psi = -\mathrm{i}\psi' \tag{6.48}$$

for $\psi$ in the domain

$$\mathfrak{D}(T) = \{\psi \in L^2([0,1]) \mid \psi' \in L^2, \psi(0) = \psi(1) = 0\}. \tag{6.49}$$

Here $\psi' \in L^2$ means that $\psi$ is differentiable (in a generalized sense) with a square-integrable derivative. Using a partial integration, we find

$$\langle \phi, -\mathrm{i}\psi' \rangle = -\mathrm{i} \int_0^1 \overline{\phi(x)}\, \psi'(x)\, dx \tag{6.50}$$

$$= -\mathrm{i}\overline{\phi(x)}\, \psi(x)\Big|_0^1 + \mathrm{i} \int_0^1 (\overline{\phi(x)})'\, \psi(x)\, dx \tag{6.51}$$

$$= \int_0^1 \overline{-\mathrm{i}\phi'(x)}\, \psi(x)\, dx = \langle -\mathrm{i}\phi', \psi \rangle. \tag{6.52}$$

The calculation above holds for all $\psi \in \mathfrak{D}(T)$ and all $\phi$ in the domain

$$\mathfrak{D}(T^\dagger) = \{\psi \in L^2([0,1]) \mid \psi' \in L^2([0,1])\}. \tag{6.53}$$

The boundary condition on $\psi$ is sufficient for the vanishing of the boundary term in (6.51), no additional conditions are required for $\phi$. The domain $\mathfrak{D}(T^\dagger)$ is strictly larger than $\mathfrak{D}(T)$. Hence the adjoint operator $T^\dagger$ is a nontrivial extension of $T$. We conclude that $T$ is symmetric, but not self-adjoint.

EXAMPLE 6.6.2. The operator $H_D = -\frac{1}{2}\frac{d^2}{dx^2}$ on the domain of twice differentiable functions with Dirichlet boundary conditions is symmetric, as was shown by the calculation in the previous section. Here the partial integration introduces boundary terms in (6.26) which vanish only if both $\phi$ and $\psi$ satisfy the Dirichlet boundary condition. Hence in this case the domain of $H_D^\dagger$ is exactly the same as $\mathfrak{D}(H_D)$ and therefore $H_D$ is self-adjoint on this domain.

EXAMPLE 6.6.3. The operator of multiplication by a real-valued function $f(\mathbf{x})$,

$$(T\psi)(\mathbf{x}) = f(\mathbf{x})\, \psi(\mathbf{x}), \tag{6.54}$$

is self-adjoint on the domain

$$\mathfrak{D}(T) = \left\{ \psi \in L^2(\mathbb{R}^n) \,\Big|\, \int |f(\mathbf{x})\psi(\mathbf{x})|\, d^n x < \infty \right\}. \tag{6.55}$$

This example includes the position operator and the operator of potential energy.

If $T$ is symmetric but not self-adjoint, then it may happen that $T^\dagger$ is not symmetric. In this case the operator $T^\dagger$ may even have complex eigenvalues, as it is the case for the momentum operator with Dirichlet boundary conditions (Example 6.6.1). The function $\phi(x) = e^x$ is differentiable on $[0, 1]$ and

$$T^\dagger \phi = -\mathrm{i}\phi' = -\mathrm{i}\phi \tag{6.56}$$

shows that $\phi$ is an eigenfunction of $T^\dagger$ with eigenvalue $-\mathrm{i}$.

If the adjoint operator $T^\dagger$ is also symmetric, then one can show that $T^\dagger$ is in fact self-adjoint (i.e., $T^{\dagger\dagger} = T^\dagger$). In this case the operator $T$ is called *essentially self-adjoint.*

One of the reasons why the self-adjoint operators are so important is Stone's theorem, which states that the one-parameter unitary groups are in one-to-one correspondence with the self-adjoint operators.

THEOREM 6.2. *If $U(t)$ is a one-parameter unitary group, then the generator $H$ is a self-adjoint operator. Conversely, if $H$ is a self-adjoint operator, then $H$ is the generator of is a unique one-parameter unitary group $U(t)$.*

PROOF. A mathematically rigorous proof of Stone's theorem is beyond the scope of this book. We refer the reader to books on functional analysis or mathematical physics. □

Usually, one writes the unitary group generated by $H$ as

$$U(t) = \mathrm{e}^{-\mathrm{i}Ht}. \tag{6.57}$$

We found already that this notation is meaningful in the case of the free Schrödinger operator. For arbitrary self-adjoint operators, $U(t)$ can be defined as the exponential function of $H$ with the help of the spectral theorem. The notation also makes sense in view of the evolution equation associated to $U(t)$, which is just a differentiation rule for the exponential function of $H$:

$$\mathrm{i}\,\frac{d}{dt}\,\mathrm{e}^{-\mathrm{i}Ht}\,\psi = H\,\mathrm{e}^{-\mathrm{i}Ht}\,\psi, \qquad \text{for all } \psi \in \mathfrak{D}(H). \tag{6.58}$$

If $\psi$ is an eigenfunction of $H$ belonging to the eigenvalue $E$, then

$$\mathrm{e}^{-\mathrm{i}Ht}\,\psi = \mathrm{e}^{-\mathrm{i}Et}\,\psi. \tag{6.59}$$

# 6.7. Translation Group

The energy observable $H$ generates the time evolution. The unitary groups generated by other observables are of similar importance.

## 6.7.1. Translations

For a one-dimensional wave function $\psi$ and for real numbers $a$,

$$\psi_a(x) = \psi(x - a) \tag{6.60}$$

is a function obtained from $\psi$ by a translation (if $a > 0$, the function is shifted to the right). The mapping

$$\tau_a : \psi \to \psi_a \qquad \text{translation by } a \tag{6.61}$$

is unitary. In fact, $a \to \tau_a$ is a one-parameter unitary group in the sense of Definition 6.2.

We now calculate the generator of this group. To this purpose we consider a differentiable wave function and form the expression

$$\mathrm{i}\,\frac{\tau_a\psi(x) - \psi(x)}{a} = \mathrm{i}\frac{\psi(x-a) - \psi(x)}{a} \longrightarrow -\mathrm{i}\frac{d}{dx}\psi(x). \tag{6.62}$$

Thus, the generator of translations is the momentum operator $p$ and we may write

$$\tau_a = \mathrm{e}^{-\mathrm{i}pa}. \tag{6.63}$$

If the function $\psi$ can be differentiated infinitely often, we can expand $\psi(x-a)$ in a Taylor series around $x$:

$$\begin{aligned}
\psi(x-a) &= \psi(x) - a\psi'(x) + \frac{a^2}{2}\psi''(x) - \cdots \\
&= \sum_{n=0}^{\infty} \frac{a^n}{n!}\Bigl(-\frac{d}{dx}\Bigr)^n \psi(x) \\
&= \sum_{n=0}^{\infty} \frac{a^n}{n!}(-\mathrm{i}p)^n\psi(x).
\end{aligned}$$

This is just the power series of the exponential function $\exp(-\mathrm{i}pa)$. Of course, the unitary translation group $\tau_a$ can be applied to all functions in the Hilbert space, while the power series is meaningless for functions which are not differentiable. Nevertheless, this consideration indicates that the power series

$$\mathrm{e}^{-\mathrm{i}pa}\psi = \sum_{n=0}^{\infty} \frac{(-\mathrm{i}a^n)}{n!}\,p^n\psi \tag{6.64}$$

does make sense on suitable (dense) linear subsets of the Hilbert space.

Ψ A wave function $\psi$ is differentiable (in the ordinary way) if the limit in Eq. (6.62) exists in a pointwise sense for each $x$ (as a limit of complex numbers). The momentum operator has a larger domain. For $\psi$ in this domain it is only required that the limit exists in $L^2$: This simply means that there is a $\phi \in L^2$ such that

$$\lim_{a\to 0} \int |\frac{\psi(x-a)-\psi(x)}{a} - \phi(x)|^2 \, dx = 0. \tag{6.65}$$

The momentum operator has been defined in Section 4.2.2 as the inverse Fourier transform of the multiplication operator in momentum space,

$$p = -\mathrm{i}\frac{d}{dx} = \mathcal{F}^{-1} k \mathcal{F}. \tag{6.66}$$

Because the multiplication operator is self-adjoint on the maximal domain (Example 6.6.3), and because the Fourier transform is unitary, the momentum operator is self-adjoint on the domain of functions $\psi$ for which $k\hat{\psi}(k)$ is square-integrable. These are functions that are differentiable in the generalized sense of Eq. (2.90). The calculation above suggests that equivalently the generalized derivative can be defined by the limit in (6.65).

The relation between multiplication operator and momentum operator is a special case of the following useful theorem:

THEOREM 6.3. *If $U$ is unitary and $T$ is self-adjoint on $\mathfrak{D}(T)$, then the operator $S = UTU^\dagger$ is self-adjoint on $\mathfrak{D}(S) = U\mathfrak{D}(T)$. Moreover, the unitary groups are related by*

$$U \mathrm{e}^{-\mathrm{i}Tt} U^\dagger = \mathrm{e}^{-\mathrm{i}St}. \tag{6.67}$$

### 6.7.2. Translations in momentum space

For a real number $b$ we define the operator of multiplication by $\exp(\mathrm{i}xb)$,

$$\mu_b \, \psi(x) = \mathrm{e}^{\mathrm{i}xb} \, \psi(x). \tag{6.68}$$

The operators $\mu_b$ form a unitary group with parameter $b$. The generator of this group is $-x$. We already know (see Section 2.6.1) that $\mu_b$ describes a translation in momentum space,

$$\mathrm{e}^{\mathrm{i}xb}\psi(x) \longleftrightarrow \hat{\psi}(k-b). \tag{6.69}$$

Thus, the unitary groups $\tau_a$ and $\mu_b$ are related by a Fourier transform,

$$\mathcal{F} \, \mathrm{e}^{\mathrm{i}xb} \, \mathcal{F}^{-1} = \mathrm{e}^{-\mathrm{i}pb}. \tag{6.70}$$

We finally note that the unitary groups $\exp(-ipa)$ and $\exp(ixb)$ have straightforward generalizations to higher dimensions,

$$e^{-i\mathbf{p}\cdot\mathbf{a}} = e^{-i(p_1 a_1 + p_2 a_2 + \ldots + p_n a_n)} = e^{-ip_1 a_1} e^{-ip_2 a_2} \cdots e^{-ip_n a_n} \tag{6.71}$$

$$e^{i\mathbf{x}\cdot\mathbf{b}} = e^{i(x_1 b_1 + x_2 b_2 + \ldots + x_n b_n)} = e^{ix_1 b_1} e^{ix_2 b_2} \cdots e^{ix_n b_n} \tag{6.72}$$

CD 2.7 lets you play with the translations in position space and momentum space. You may change the parameters $a$ and $b$ manually and watch the effect on the wave functions both in position and momentum space.

## 6.8. Weyl Relations

Two bounded linear operators $A$ and $B$ are said to *commute* if

$$AB\psi = BA\psi, \qquad \text{for all } \psi \in \mathfrak{H}. \tag{6.73}$$

It is easy to see that the translations in position space do not commute with the translations in momentum space,

$$\tau_a\, \mu_b\, \psi(x) = \tau_a\, e^{ibx}\, \psi(x) = e^{ib(x-a)}\, \psi(x-a), \tag{6.74}$$

$$\mu_b\, \tau_a\, \psi(x) = \mu_b\, \psi(x-a) = e^{ibx}\, \psi(x-a). \tag{6.75}$$

This implies the Weyl relations.

**Weyl relations:**

The unitary groups $\tau_a$ (translations in position space) and $\mu_b$ (translations in momentum space) satisfy the relations

$$\tau_a\, \mu_b = e^{-iab}\, \mu_b\, \tau_a, \qquad \text{for all } a \text{ and } b \text{ in } \mathbb{R}, \tag{6.76}$$

that is,

$$e^{-ipa}\, e^{ixb} = e^{-iab}\, e^{ixb}\, e^{-ipa} \qquad \text{for all } a \text{ and } b \text{ in } \mathbb{R}. \tag{6.77}$$

The noncommutativity of the translations in position and momentum space is visualized in CD 2.8. A wave packet is first translated by $a$ in position space, then shifted by $b$ in position space. After that the inverse operations are applied, first the translation by $-a$, then the momentum shift by $-b$. It can be seen that the final result differs from the original wave packet by a phase factor.

We may deduce from this a formula which will be useful later on. First we show that the operators

$$W(s) = \tau_s \, \mu_s \, \mathrm{e}^{\mathrm{i}s^2/2} \tag{6.78}$$

form a one-parameter unitary group. Using the Weyl relations, it is easy to prove the group property

$$W(s)\, W(t) = \tau_s \, \mu_s \, \tau_t \, \mu_t \, \mathrm{e}^{\mathrm{i}(s^2+t^2)/2} \tag{6.79}$$

$$= \tau_s \, \mathrm{e}^{\mathrm{i}st} \, \tau_t \, \mu_s \, \mu_t \, \mathrm{e}^{\mathrm{i}(s^2+t^2)/2} \tag{6.80}$$

$$= \tau_{s+t} \, \mu_{s+t} \, \mathrm{e}^{\mathrm{i}(s^2+2st+t^2)/2} = W(s+t) \tag{6.81}$$

and the unitarity

$$W(s)^\dagger = \mu_{-s} \, \tau_{-s} \, \mathrm{e}^{-\mathrm{i}s^2/2} = \mathrm{e}^{\mathrm{i}s^2} \, \tau_{-s} \, \mu_{-s} \, \mathrm{e}^{-\mathrm{i}s^2/2} = W(-s). \tag{6.82}$$

The generator of $W(s)$ can be determined by a differentiation. Because the groups $\tau_a$ and $\mu_b$ are generated by $p$ and $x$,

$$\mathrm{i}\frac{d}{ds}\, \tau_s \psi = p \, \tau_s \psi, \qquad \mathrm{i}\frac{d}{ds}\, \mu_s \psi = (-x)\, \mu_s \psi, \tag{6.83}$$

we find (using the product rule for the differentiation) that

$$\mathrm{i}\frac{d}{ds}\, W(s)\psi = p \, \tau_s \, \mu_s \, \mathrm{e}^{\mathrm{i}s^2/2}\psi - \tau_s \, x \, \mu_s \, \mathrm{e}^{\mathrm{i}s^2/2}\psi + \tau_s \, x \, \mu_s \, (-s)\, \mathrm{e}^{\mathrm{i}s^2/2}\psi. \tag{6.84}$$

With $\tau_s \, x = (x-s)\tau_s$ this simplifies to

$$\mathrm{i}\frac{d}{ds}\, W(s)\psi = (p - x)\, W(s)\psi. \tag{6.85}$$

Hence we find that the infinitesimal generator of the unitary group $W(s)$ is the operator $p - x$ (which is self-adjoint on a suitable domain). We may write

$$\mathrm{e}^{-\mathrm{i}(p-x)s} = \mathrm{e}^{-\mathrm{i}ps} \, \mathrm{e}^{\mathrm{i}xs} \, \mathrm{e}^{\mathrm{i}s^2/2} = \mathrm{e}^{\mathrm{i}xs} \, \mathrm{e}^{-\mathrm{i}ps} \, \mathrm{e}^{\mathrm{i}s^2/2}. \tag{6.86}$$

A slight generalization of the calculation above proves the following useful formula.

**Another form of the Weyl relations:**

$$\mathrm{e}^{-\mathrm{i}(ap-bx)} = \mathrm{e}^{-\mathrm{i}pa} \, \mathrm{e}^{\mathrm{i}xb} \, \mathrm{e}^{\mathrm{i}ab/2} = \mathrm{e}^{\mathrm{i}xb} \, \mathrm{e}^{-\mathrm{i}pa} \, \mathrm{e}^{-\mathrm{i}ab/2}. \tag{6.87}$$

# 6.9. Canonical Commutation Relations

Two self-adjoint operators $P$ and $X$ are said to satisfy the *canonical commutation relations* if

$$[P, X] = PX - XP = -\mathrm{i}\,\mathbf{1} \tag{6.88}$$

holds on a suitable (dense) domain. You have learned in Section 4.5 that the position and momentum operators satisfy the canonical commutation relations.

The Weyl relations imply the canonical commutation relations. Let $\exp(-\mathrm{i}At)$ and $\exp(-\mathrm{i}Bs)$ be unitary groups satisfying the Weyl relations

$$\mathrm{e}^{-\mathrm{i}At}\,\mathrm{e}^{-\mathrm{i}Bs} = \mathrm{e}^{-\mathrm{i}Bs}\,\mathrm{e}^{-\mathrm{i}At}\,\mathrm{e}^{\mathrm{i}st}. \tag{6.89}$$

Then the infinitesimal generators satisfy the canonical commutation relations,

$$[A, B] = -\mathrm{i}. \tag{6.90}$$

This can be shown by applying $d^2/dsdt$ to both sides of Eq. (6.89) and then setting $s = t = 0$.

If two operators $P$ and $X$ satisfy the canonical commutation relations, then they cannot both be bounded. Indeed, the momentum and position operators are both unbounded. In particular, the relation $[P, X] = -\mathrm{i}\,\mathbf{1}$ cannot hold for matrices in a finite dimensional vector space. This can be seen by taking the trace on both sides of this equation, which leads to a contradiction.

$\boxed{\Psi}$ The canonical commutation relations follow from the Weyl relations, but the converse is not true: The canonical commutation relations do not imply the Weyl relations. There are examples of self-adjoint operators $A$ and $B$, that satisfy the canonical commutation relations on an invariant dense domain $\mathfrak{D}$. (The invariance means that $A : \mathfrak{D} \to \mathfrak{D}$ and $B : \mathfrak{D} \to \mathfrak{D}$, so that the products $AB$ and $BA$ and the commutator are well defined on $\mathfrak{D}$.) The operators $A$ and $B$ in these examples are both essentially self-adjoint on $\mathfrak{D}$, and still the unitary groups do not fulfill the Weyl relations. An additional condition ($A^2 + B^2$ be essentially self-adjoint on $\mathfrak{D}$) is required in order to derive the Weyl relations from the canonical commutation relations.

CD 3.8 shows the quantum-mechanical motion of a free particle in phase space (the space formed by position and momentum coordinates of a particle). The animation shows the function $f(x,k,t) = \psi(x,t)\hat{\psi}(k,t)$ with the position coordinate on the horizontal axis and the momentum coordinate on the vertical axis. In classical mechanics, the state would be described by a point in phase space that moves in horizontal direction with constant velocity. The quantum particle is only approximately localized around the classical position in phase space and spreads with time in the horizontal direction.

## 6.10. Commutator and Uncertainty Relation

You know already that the uncertainty of an observable $A$ in a state $\psi$ (with $\|\psi\| = 1$) is defined as

$$\Delta_\psi A = \|(A - \langle A\rangle_\psi)\psi\| = \sqrt{\langle (A - \langle A\rangle_\psi)^2\rangle_\psi}. \tag{6.91}$$

It describes how the measured values of $A$ are scattered around their mean value $\langle A\rangle_\psi = \langle \psi, A\psi\rangle$. In Section 2.8.1 Heisenberg's uncertainty relation was derived by making use of the fact that the commutator of momentum and position operators is $\mathrm{i}[(-\mathrm{i}\nabla), \mathbf{x}] = n$ (see Eq. (2.117)). It is a quite general observation that the commutator $[A, B]$ poses a restriction on the uncertainties of the two observables $A$ and $B$ for the same state $\psi$.

**General uncertainty relation**:

Suppose $A$ and $B$ are self-adjoint operator, and let $\psi$ be in the domain of the commutator $\mathrm{i}[A, B]$. Then

$$\Delta_\psi A\ \Delta_\psi B \geq \tfrac{1}{2}\,|\,\langle [A,B]\,\rangle_\psi\,|. \tag{6.92}$$

Thus, if the observables $A$ and $B$ do not commute, the uncertainties of $A$ and $B$ cannot both be small simultaneously.

PROOF. For an arbitrary real number $x$ consider the quantity

$$\|(A - \mathrm{i}xB)\psi\|^2 = \|B\psi\|^2 x^2 - x\,\langle \psi, \mathrm{i}[A,B]\,\psi\rangle + \|A\psi\|^2 \tag{6.93}$$

which is always non-negative. The well-known condition for $ax^2 + bx + c \geq 0$ is $b^2 - 4ac \leq 0$. Hence we must have

$$\langle \psi, \mathrm{i}[A,B]\psi\rangle^2 \leq 4\,\|B\psi\|^2\,\|A\psi\|^2. \tag{6.94}$$

If you replace $A$ by $A - \langle A\rangle_\psi$ and $B$ by $B - \langle B\rangle_\psi$, you will find that the commutator is not changed because

$$[A - a, B - b] = [A, B], \quad \text{for arbitrary real numbers } a \text{ and } b. \tag{6.95}$$

Notice that for $A$ and $B$ self-adjoint, the expression $\mathrm{i}[A,B]$ is symmetric and has only real expectation values. Therefore, we obtain

$$\|(A - \langle A\rangle_\psi)\psi\|^2 \, \|(B - \langle B\rangle_\psi)\psi\|^2 \geq \tfrac{1}{4} \,|\, \langle \psi, [A,B]\, \psi\rangle \,|^2, \tag{6.96}$$

which in view of Eq (6.91) is equivalent with the general uncertainty relation. □

EXERCISE 6.5. *For symmetric operators $A$ and $B$ the expression* $\mathrm{i}\,[A,B]$ *is also symmetric.*

EXERCISE 6.6. *The function $f(\lambda) = \|(A-\lambda)\psi\|^2$ is minimal for $\lambda = \langle A\rangle_\psi$.*

## 6.11. Symmetries and Conservation Laws

Two (possibly unbounded) operators $A$ and $B$ are said to *commute*, if the corresponding unitary groups commute,[2] that is, if

$$[\mathrm{e}^{-\mathrm{i}sA}, \mathrm{e}^{-\mathrm{i}tB}] = 0, \quad \text{for all } s \text{ and } t. \tag{6.97}$$

The observables corresponding to commuting operators are called *compatible*. For compatible observables there is no a priori bound on the accuracy of simultaneously determined values.

When the Hamiltonian $H$ of a physical system commutes with some self-adjoint operator $T$, then the Schrödinger equation has the following invariance property: For any solution $\psi(t) = \exp(-\mathrm{i}Ht)\,\psi_0$, the expression $\phi_a(t) = \exp(-\mathrm{i}Ta)\psi(t)$ (for arbitrary $a \in \mathbb{R}$) is again a solution. The proof is very simple:

$$\phi_a(t) = \mathrm{e}^{-\mathrm{i}Ta}\,\psi(t) = \mathrm{e}^{-\mathrm{i}Ta}\,\mathrm{e}^{-\mathrm{i}Ht}\,\psi_0 = \mathrm{e}^{-\mathrm{i}Ht}\,\mathrm{e}^{-\mathrm{i}Ta}\,\psi_0.$$

This shows that $\phi_a(t)$ is the solution corresponding to the initial condition $\phi_a(0) = \exp(-\mathrm{i}Ta)\psi_0$. A unitary group that commutes with the time evolution is called a *symmetry* or *invariance* of the Schrödinger equation.

The compatibility of the Hamiltonian with the observable $T$ means that the transformation generated by $T$ commutes with the time evolution. Under certain conditions it follows that $T$ is a conserved quantity, a *constant of*

[2]Notice that the definition of commutativity in terms of the unitary groups avoids domain questions because the unitary groups are bounded and defined everywhere in the Hilbert space. If $A$ and $B$ are bounded, then (6.97) is equivalent to (6.73).

*motion*. Formally (i.e., without worrying about domains) we can differentiate the commutator of the commuting groups,

$$\begin{aligned} 0 &= \frac{d}{da}\Big(\mathrm{e}^{-\mathrm{i}Ta}\,\mathrm{e}^{-\mathrm{i}Ht} - \mathrm{e}^{-\mathrm{i}Ht}\,\mathrm{e}^{-\mathrm{i}Ta}\Big) \\ &= (-\mathrm{i}T)\,\mathrm{e}^{-\mathrm{i}Ta}\,\mathrm{e}^{-\mathrm{i}Ht} - \mathrm{e}^{-\mathrm{i}Ht}\,(-\mathrm{i}T)\,\mathrm{e}^{-\mathrm{i}Ta} \\ &= -\mathrm{i}\,\mathrm{e}^{-\mathrm{i}Ta}\Big(T\,\mathrm{e}^{-\mathrm{i}Ht} - \mathrm{e}^{-\mathrm{i}Ht}\,T\Big) \end{aligned} \tag{6.98}$$

and obtain

$$T\,\mathrm{e}^{-\mathrm{i}Ht} = \mathrm{e}^{-\mathrm{i}Ht}\,T. \tag{6.99}$$

This relation says in particular that the expectation values of $T$ are constant in time:

$$\begin{aligned} \langle T\rangle_{\psi(t)} &= \langle \psi(t)\,, T\,\psi(t)\rangle = \langle \mathrm{e}^{-\mathrm{i}Ht}\psi_0\,, T\,\mathrm{e}^{-\mathrm{i}Ht}\psi_0\rangle \\ &= \langle \mathrm{e}^{-\mathrm{i}Ht}\psi_0\,, \mathrm{e}^{-\mathrm{i}Ht}\,T\,\psi_0\rangle = \langle \psi_0\,, T\,\psi_0\rangle = \langle T\rangle_{\psi_0}. \end{aligned} \tag{6.100}$$

Probably you know Noether's theorem from classical mechanics. It states that a symmetry of the system is related to a conservation law. Here we have obtained the quantum-mechanical version.

**Noether's theorem:**

If the Schrödinger equation has a symmetry with respect to some unitary group, then the observable corresponding to the self-adjoint generator of the group is a constant of motion.

EXAMPLE 6.11.1. The free Schrödinger equation is invariant with respect to translations. If $\psi(x,t)$ is the solution belonging to some initial function $\psi_0(x)$, then the translated function $\psi(x-a,t)$ is a solution corresponding to the translated initial state $\psi_0(x-a)$. The generator $p = -\mathrm{i}d/dx$ of the translations is the momentum operator. For free particles, the momentum is a constant of motion.

Another important point is the connection between symmetry and degeneracy of eigenvalues. Consider a system that has a symmetry with respect to some unitary group $\exp(-\mathrm{i}Ta)$. Assume that the system is in an eigenstate $\psi$ of the Hamiltonian, that is, $H\psi = E\psi$, where $E$ is the eigenvalue. Because $H$ commutes with the unitary group, you can see that $\psi_a = \exp(-\mathrm{i}Ta)\psi$ is also an eigenvector of $H$ belonging to the same eigenvalue $E$:

$$H\,\psi_a = H\,\mathrm{e}^{-\mathrm{i}Ta}\,\psi = \mathrm{e}^{-\mathrm{i}Ta}\,H\,\psi = \mathrm{e}^{-\mathrm{i}Ta}\,E\,\psi = E\,\psi_a. \tag{6.101}$$

Hence, if you have found an eigenvector $\psi$ of $H$, the symmetry will allow you to find many other eigenvectors $\psi_a$, $a \in \mathbb{R}$, all belonging to the same eigenvalue. If the newly found eigenvectors are orthogonal to the first one, the multiplicity of the eigenvalue will be greater than one, and one could say that symmetry implies degeneracy. But this need not always be the case. Consider a situation where $\psi$ is not only an eigenvector of $H$, but also an eigenvector of $T$, the generator of the symmetry transformation. We have $T\psi = \lambda\psi$ with some real number $\lambda$. Then the transformed eigenvector $\psi_a = \exp(-\mathrm{i}\lambda a)\psi$ differs only by a phase factor from $\psi$, that is, $\psi$ and $\psi_a$ are the same state and the degeneracy is not increased by the symmetry.

$\boxed{\Psi}$ When you write down Eq. (6.100) you have to assume that both the initial value $\psi_0$ and the corresponding solution $\psi(t)$ are in the domain of $T$. Assuming that $H$ and $T$ commute, see Eq. (6.97), one can indeed prove that there is a dense invariant subspace $\mathfrak{D}$ contained in the domains of the operators $H$ and $T$, which is invariant with respect to the unitary groups,

$$e^{-\mathrm{i}Ht}\,\mathfrak{D} \subset \mathfrak{D}, \quad e^{-\mathrm{i}Ta}\mathfrak{D} \subset \mathfrak{D}, \quad \text{for all real numbers } a \text{ and } t. \tag{6.102}$$

Here, for example, the set $\exp(-\mathrm{i}Ht)\mathfrak{D}$ is defined as $\{\exp(-\mathrm{i}Ht)\psi \mid \psi \in \mathfrak{D}\}$. On the invariant domain $\mathfrak{D}$ the commutator of the operators $H$ and $T$ vanishes,

$$[H, T] = 0 \quad \text{on } \mathfrak{D}, \tag{6.103}$$

and it makes sense to write

$$T\,e^{-\mathrm{i}Ht} = e^{-\mathrm{i}Ht}\,T \quad \text{on } \mathfrak{D}. \tag{6.104}$$

However, the converse is *not* true. The vanishing of the commutator $[H, T]$ on some dense domain does *not* imply that the operators $H$ and $T$ commute in the sense of Eq. (6.97) (there are counterexamples). For this reason, the vanishing of the commutator $[H, T]$ and the commutativity of the operators $H$ and $T$ are not synonymous in the mathematical literature.

*Chapter 7*

# Harmonic Oscillator

**Chapter summary**: The harmonic oscillator is among the most important examples of explicitly solvable problems, whether in classical or quantum mechanics. It appears in every textbook in order to demonstrate some general principles by explicit calculations. In some respect, the classical and quantum properties of the harmonic oscillator are similar. For example, the classical observation that the oscillation frequency does not depend on the amplitude corresponds to the fact that all quantum states are periodic in time with the same period.

Like the particle in a box, the harmonic oscillator is a system that has only bound states. There is a discrete set of allowed energies and an associated basis of energy eigenfunctions. All energy eigenstates can be generated from the ground state by repeated application of a creation operator $A^\dagger$. This approach to the solution of the eigenvalue problem is related to the supersymmetry of the harmonic oscillator.

It is rewarding to calculate the time dependence of position and momentum observables, which leads to the conclusion that their expectation values always follow the laws of classical mechanics. Moreover, all eigenstates of the harmonic oscillator are also eigenstates of the Fourier transformation $\mathcal{F}$, hence the motion in momentum space looks exactly like the motion in position space. If the initial state is a translated eigenstate, then the position and momentum distributions oscillate back and forth without changing their shape.

Among the more mathematical results obtained in this chapter is a proof of the completeness of the eigenfunctions and the calculation of the integral kernel of the unitary time-evolution operator (Mehler kernel). As a result, the Schrödinger equation for an arbitrary initial function can be solved with an integration. Of particular interest is the behavior of Gaussian wave packets, because their motion is very similar to the motion of a classical particle. The most particlelike states are the coherent states—Gaussian functions that optimize the uncertainty relation for all times. The corresponding initial state is a translated ground state and can also be defined as an eigenvector of the annihilation operator $A$.

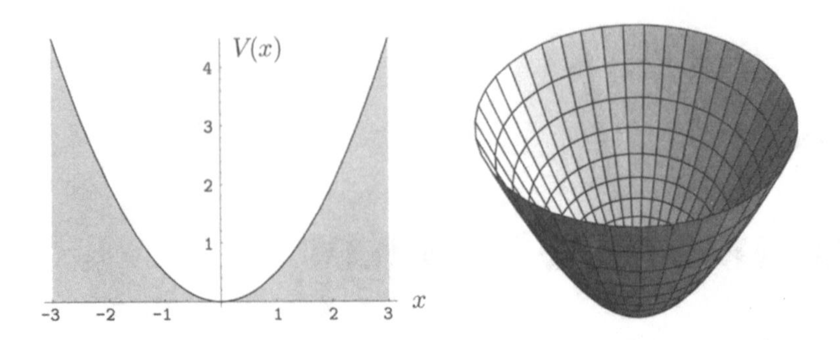


FIGURE 7.1. The harmonic oscillator potential in one and two space dimensions with spring constant $k = 1$.

## 7.1. Basic Definitions and Properties

### 7.1.1. Classical mechanics

The harmonic oscillator is defined as a particle subject to a linear force field. In classical mechanics, the harmonic oscillator can be pictured as a pointlike mass attached to a spring. The spring is idealized in the sense that it has no mass and can be stretched infinitely in both directions. The force of the spring is directed toward the equilibrium point and is proportional to the distance of the particle from the equilibrium position and to a material constant $k$.

The force $F$ can be expressed in terms of a potential function $V$,

$$F(x) = -kx = -\frac{d}{dx}V(x), \qquad V(x) = \frac{kx^2}{2}. \tag{7.1}$$

In classical mechanics, the constant $k$ is called the *spring constant*. It describes the stiffness of the spring.

The generalization to higher dimensions is straightforward. With $\mathbf{x} = (x_1, \dots, x_n)$ we have (see Fig. 7.1)

$$F(\mathbf{x}) = -k\mathbf{x} = -\nabla V(\mathbf{x}), \qquad V(\mathbf{x}) = k\,\frac{\mathbf{x}\cdot\mathbf{x}}{2}. \tag{7.2}$$

The spring constant $k$ determines the *oscillator frequency* $\omega$,

$$\omega = \sqrt{\frac{k}{m}}, \qquad k = m\omega^2. \tag{7.3}$$

The motion of a classical particle in a harmonic oscillator potential is given by

$$x(t) = a\,\sin(\omega t - b), \qquad p(t) = m\,\dot{x}(t) = m\omega a\,\cos(\omega t - b), \tag{7.4}$$

where the constants $a$ and $b$ have to be determined from given initial conditions $x(0) = x_0$ and $p(0) = p_0$. Hence the classical motion is an oscillation with frequency $\nu = \omega/2\pi$, which is independent of the amplitude. The motion takes place between the classical turning points $x = \pm a$. The interval $[-a, a]$ will be called the *classically allowed region.*

CD 5.1 investigates the classical motion of a harmonic oscillator.

The Hamiltonian function of the harmonic oscillator in classical mechanics is the sum of the kinetic energy (expressed as a function of the momentum) and the potential energy $V(x)$, that is,

$$H(x,p) = \frac{p^2}{2m} + \frac{kx^2}{2}. \tag{7.5}$$

### 7.1.2. Quantum mechanics

We can now apply the substitution rule $p \to -\mathrm{i}\hbar\, d/dx$, $x \to$ multiplication by $x$, to the classical Hamiltonian function. We obtain the quantum-mechanical Hamiltonian operator

$$H = -\frac{\hbar^2}{2m}\frac{d^2}{dx^2} + \frac{kx^2}{2}, \tag{7.6}$$

which acts on square-integrable wave functions $\psi$ in the Hilbert space $L^2(\mathbb{R})$. The time evolution of a state of a quantum harmonic oscillator is then described by a solution of the (time-dependent) Schrödinger equation

$$\mathrm{i}\hbar\frac{d}{dt}\psi(x,t) = H\,\psi(x,t). \tag{7.7}$$

The harmonic oscillator is the prototypical case of a system that has only bound states: All states remain under the influence of the force field for all times; no state can escape toward infinity. Although such a system does not exist in nature, the harmonic oscillator is often used to approximate the motion of more realistic systems in the neighborhood of a stable equilibrium point (see Fig. 7.2).

A system has a stable equilibrium at $x = x_0$ if the potential function $V(x)$ of the system has a minimum at $x = x_0$. If the minimum is characterized by $V'(x_0) = 0$ and $V''(x_0) > 0$, we can approximate the potential near $x = x_0$ by the first coefficients of its Taylor series:

$$V(x) = V(x_0) + \frac{(x-x_0)^2}{2}V''(x_0) + O((x-x_0)^3). \tag{7.8}$$

The harmonic oscillator potential of Eq. (7.1) is obtained if we set $x_0 = 0$, $V(0) = 0$, and $V''(0) = k$. (Counterexample: The potential function

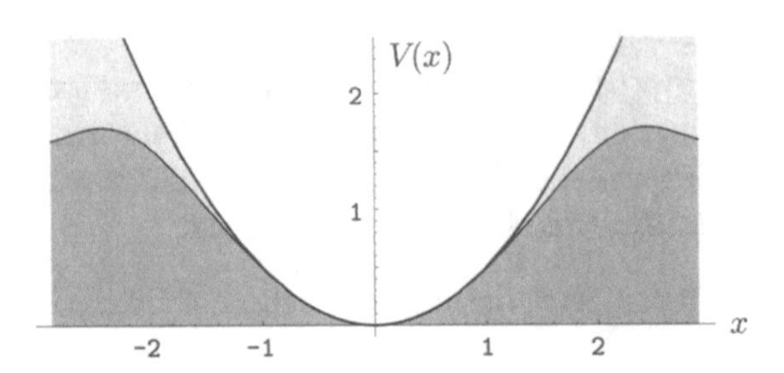


FIGURE 7.2. Approximation of a potential function near an equilibrium by a harmonic oscillator potential

$V(x) = x^4$ cannot be approximated by an oscillator potential). If the local minimum of the potential is sufficiently deep, some solutions of the Schrödinger equation with potential $V$ can be approximated (at least for some time) by the motion in a harmonic oscillator potential. In this sense, for example, the motion of a pendulum near the equilibrium is described by a harmonic oscillator. It should be noted, however, that many mathematical properties of the Schrödinger equation depend on the global rather than local behavior of the potential function.

EXERCISE 7.1. *Investigate the classical motion of the harmonic oscillator for various initial conditions. The initial condition $x_0 = 0$, $p_0 = 0$ corresponds to the stable equilibrium: The particle remains at rest at the minimum of the harmonic oscillator potential. Find the solution $(x(t), p(t))$ corresponding to the initial condition $(x_0, p_0)$. Describe the classical motion of the harmonic oscillator in two dimensions.*

EXERCISE 7.2. *Describe the classical motion in the potential*

$$V(x) = k\frac{(x - x_0)^2}{2} + V_0. \tag{7.9}$$

EXERCISE 7.3. *Show that $H(x(t), p(t)) = E$ is a constant of motion. It is a unique feature of the harmonic oscillator that the oscillation frequency of the motion is independent of the energy $E$ of the oscillator. But the amplitude $a$ of the classical motion does depend on the energy. Show that the amplitude is given by*

$$a = \sqrt{2E/k}. \tag{7.10}$$

### 7.1.3. Scaling transformation of the Hamiltonian

For the graphical representation as well as for the mathematical investigation it is useful to change the units of length, time, and energy in order to simplify the expressions. This can be done using an appropriate scaling transformation.

We consider a scaling transformation $x \to \tilde{x} = \lambda x$ with some $\lambda > 0$. Given a scalar function $\psi(x)$, the scaled function $\tilde{\psi}$ is defined by

$$\tilde{\psi}(\tilde{x}) = \psi(x). \tag{7.11}$$

The derivative of $\tilde{\psi}$ with respect to $\tilde{x}$ can be calculated as

$$\frac{d}{dx}\,\psi(x) = \frac{d}{dx}\,\tilde{\psi}(\tilde{x}) = \frac{d}{d\tilde{x}}\,\tilde{\psi}(\tilde{x})\,\frac{d\tilde{x}}{dx} = \lambda\,\frac{d}{d\tilde{x}}\,\tilde{\psi}(\tilde{x}). \tag{7.12}$$

For the second derivative (operator of kinetic energy) we get

$$\frac{d^2}{dx^2}\,\psi(x) = \lambda^2\,\frac{d^2}{d\tilde{x}^2}\,\tilde{\psi}(\tilde{x}). \tag{7.13}$$

The behavior of the position operator under a scaling transformation is given by

$$x\,\psi(x) = \frac{\tilde{x}}{\lambda}\,\tilde{\psi}(\tilde{x}), \quad x^2\,\psi(x) = \frac{\tilde{x}^2}{\lambda^2}\,\tilde{\psi}(\tilde{x}). \tag{7.14}$$

The action of the Schrödinger operator on the scaled function can thus be described as follows.

$$H\psi(x) = \Bigl(-\frac{\hbar^2}{2m}\,\lambda^2\,\frac{d^2}{d\tilde{x}^2} + \frac{k}{2}\,\frac{1}{\lambda^2}\,\tilde{x}^2\Bigr)\tilde{\psi}(\tilde{x}). \tag{7.15}$$

### 7.1.4. Dimensionless units

Because of the opposite scaling behavior of the two summands in Eq. (7.15) we can choose the scaling parameter $\lambda = \lambda_0$ such that

$$\frac{\hbar^2}{2m}\,\lambda_0^2 = \frac{k}{2}\,\frac{1}{\lambda_0^2}, \tag{7.16}$$

that is,

$$\lambda_0^2 = \sqrt{\frac{km}{\hbar^2}} \quad \text{or} \quad \lambda_0 = \sqrt{\frac{m\omega}{\hbar}}, \tag{7.17}$$

where $\omega$ is the oscillator frequency defined in Eq. (7.3). Thus, we define the new position variable by

$$\tilde{x} = x\sqrt{\frac{m\omega}{\hbar}}. \tag{7.18}$$

This means that the displacement will be measured in multiples of the length $\sqrt{\hbar/m\omega}$. According to Eq. (7.15) the Schrödinger operator becomes

$$H\psi(x) = \frac{\hbar\omega}{2}\Bigl(-\frac{d^2}{d\tilde{x}^2} + \tilde{x}^2\Bigr)\,\tilde{\psi}(\tilde{x}). \tag{7.19}$$

The Schrödinger equation (7.7) can be further simplified if we also scale the time variable and define $\tilde{t} = \omega t$ such that

$$\omega \frac{d}{d\tilde{t}} \tilde{\psi}(\tilde{x}, \tilde{t}) = \frac{d}{dt} \psi(x,t). \tag{7.20}$$

The Schrödinger equation in the new units finally reads (after division by $\hbar\omega$)

$$\mathrm{i} \frac{d}{d\tilde{t}} \tilde{\psi}(\tilde{x}, \tilde{t}) = \frac{1}{2}\Bigl(-\frac{d^2}{d\tilde{x}^2} + \tilde{x}^2\Bigr) \tilde{\psi}(\tilde{x}, \tilde{t}). \tag{7.21}$$

If we redefine the scale of the energy such that $\tilde{E} = E/\hbar\omega$, then the Hamiltonian operator

$$\tilde{H} = -\frac{1}{2}\frac{d^2}{d\tilde{x}^2} + \frac{\tilde{x}^2}{2} \tag{7.22}$$

is the operator for the energy of the harmonic oscillator measured in the new units.

The new variables $\tilde{t} = \omega t$ and $\tilde{x} = (m\omega/\hbar)^{(1/2)} x$ have dimensionless units because $\omega = \sqrt{k/m}$ has dimension 1/[time], and $\hbar$ has the dimension of an action (= [energy]×[time] = [mass]×[length]$^2$/[time]).

From now on, the tilde on the symbols for quantities in dimensionless units will be omitted.

### 7.1.5. Orders of magnitude

In order to estimate the expected order of magnitude of the phenomena encountered here, it is useful to insert the actual values for the physical constants. We assume that $m$ is the mass of an electron moving in an oscillator potential with spring constant $k = 1$. The numerical values of the physical constants are

$$m = 0.9109 \cdot 10^{-30}\mathrm{kg}, \qquad \hbar = 1.0546 \cdot 10^{-34}\mathrm{Joule}\cdot\mathrm{sec}, \tag{7.23}$$

and the scales for time, length, and energy are set by

$$\omega = 1.048 \cdot 10^{15}/\mathrm{sec}, \qquad \sqrt{\frac{m\omega}{\hbar}} = 3.008 \cdot 10^{9}/\mathrm{meter}, \tag{7.24}$$

$$\frac{1}{\hbar\omega} = 9.050 \cdot 10^{18}/\mathrm{Joule}. \tag{7.25}$$

In the following we shall use the dimensionless units. This should be remembered in particular when viewing the pictures: For electrons in the field of a harmonic oscillator with spring constant $k = 1$ the connection

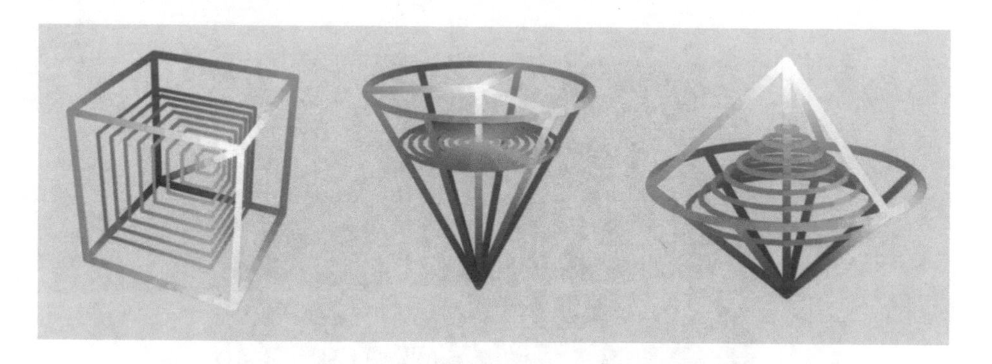

COLOR PLATE 1. Color manifold in various representations: The RGB color cube, the HSB cone, and the HLS double-cone. Each figure indicates the shape of the manifold and shows a collection of lines with constant saturation on a surface with brightness 0.75. (Section 1.2.2.)

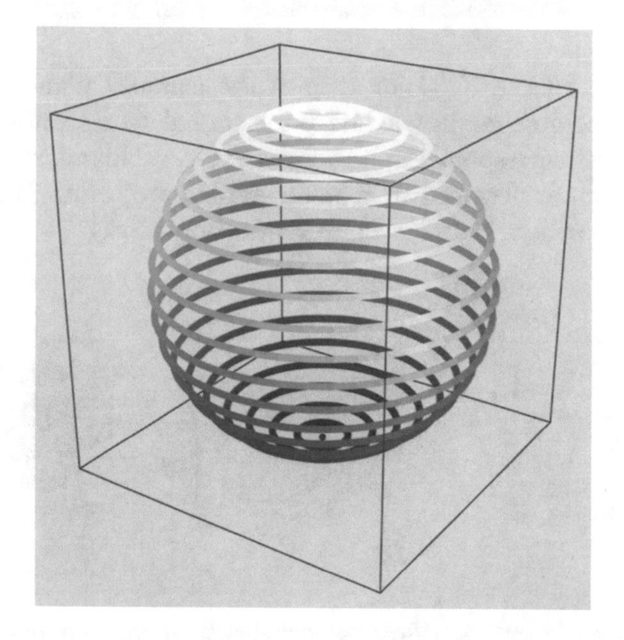

COLOR PLATE 2. The color map of the sphere is defined by mapping the surface of the color manifold in the HLS system to the sphere. With the help of a stereographic projection, the colored sphere can be used to define a color map of the complex plane. (Section 1.2.3.)

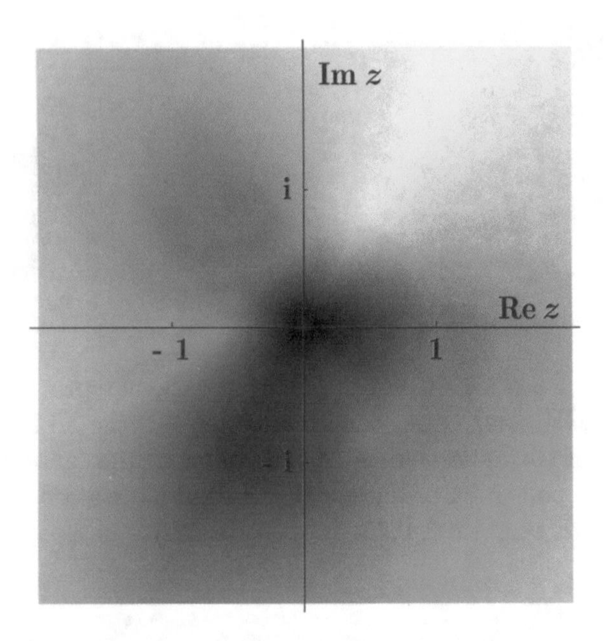


Color Plate 3. Color map of the complex plane. Each complex number has a hue proportional to its phase, the lightness corresponds to its absolute value. This color map of the plane is obtained by a stereographic projection from the colored sphere in Color Plate 2. (Section 1.2.3.)

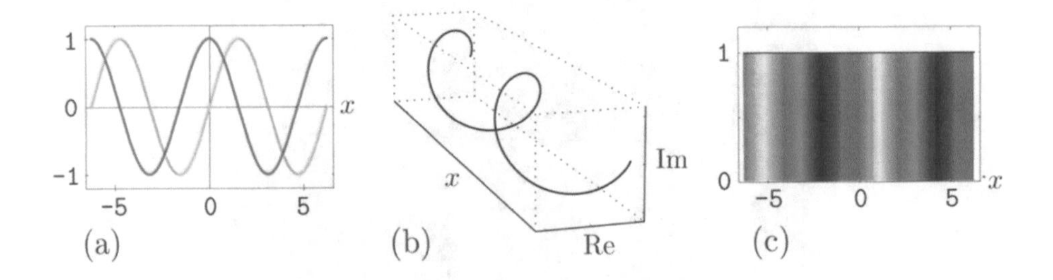


Color Plate 4. Various visualizations of the complex-valued function $x \to \exp(ix)$. See CD 1.8 for other examples. (a) Separate plots of the real part (red) and the imaginary part (yellow-green), (b) representation as a space curve, (c) plot of the absolute value with a color code for the phase. (Section 1.3.1.)

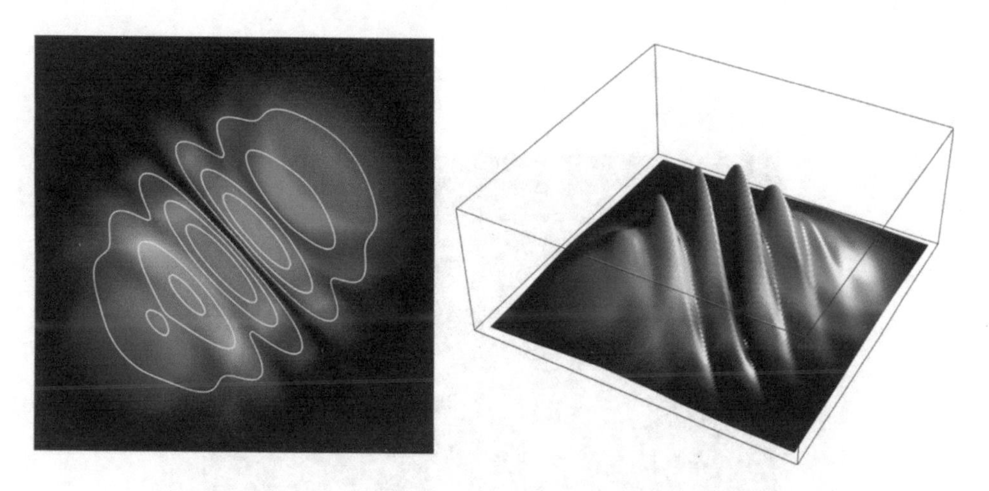

Color Plate 5. Visualizations of a wave function in two dimensions. The left graphic shows the function as a "density plot" with additional contour lines for the absolute value. In the three-dimensional surface plot the height of the surface gives the absolute value of the wave function. In both cases, the color describes the complex values according to Color Plate 3. (Section 1.3.2.)

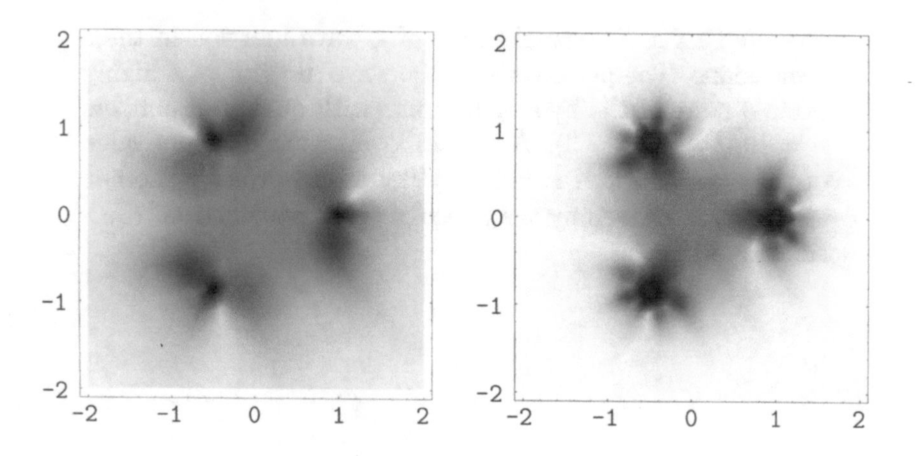


Color Plate 6. Visualization of the function $\psi(x, y) = (x + \mathrm{i}y)^3 - 1$ using the color map of Color Plate 3. The left graphic shows the function $\psi$ and the right graphic shows its square $\psi^2$. The zeros of $\psi^2$ are of second order. This can be easily recognized because all colors appear twice on a small circle around each zero. (Section 1.3.2.)

COLOR PLATE 7. Visualization of a wave function in three dimensions. The picture shows the wave function of a highly excited state of the hydrogen atom (with quantum numbers $n = 10$, $l = 5$, $m = 3$). A certain level of the absolute value of the wave function is indicated by an isosurface. The hue of the color is given by the phase of the wave function. (Section 1.3.2.)

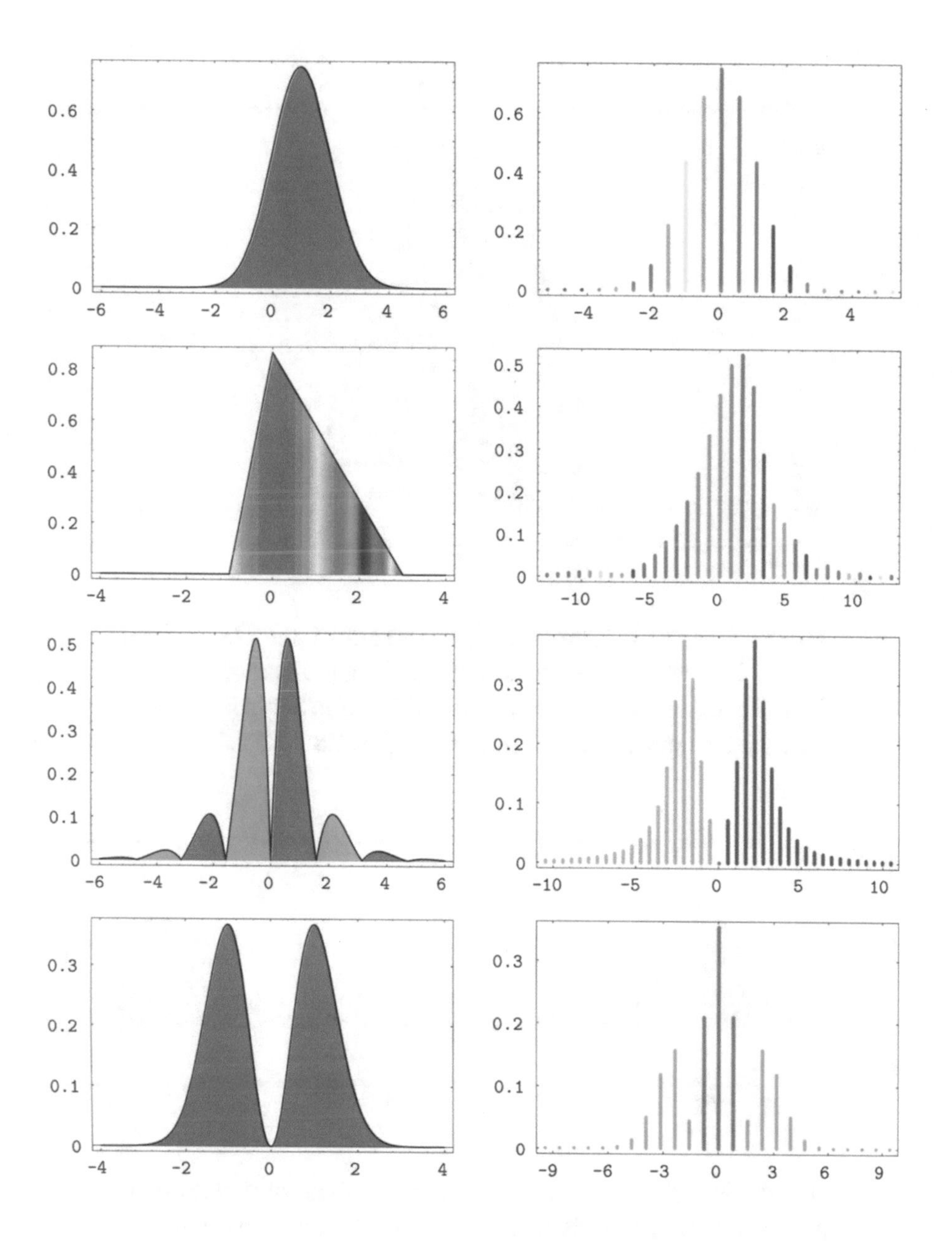


COLOR PLATE 8. Various functions and the spectrum of the Fourier amplitudes $\hat{\psi}(k_n^{(L)})$ defined in Eq. (2.13), with $k_n^{(L)} = n\pi/L$. The lines describe the absolute values (length) and the phases (color) of the Fourier amplitudes. (Section 2.1.2.)

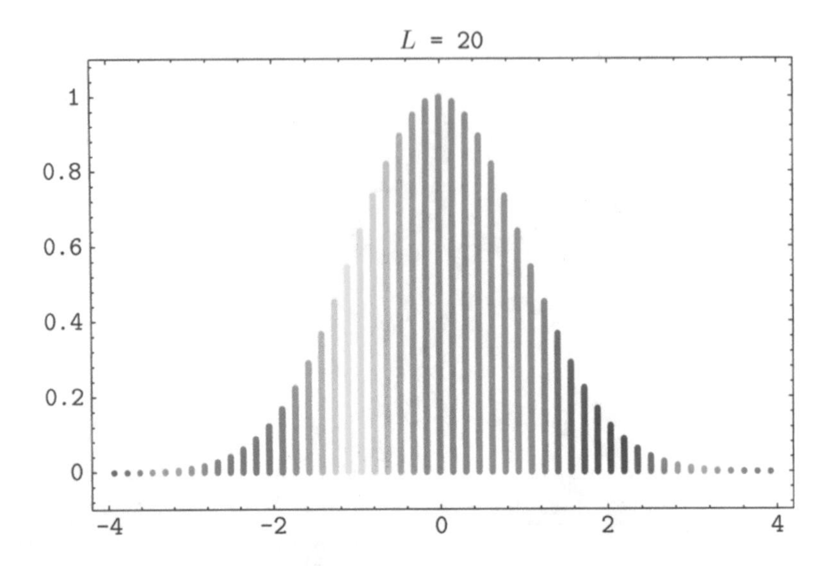


COLOR PLATE 9. The Fourier spectrum of the Gaussian in Fig. 8, but with respect to a much larger interval $[-L, L]$. This illustrates the transition from the Fourier spectrum to the Fourier transformed function $\hat{\psi}$. (Section 2.3.1.)

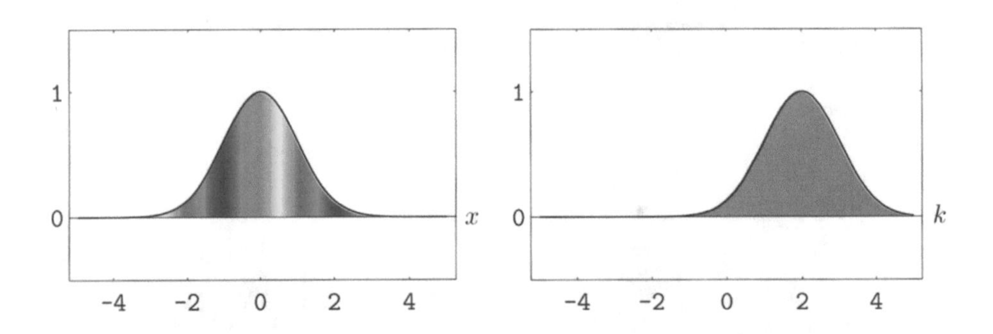


COLOR PLATE 10. The Fourier transform of a Gaussian function $\exp(-x^2/2)$ is again a Gaussian function. The picture shows the function $\exp(2ix - x^2/2)$ (left) and its Fourier transform $\exp(-(k-2)^2/2)$ (right). The translation by 2 in momentum space corresponds to a phase shift by $2ix$ in position space. (Section 2.6.1.)

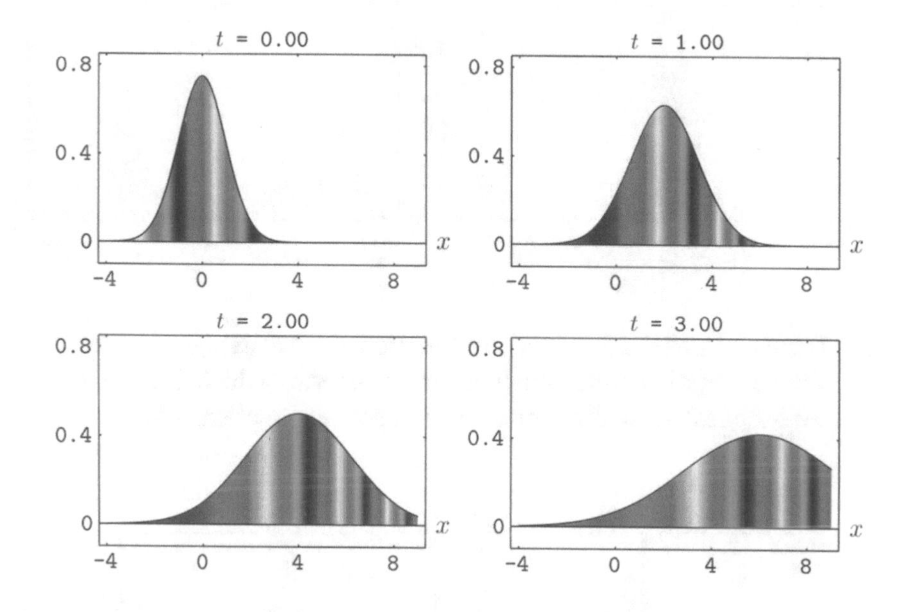


COLOR PLATE 11. Time evolution of a Gaussian wave packet with average momentum 2. It can be seen that the maximum of the wave packet moves according to classical physics with velocity 2. During the time evolution the wave packet spreads and contributions of higher momenta accumulate in front of the maximum. (Section 3.3.2.)

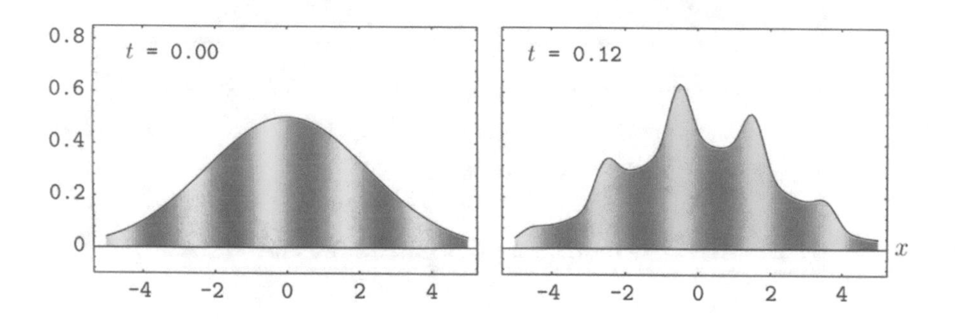


COLOR PLATE 12. There is always a flow in the direction of increasing phase. Hence if a yellow region of a wave packet is surrounded by red, then the wave function will increase in the yellow region. (Section 3.5.)

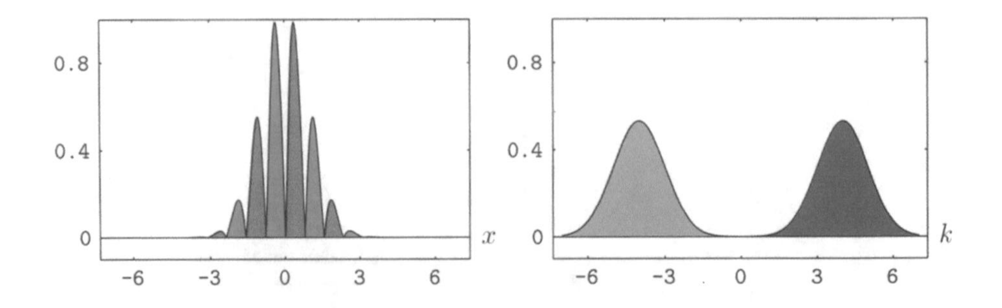


COLOR PLATE 13. Here the function $\sin(4x)\exp(-x^2/2)$ is shown together with its Fourier transform, which has two well-separated peaks in momentum space. (Section 3.6.)

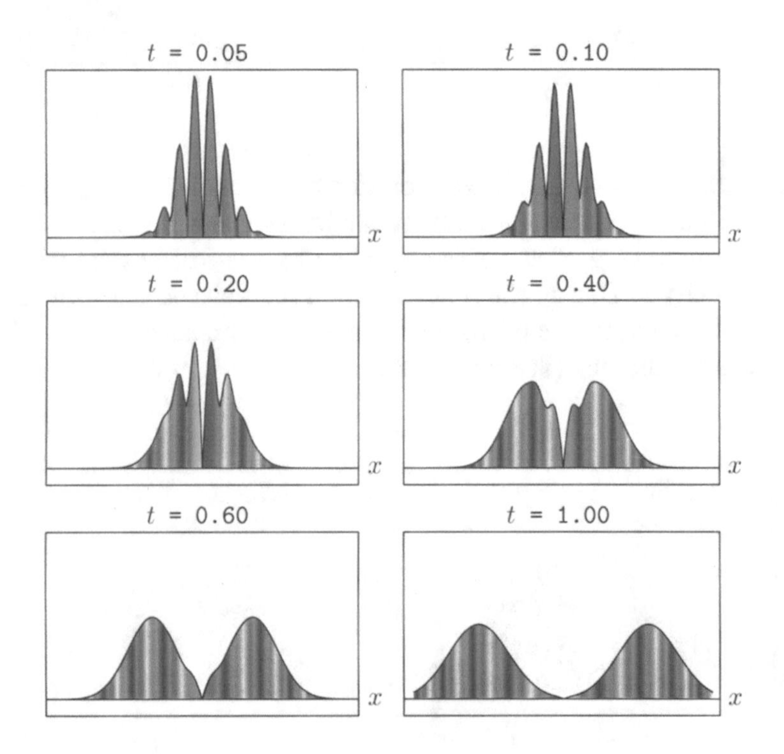


COLOR PLATE 14. Some snapshots from the time evolution of the function in Color Plate 13. For sufficiently large times the localization in position space can be understood from the distribution of momenta in the Fourier transform of the initial wave packet. (Section 3.6. See also CD 3.14.)

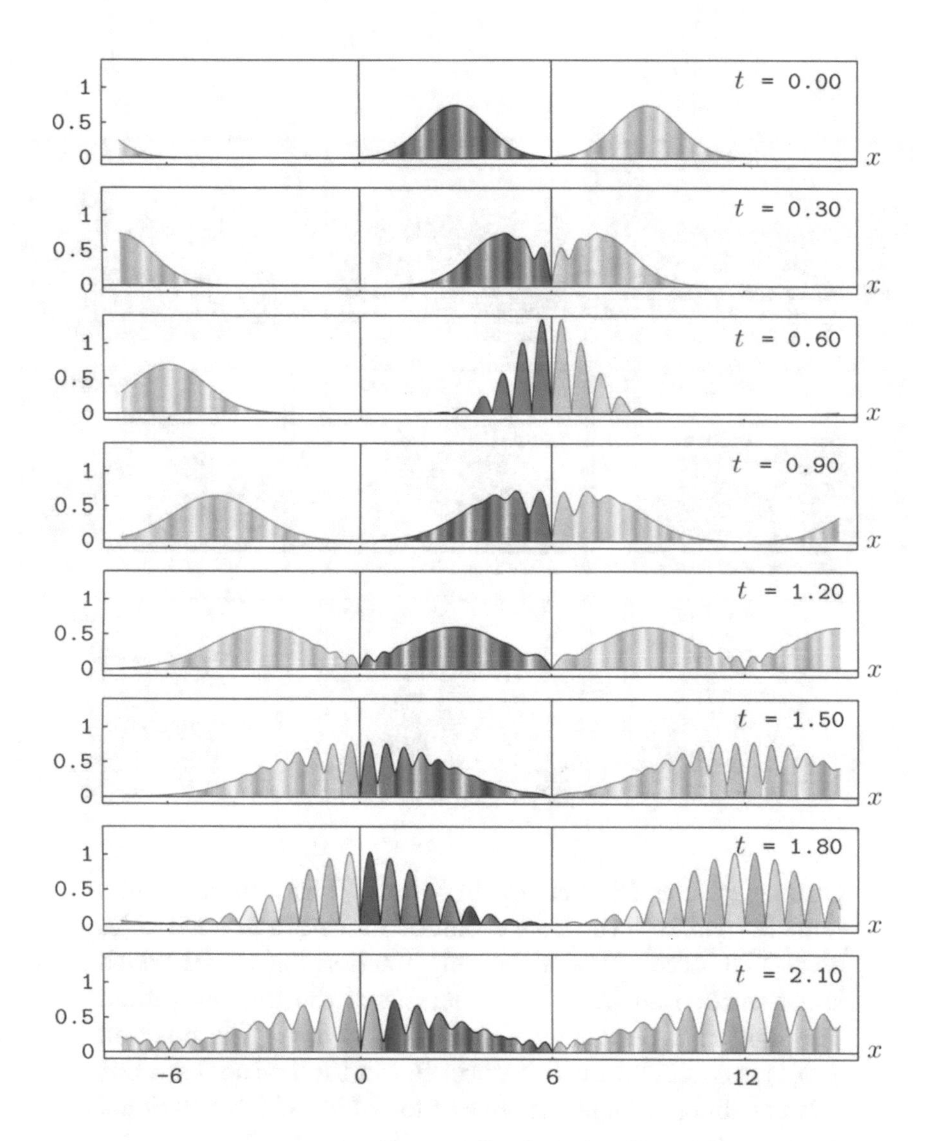


COLOR PLATE 15. Particle in a box by the method of mirrors. The initial function is a centered Gaussian function. The motion between the walls can be understood as a superposition of mirror wave packets. Initially, only the mirror waves that move toward the box are shown. The part of the wave function in the physical region (inside the box) is drawn with higher saturation. (Section 5.3.2.)

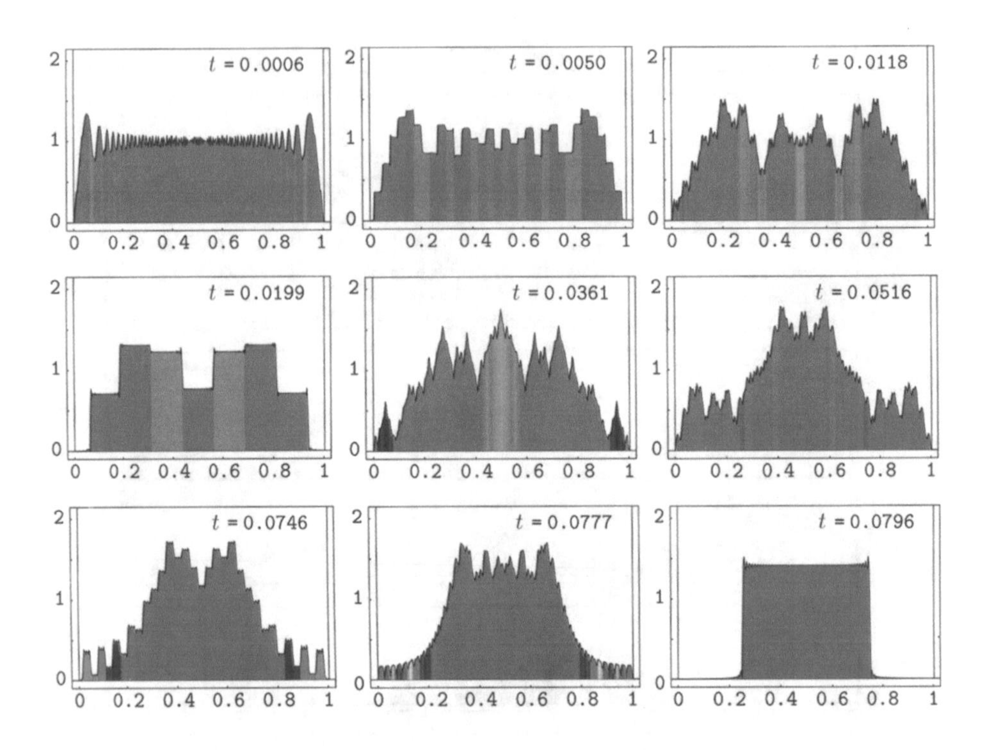


COLOR PLATE 16. Particle in a box. This graphic shows some frames from the time evolution of a state in a Dirichlet box. The initial state is the unit function $\psi_0(x) = 1$ which is not in the domain of the generator of the time evolution. The motion is periodic with period $T = 4/\pi$. The function $\psi(x,t)$ is continuous with respect to $t$ in the $L^2$-topology, but it is not differentiable. At times $t$ for which $t/T$ is a rational number, $\psi(x,t)$ is a step function. (Section 5.5.)

COLOR PLATE 17. Aharonov-Bohm effect. Scattering from the left at an obstacle with a magnetic field inside. The node line behind the obstacle is due to the influence of the magnetic vector potential. (Section 8.3.)

COLOR PLATE 18. Potential barrier with two holes (double-slit experiment). The initial state has an uncertainty of position which is larger than the distance between the two holes. So one cannot predict through which of the holes the particle will actually go. (Section 5.7.2.)

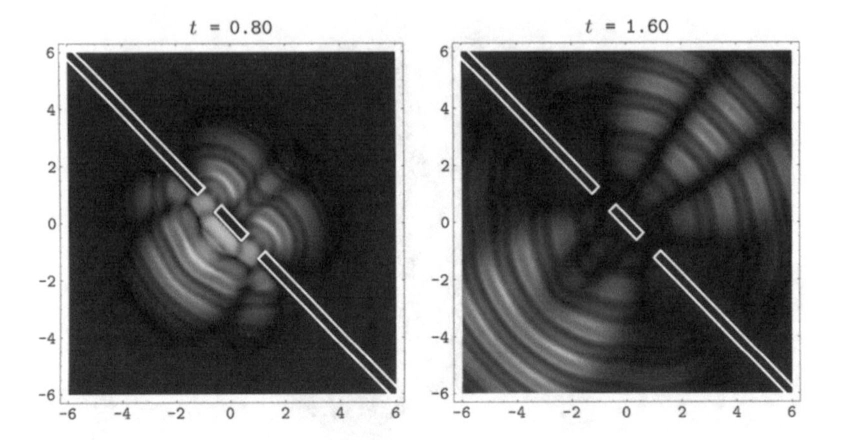


Color Plate 19. Time evolution of the initial function shown in Color Plate 18. A part of the wave function penetrates through the holes in the screen. Behind the screen emerges an interference pattern which shows that in certain directions there is only a small probability of observing the particle. In this example, the probability of being scattered to angles about $\pm 15^\circ$ off the forward direction has a minimum. (Section 5.7.4.)

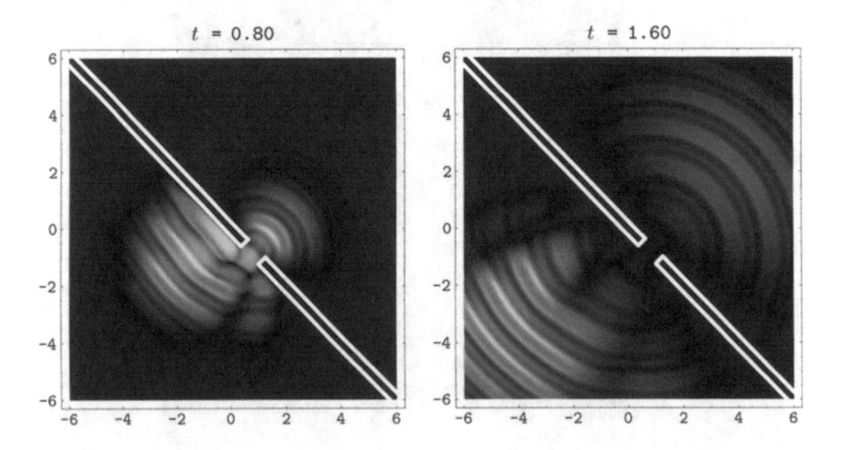


Color Plate 20. The double slit experiment with one hole closed. The wave emerging from the hole behind the screen is an approximately spherical wave without visible interference pattern, as one might expect from Huygens' principle. (Section 5.7.4.)

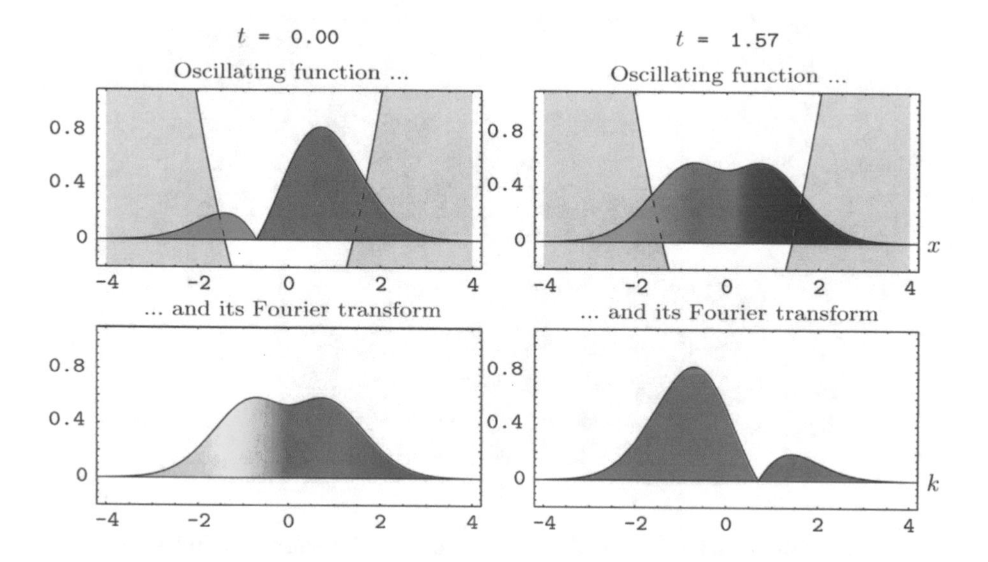


Color Plate 21. Two snapshots of an oscillating state in a harmonic-oscillator potential. The wave function and its Fourier transform are shown at times $t = 0$ and $t = \pi/2$. The initial state is a superposition of $\phi_0$ and $\phi_1$. (Section 7.3.)

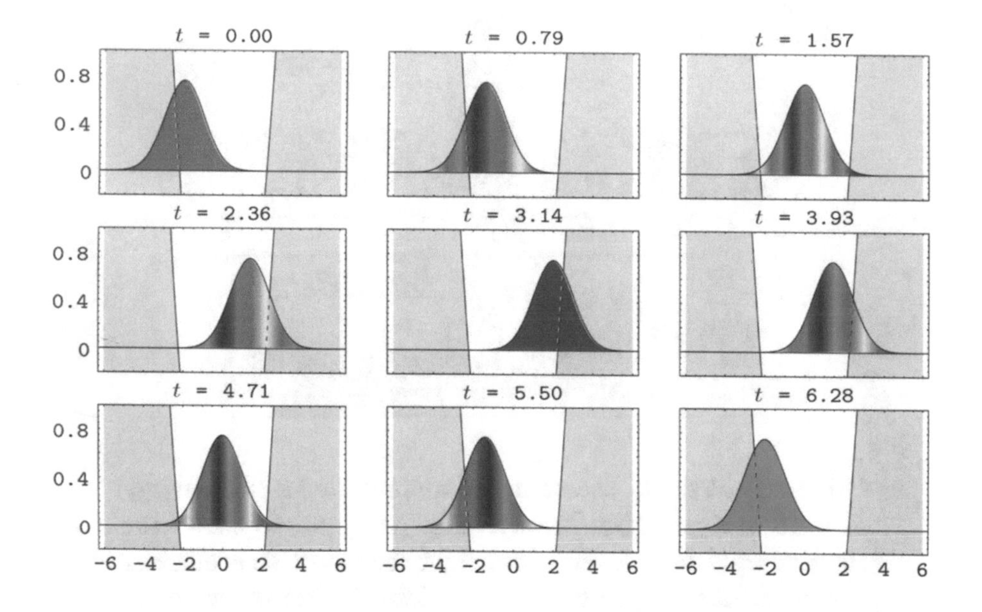


Color Plate 22. The time evolution of a coherent state of the harmonic oscillator. (Section 7.4.3.)

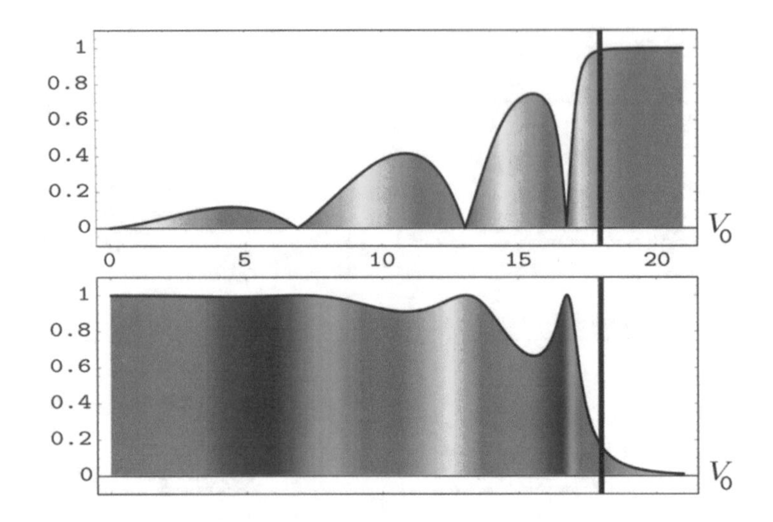


COLOR PLATE 23. Reflection and transmission coefficients for the scattering at a potential barrier. The barrier has a width $R = 1$ and the energy is $E = 18$. (Section 9.6.3.)

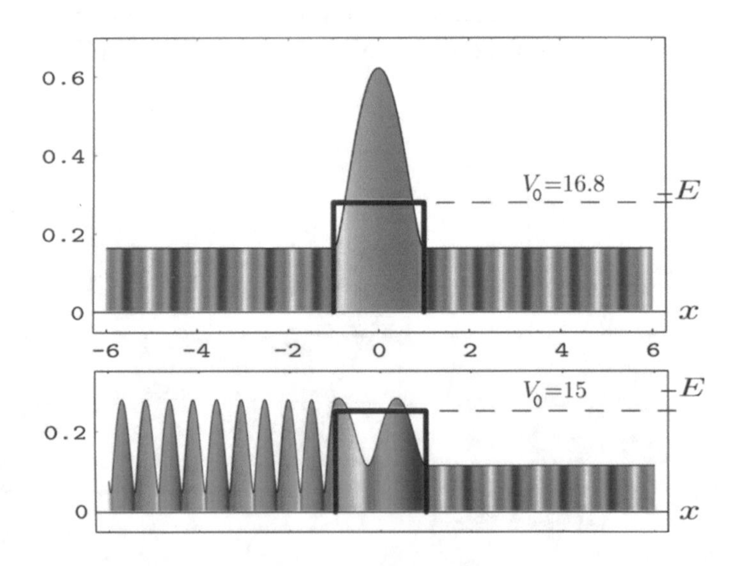


COLOR PLATE 24. Scattering of a plane waves with energy $E = 18$ at a potential barrier. The image above shows the solution at $V_0 = 16.8$. For this height the reflection coefficient is zero, as one can see from Color Plate 23. In the image below, at $V_0 = 15$, the part of the solution on the left-hand side shows the interference between an incoming and a reflected plane wave. (Section 9.6.3.)

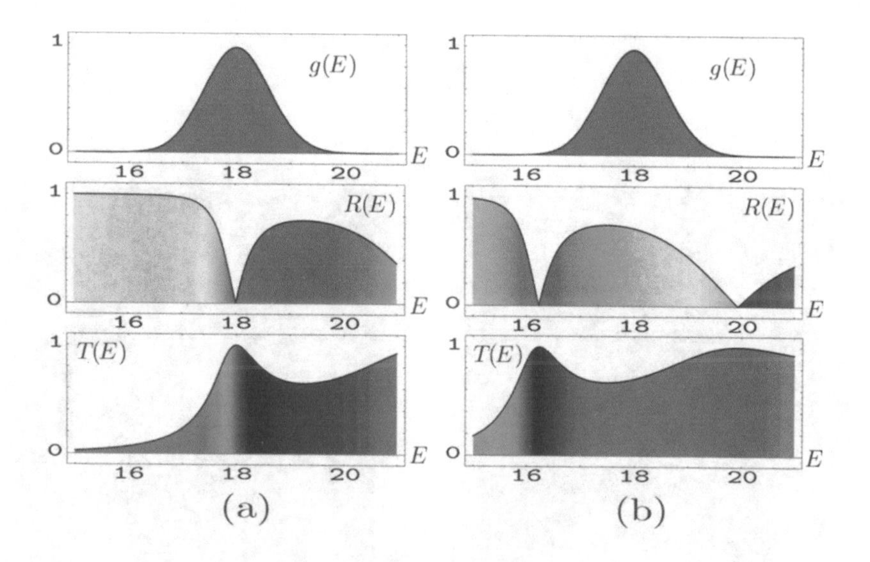


COLOR PLATE 25. Energy representation $g(E)$ of an incoming wave packet, and the scattering coefficients at a barrier. (a) $V_0 = 16.8$. (b) $V_0 = 15$. (Section 9.6.3.)

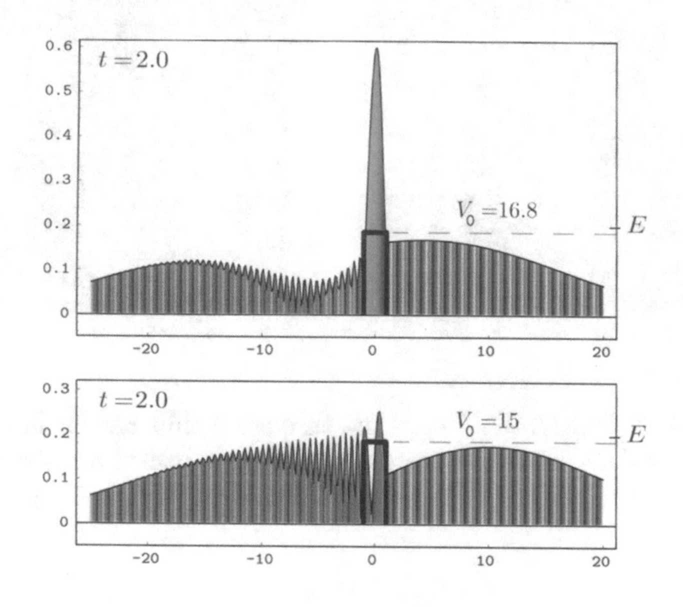


COLOR PLATE 26. Scattering with energies strictly higher than the potential barrier for the two situations of Color Plate 25. (Section 9.6.3.)

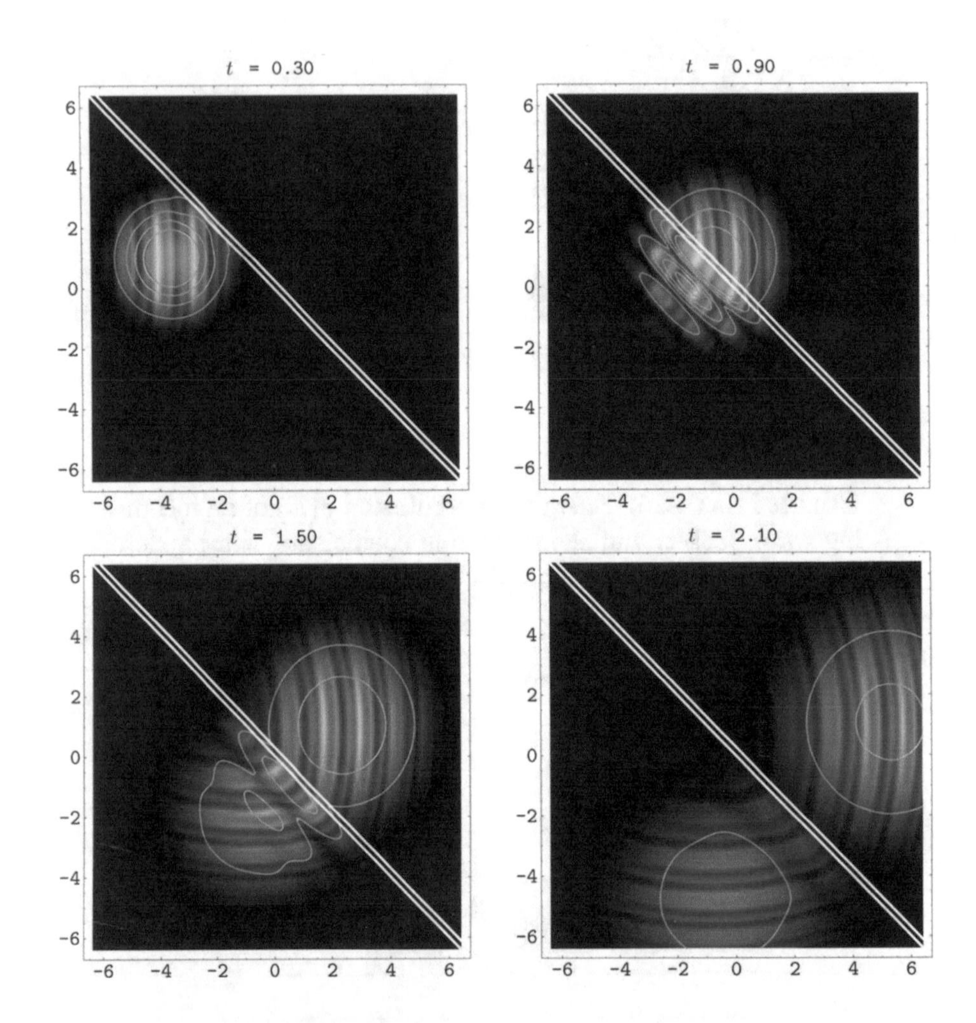


Color Plate 27. Tunneling through a thin barrier in two dimensions. Here the energies in the wave packet are strictly below the height of the potential barrier. (Section 9.7.)

of the dimensionless units with the units of the international system is as follows:

$$\text{length unit} \cong 3.324 \cdot 10^{-10} \text{ meters,} \tag{7.26}$$

$$\text{time unit} \cong 0.954 \cdot 10^{-15} \text{ seconds,} \tag{7.27}$$

$$\text{energy unit} \cong 1.105 \cdot 10^{-19} \text{ Joule.} \tag{7.28}$$

Bohr's radius of a hydrogen atom ($r_0 = 0.53 \cdot 10^{-10}$ meters) is 0.16 length units. The speed of light ($c = 2.9979 \cdot 10^8$ meters/second) has the value 861 in dimensionless units.

EXERCISE 7.4. *Assume that the spring constant $k$ is equal to the electromagnetic coupling constant $e$ (the elementary charge). Find the relations between the dimensionless units of length, time, and energy and the corresponding units of the international system.*

## 7.2. Eigenfunction Expansion

It can be expected from the classical mechanical analogy that the quantum time evolution of the harmonic oscillator will be more similar to the motion of a particle bouncing between two walls than to the free motion on the line $\mathbb{R}$. Therefore, our strategy to solve the Schrödinger equation for the harmonic oscillator will follow the method of Section 5.4.2. Hence the first step toward a solution of the initial-value problem is to solve the eigenvalue problem.

### 7.2.1. Eigenvalues of the Hamiltonian

The eigenvalue problem for the energy operator

$$H = -\frac{1}{2}\frac{d^2}{dx^2} + \frac{x^2}{2}. \tag{7.29}$$

consists in finding all numbers $E$ (eigenvalues) for which the stationary Schrödinger equation

$$H\psi = E\psi \tag{7.30}$$

has a nonzero square-integrable solution $\psi(x)$. For the one-dimensional situation considered here, Eq. (7.30) is an ordinary differential equation for a function $\psi$ depending only on the space variable $x$.

CD 5.2 lets you explore the first few energy-eigenstates of the quantum-mechanical harmonic oscillator.

When an eigenvalue $E$ is known together with a square-integrable solution $\phi(x)$ of the stationary Schrödinger equation, a solution $\psi(x,t)$ of Eq. (7.48) with initial value $\phi$ is given by

$$\psi(t,x) = \mathrm{e}^{-\mathrm{i}Et}\phi(x). \tag{7.31}$$

In fact, this is the unique solution, which is equal to $\phi(x)$ at $t=0$.

For the harmonic oscillator the complete solution of the eigenvalue problem is given in the box below. The mathematically interested reader will learn in Section 7.7 below how this solution can be derived. Here we discuss only the result and its implications.

**Eigenvalue problem for the harmonic oscillator**:

The equation

$$\left(-\frac{1}{2}\frac{d^2}{dx^2} + \frac{x^2}{2}\right)\phi(x) = E\,\phi(x) \tag{7.32}$$

has square-integrable solutions only for $E = E_n$, where

$$E_n = n + \frac{1}{2}, \quad n = 0,1,2,3,\ldots. \tag{7.33}$$

The unique solution belonging to $E_n$ is given by

$$\phi_n(x) = \left(\frac{1}{\pi}\right)^{1/4} \frac{1}{\sqrt{2^n\,n!}}\,H_n(x)\,\mathrm{e}^{-x^2/2}, \tag{7.34}$$

where $H_n(x)$ is the *Hermite polynomial* of order $n$. The *Hermite functions* $\phi_n$, $n = 0,1,2,\ldots$, form an orthonormal basis in the Hilbert space $L^2(\mathbb{R})$.

The Hermite polynomials are defined as

$$H_n(x) \equiv \sum_{j=0}^{[n/2]} (-1)^j \frac{n!}{(n-2j)!\,j!}(2x)^{n-2j}, \quad n = 0,1,2,3,\ldots \tag{7.35}$$

(the symbol $[n/2]$ denotes the greatest integer $\leq n/2$). For example,

$$H_0(x) = 1, \quad H_1(x) = 2x, \quad H_2(x) = 4x^2 - 2, \quad \ldots. \tag{7.36}$$

Figure 7.3 shows plots of $|\phi_n(x)|^2$ and the eigenvalues $E_n$.

EXERCISE 7.5. *Verify that the Hermite polynomials have the property*

$$H_{n+1}(x) = -\frac{d}{dx}H_n(x) + 2x\,H_n(x). \tag{7.37}$$

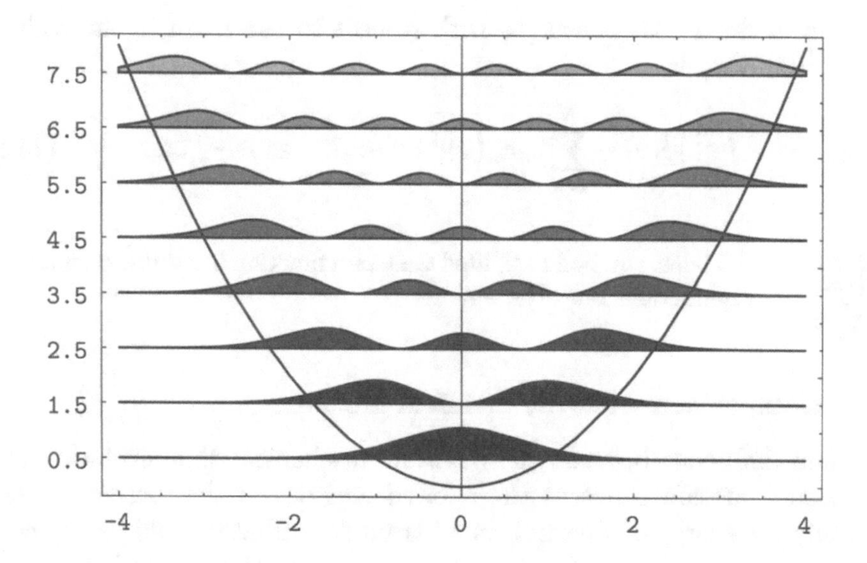


FIGURE 7.3. The position probability densities $|\phi_n(x)|^2$ of the lowest oscillator eigenstates drawn at a height corresponding to their energy. The horizontal lines show the values of the energy of the eigenstates. The part of a horizontal line inside the potential curve is the classically allowed region for a classical particle with that energy.

EXERCISE 7.6. *The Hermite polynomials can also be defined by*

$$H_n(x) = (-1)^n \exp(x^2) \frac{d^n}{dx^n} \exp(-x^2). \tag{7.38}$$

*Compare this formula with Eq.* (7.35).

### 7.2.2. Expansion into eigenfunctions

Every square-integrable wave function $\phi$ can be written as a linear combination

$$\phi(x) = \sum_{n=0}^{\infty} c_n \phi_n, \tag{7.39}$$

because the eigenfunctions $\phi_n$ of the harmonic oscillator form an orthonormal basis in the Hilbert space; see Section 5.4.4. The coefficients are given by

$$c_n = \langle \phi_n, \phi \rangle = \int_{-\infty}^{\infty} \phi_n(x)\, \phi(x)\, dx. \tag{7.40}$$

The sum in Eq. (7.39) converges with respect to the norm in the Hilbert space $L^2(\mathbb{R})$, that is,

$$\int_{-\infty}^{\infty} |\phi(x) - \sum_{n=0}^{N} c_n \phi_n(x)|^2 \, dx \to 0, \quad \text{as } N \to \infty. \tag{7.41}$$

CD 5.9 lets you build a shifted Gaussian function by adding oscillator eigenfunctions one after another.

### 7.2.3. Comparison with the classical motion

The main difference between the quantum-mechanical time evolution and the classical motion concerns the allowed energies of the system. While classically the energy of a particle in a harmonic oscillator could be any non-negative number $E \geq 0$, the energy of a quantum particle can only be one of the numbers $E_n$. The lowest energy accessible to the quantum oscillator is $E_0 = 1/2$. This can be understood with the help of the uncertainty relation: In order to have a small kinetic energy, the wave function of the ground state has to be well localized near $p = 0$ in momentum space. The potential energy can only be made small by concentrating the wave packet near the minimum of the potential. But, according to the uncertainty relation, this would extend the momentum distribution and thus increase the average kinetic energy. Hence the sum of kinetic and potential energy cannot be made arbitrarily small—even if the ground state minimizes the uncertainty relation, as it is the case for the harmonic oscillator. Indeed, the expectation value of the energy of a harmonic oscillator is

$$\begin{aligned}
\langle H \rangle_\phi &= \langle \phi, H\phi \rangle \\
&= \left\langle \phi, -\frac{1}{2}\frac{d^2}{dx^2}\phi \right\rangle + \left\langle \phi, \frac{x^2}{2}\phi \right\rangle \\
&= \frac{1}{2}\left\langle -\mathrm{i}\frac{d}{dx}\phi, -\mathrm{i}\frac{d}{dx}\phi \right\rangle + \frac{1}{2}\langle x\,\phi, x\,\phi \rangle \\
&= \frac{1}{2}\left\| -\mathrm{i}\frac{d}{dx}\phi \right\|^2 + \frac{1}{2}\|x\phi\|^2
\end{aligned} \tag{7.42}$$

Using the relation $a^2 + b^2 \geq 2ab$ (for all $a, b \in \mathbb{R}$), and the uncertainty principle (2.109), we can estimate this further as

$$\langle H \rangle_\phi \geq \left\| -\mathrm{i}\frac{d}{dx}\phi \right\| \|x\phi\| \geq \frac{1}{2}\|\phi\|^2. \tag{7.43}$$

Hence for a normalized state $\phi$ the expectation value of the energy must always be $\geq 1/2 = E_0$.

For all eigenstates $\phi_n$ of $H$ the expectation values of position and momentum are zero because the densities $|\phi_n(x)|^2$ and $|\hat{\phi}_n(k)|^2$ are symmetric with respect to the origin,

$$\langle x \rangle_{\phi_n} = 0, \qquad \langle k \rangle_{\phi_n} = 0. \tag{7.44}$$

This is also true for the classical motion, if we interpret the expectation value as the time average. We can evaluate a classical position probability by computing the fraction $dt/\pi$ of time per semi-period spent in an infinitesimal region $dx$ around $x$. From $x(t) = x_0 \sin(t)$ we find

$$\frac{dx}{dt} = x_0 \cos(t) = x_0 \sqrt{1 - \sin^2(t)} = \sqrt{x_0^2 - x(t)^2}, \tag{7.45}$$

and hence (with $T = 2\pi$)

$$\frac{dt}{\pi} = \frac{dx}{\pi \sqrt{x_0^2 - x^2}} = \rho(x)\, dx. \tag{7.46}$$

The classical density has singularities at $x = \pm x_0$, that is, at the classical turning points. For a classical state with energy $E_n$ we have $x_0^2 = 2E_n$.

Plotting the quantum probability density $|\psi_n(x)|^2$ together with the classical probability density

$$\rho(x) = \frac{1}{\pi \sqrt{2E_n - x^2}}, \tag{7.47}$$

we find that the quantum density oscillates around the classical density (see Fig. 7.4).

CD 5.2 contains a comparison between the classical and quantum-mechanical position probability densities for the eigenfunctions $\phi_0, \dots, \phi_{20}$.

## 7.3. Solution of the Initial-Value Problem

### 7.3.1. The time evolution

We want to solve the time-dependent Schrödinger equation (in dimensionless units) for a given initial function $\psi^{(0)}$

$$\mathrm{i} \frac{d}{dt} \psi(x,t) = \frac{1}{2} \Big( -\frac{d^2}{dx^2} + x^2 \Big) \psi(x,t), \quad \psi(x, t=0) = \psi^{(0)}(x). \tag{7.48}$$

At every instance of time $t$ the solution $\psi(x,t)$ has to be a square-integrable function of the space variable $x \in \mathbb{R}$ in order to have a useful interpretation (see Section 3.4).

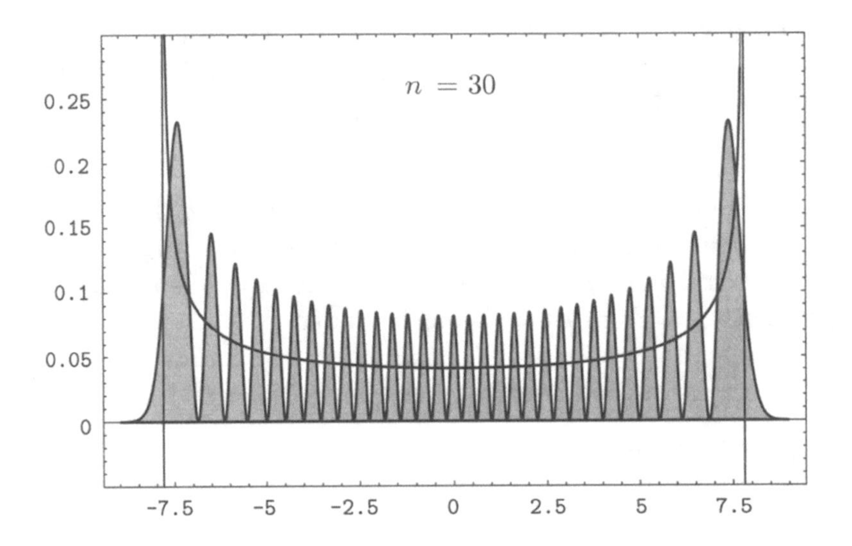


FIGURE 7.4. The classical probability density versus the quantum probability density for the state $\phi_{30}$. The vertical lines indicate the position of the classical turning points.

The time evolution of an arbitrary initial state $\psi$ follows from the time dependence of the eigenstates $\phi_n$,

$$\phi_n(x,t) = \mathrm{e}^{-\mathrm{i}E_n t}\,\phi_n(x). \tag{7.49}$$

Expanding the initial state $\psi$ into oscillator eigenfunctions we get the representation

$$\psi^{(0)}(x) = \sum_{n=0}^{\infty} c_n \phi_n(x) \tag{7.50}$$

with suitable coefficients $c_n$. Inserting the time evolution of the eigenstates we obtain immediately the solution

$$\psi(x,t) = \sum_{n=0}^{\infty} c_n\,\mathrm{e}^{-\mathrm{i}E_n t}\,\phi_n(x) = \mathrm{e}^{-\mathrm{i}t/2} \sum_{n=0}^{\infty} c_n\,\mathrm{e}^{-\mathrm{i}nt}\,\phi_n(x). \tag{7.51}$$

The superposition of two or more eigenstates depends on time in a nontrivial way. See CD 5.3–CD 5.9 for a few typical examples. These movies show oscillating states which are obtained from superpositions of only a few eigenstates. See also Color Plate 21.

### 7.3.2. Periodic time dependence

From Eq. (7.51) above we obtain the following result.

**Periodic time dependence of oscillating states:**

For every initial state $\phi$ the solution $\psi(x,t)$ of Eq. (7.48) is periodic with period $T = 4\pi$. But while the *wave function* $\psi$ has period $4\pi$, the quantum-mechanical *state* (and in particular the position probability density $|\psi(x,t)|^2$) has period $2\pi$ (which is the period of the corresponding classical system in dimensionless units).

EXERCISE 7.7. *Prove the statement above.*

In fact, it is only necessary to know the wave function in the time interval from 0 to $\pi$. From this, the time evolution for other times follows easily:

$$\begin{aligned}\psi(x,t+\pi) &= \mathrm{e}^{-\mathrm{i}\pi/2}\,\mathrm{e}^{-\mathrm{i}t/2}\sum_{n=0}^{\infty} c_n\,\mathrm{e}^{-int}\,\mathrm{e}^{-in\pi}\,\phi_n(x)\\ &= \mathrm{e}^{-\mathrm{i}\pi/2}\,\mathrm{e}^{-\mathrm{i}t/2}\sum_{n=0}^{\infty} c_n\,\mathrm{e}^{-int}\,(-1)^n\,\phi_n(x).\end{aligned}$$

Now we use the following symmetry property of the oscillator eigenfunctions,

$$\phi_n(-x) = (-1)^n\,\phi_n(x), \tag{7.52}$$

which follows from the explicit form of the Hermite polynomials, Eq. (7.35). Hence

$$\psi(x,t+\pi) = \mathrm{e}^{-\mathrm{i}\pi/2}\,\mathrm{e}^{-\mathrm{i}t/2}\sum_{n=0}^{\infty} c_n\,\mathrm{e}^{-int}\,\phi_n(-x) = \mathrm{e}^{-\mathrm{i}\pi/2}\,\psi(-x,t).$$

**Time symmetry of oscillating states:**

Any solution of the Schrödinger equation with a harmonic oscillator potential has the property

$$\psi(x,t+\pi) = \mathrm{e}^{-\mathrm{i}\pi/2}\,\psi(-x,t), \tag{7.53}$$

and hence

$$\psi(x,t+2\pi) = -\psi(x,t). \tag{7.54}$$

In addition, whenever the initial function is real-valued, we find

$$\psi(x,\pi-t) = \mathrm{e}^{-\mathrm{i}\pi/2}\,\overline{\psi(-x,t)} \qquad (\psi^{(0)}\text{ real-valued}). \tag{7.55}$$

PROOF. The result Eq. (7.53) has been shown above. For Eq. (7.55) we observe that

$$\psi(x, \pi - t) = e^{-i\pi/2}\, \psi(-x, -t). \tag{7.56}$$

Now, Eq. (7.51) shows that

$$\overline{\psi(x,t)} = e^{it/2} \sum_{n=0}^{\infty} \overline{c_n}\, e^{int}\, \phi_n(x).$$

But for a real-valued initial function also the expansion coefficients $c_n$ are real-valued, see Eq. (7.40). Hence $\overline{\psi(x,t)} = \psi(x,-t)$ and Eq. (7.56) implies the result Eq. (7.55). □

For real-valued initial functions it is thus sufficient to know the time evolution for the interval $[0, \pi/2]$.

EXERCISE 7.8. *The time period of the wave function is not gauge-invariant. Show that the time evolution in the potential* $V(x) = \frac{1}{2}x^2 - \frac{1}{2}$ *has the property*

$$\psi(x, t + 2\pi) = \psi(x, t). \tag{7.57}$$

### 7.3.3. Fourier transform of oscillator states

If you remember Section 2.6.2, you will have no difficulties to calculate the Fourier transform of the eigenvalue equation

$$\frac{1}{2}\Bigl(-\frac{d^2}{dx^2} + x^2\Bigr)\, \phi_n(x) = E_n\, \phi_n(x). \tag{7.58}$$

You will obtain the stationary Schrödinger equation in momentum space

$$\frac{1}{2}\Bigl(k^2 - \frac{d^2}{dk^2}\Bigr)\, \hat{\phi}_n(k) = E_n\, \hat{\phi}_n(k), \tag{7.59}$$

which is identical with the Schrödinger equation in position space. The oscillator eigenfunction $\phi_n(x)$ and its Fourier transform $\hat{\phi}_n(k)$ are both eigenfunctions of the oscillator Hamiltonian. You can see that $\hat{\phi}_n(k) = \alpha\phi_n(k)$ with some constant $\alpha \in \mathbb{C}$ because the oscillator eigenfunctions belonging to the same eigenvalue are unique up to a constant factor. From the unitarity of the Fourier transform it follows that $\alpha$ is a phase factor. Indeed, one finds

$$\hat{\phi}_n(k) = (-i)^n\, \phi_n(k). \tag{7.60}$$

This result can be verified by a direct calculation: $\phi_n(x)$ is a polynomial in $x$ times $\exp(-x^2/2)$. Under a Fourier transform, $x^n \exp(-x^2/2)$ is mapped onto $(i d/dk)^n \exp(-k^2/2)$ which is equal to a polynomial in $k$ times the exponential function $\exp(-k^2/2)$. The explicit calculation gives Eq. (7.60).

From Eqs. (7.51) and (7.60) we find the time evolution of an oscillator state in momentum space,

$$\hat{\psi}(k,t) = \mathrm{e}^{-\mathrm{i}t/2} \sum_{n=0}^{\infty} c_n\, \mathrm{e}^{-\mathrm{i}nt}\, (-\mathrm{i})^n\, \phi_n(k) \tag{7.61}$$

$$= \mathrm{e}^{\mathrm{i}\pi/4} \exp\Bigl\{-\frac{\mathrm{i}}{2}\Bigl(t + \frac{\pi}{2}\Bigr)\Bigr\} \sum_{n=0}^{\infty} c_n\, \mathrm{e}^{-\mathrm{i}n(t+\pi/2)}\, \phi_n(k) \tag{7.62}$$

$$= \mathrm{e}^{\mathrm{i}\pi/4}\, \psi\bigl(k, t + \frac{\pi}{2}\bigr). \tag{7.63}$$

**Time evolution and Fourier transform:**

In the field of a harmonic oscillator, the time evolution by $\pi/2$ amounts to a Fourier transform. We have

$$\psi(x, t + \pi/2) = \mathrm{e}^{-\mathrm{i}\pi/4}\, \hat{\psi}(x,t). \tag{7.64}$$

Apart from the phase factor $\exp(\mathrm{i}\pi/4) = \mathrm{i}^{1/2}$ the motion in momentum space looks like the motion in position space shifted in time by $\pi/2$.

CD 5.3–CD 5.6 contain animations of oscillating states in momentum space. It can be seen that the wave function in momentum space is (apart from the phase factor described above) a quarter of a classical period ahead of the wave function in position space. See also Color Plate 21.

EXERCISE 7.9. *Let $(x(t), p(t))$ be any solution of the classical harmonic oscillator (in dimensionless units). Prove that*

$$p(t) = x(t + \pi/2). \tag{7.65}$$

# 7.4. Time Evolution of Observables

## 7.4.1. Time-dependence of operators

Important properties of the motion can be learned by considering the time evolution of certain observables like the position and momentum operators. We define the time evolution of an arbitrary observable $A$ by

$$A(t) = \mathrm{e}^{\mathrm{i}Ht}\, A\, \mathrm{e}^{-\mathrm{i}Ht}. \tag{7.66}$$

If one knows the operator $A(t)$, one can determine the time evolution of the expectation value of $A$ for every initial state $\psi$. The expectation value at time $t$ is given by

$$\langle A\rangle_\psi(t) = \langle \psi(t), A\,\psi(t)\rangle = \langle \psi, A(t)\,\psi\rangle. \tag{7.67}$$

Here we used $\psi(t) = \exp(-\mathrm{i}Ht)\psi$ and the fact that $\exp(\mathrm{i}Ht)$ is adjoint to $\exp(-\mathrm{i}Ht)$.

The time evolution of an operator $A$ obeys a differential equation. From Eq. (7.66) it follows immediately that

$$\begin{aligned}\frac{d}{dt}A(t) &= \left(\frac{d}{dt}\mathrm{e}^{\mathrm{i}Ht}\right)A\,\mathrm{e}^{-\mathrm{i}Ht} + \mathrm{e}^{\mathrm{i}Ht}A\left(\frac{d}{dt}\mathrm{e}^{-\mathrm{i}Ht}\right)\\ &= \mathrm{e}^{\mathrm{i}Ht}\left(\mathrm{i}H\,A - A\,\mathrm{i}H\right)\mathrm{e}^{-\mathrm{i}Ht}\\ &= \mathrm{e}^{\mathrm{i}Ht}\,\mathrm{i}[H,A]\,\mathrm{e}^{-\mathrm{i}Ht}.\end{aligned}$$

We collect these results in the following box.

**Time evolution of an observable:**

The time-dependent operator

$$A(t) = \mathrm{e}^{\mathrm{i}Ht}\,A\,\mathrm{e}^{-\mathrm{i}Ht} \tag{7.68}$$

is a solution of the *Heisenberg equation*

$$\frac{d}{dt}A(t) = \mathrm{i}\,[H, A(t)] \tag{7.69}$$

with the initial condition $A(0) = A$. When you know the solution $A(t)$ you can calculate the expectation value of the observable $A$ at time $t$ by taking the expectation value of $A(t)$ with the initial state $\psi$:

$$\langle A\rangle_{\psi(t)} = \langle A(t)\rangle_\psi. \tag{7.70}$$

In particular, if an observable commutes with the Hamiltonian $H$, then the Heisenberg equation has the trivial solution $A(t) = A(0) = A$. Such an observable $A$ is called a *constant of motion*, see Section 6.11.

### 7.4.2. Position and momentum observables

For $H$ = harmonic oscillator and $A$ = position or momentum, the time evolution can be calculated explicitly. According to Eq. (7.69) we have to calculate

$$\begin{aligned}
\mathrm{i}[H,p] &= \tfrac{\mathrm{i}}{2}\,[x^2,p] \\
&= \tfrac{\mathrm{i}}{2}\,(x\,[x,p] + [x,p]\,x) \\
&= -x, \\
\mathrm{i}[H,x] &= \tfrac{\mathrm{i}}{2}\,[p^2,x] = p,
\end{aligned}$$

from which we obtain

$$\frac{d}{dt}\,p(t) = -x(t), \qquad \frac{d}{dt}\,x(t) = p(t). \tag{7.71}$$

It can be verified easily by differentiating that the following families of operators form a solution of Eq. (7.71).

**Time evolution of position and momentum:**

In the field of a harmonic oscillator, the observables of position and momentum fulfill

$$x(t) = (\sin t)\,p + (\cos t)\,x, \tag{7.72}$$

$$p(t) = (\cos t)\,p - (\sin t)\,x. \tag{7.73}$$

This is identical to the classical time dependence of position and momentum for a particle that starts at $x$ with momentum $p$ (in dimensionless units). It follows in particular that for every quantum-mechanical initial state the expectation values of position and momentum perform a classical motion.

EXERCISE 7.10. *The classical motion* $t \to (\langle x(t)\rangle_\psi, \langle p(t)\rangle_\psi)$ *of the expectation values (with respect to an arbitrary state* $\psi$*) has a classical energy which is strictly less than the quantum-mechanical expectation value of the energy of* $\psi$*. Prove the following inequality:*

$$\tfrac{1}{2}\,(\langle p\rangle_\psi^2 + \langle x\rangle_\psi^2) < \langle \tfrac{1}{2}\,(p^2 + x^2)\rangle_\psi. \tag{7.74}$$

*Can you estimate the difference between these expressions?*

HINT: *Use Exercise 4.1 to show that the difference can be written as*

$$\tfrac{1}{2}\big[(\Delta_\psi p)^2 + (\Delta_\psi x)^2\big]. \tag{7.75}$$

### 7.4.3. Time evolution and translation

The relation

$$\mathrm{e}^{\mathrm{i}Ht}\,p\,\mathrm{e}^{-\mathrm{i}Ht} = p(t) \tag{7.76}$$

can be written for the unitary groups in the form

$$\mathrm{e}^{\mathrm{i}Ht}\,\mathrm{e}^{-\mathrm{i}x_0 p}\,\mathrm{e}^{-\mathrm{i}Ht} = \mathrm{e}^{-\mathrm{i}x_0 p(t)}. \tag{7.77}$$

Here $x_0$ is an arbitrary real number. The unitary operator $\exp(-\mathrm{i}x_0 p)$ describes a translation by $x_0$,

$$\mathrm{e}^{-\mathrm{i}x_0 p}\,\psi(x) = \psi(x - x_0). \tag{7.78}$$

Now we have occasion to use the Weyl relation for the operators $x$ and $p$,

$$\mathrm{e}^{\mathrm{i}(ax+bp)} = \mathrm{e}^{\mathrm{i}ax}\,\mathrm{e}^{\mathrm{i}bp}\,\mathrm{e}^{\mathrm{i}ab/2}, \quad \text{all } a, b \in \mathbb{R}. \tag{7.79}$$

(see Section 6.8). With the explicit form of $p(t)$ obtained in the previous section we calculate

$$\begin{aligned}\mathrm{e}^{-\mathrm{i}x_0 p(t)} &= \mathrm{e}^{-\mathrm{i}x_0\big(-(\sin t)\,x + (\cos t)\,p\big)} \\ &= \mathrm{e}^{\mathrm{i}(x_0\,\sin t)\,x}\,\mathrm{e}^{-\mathrm{i}(x_0\,\cos t)\,p}\,exp\Big(-\mathrm{i}\,\frac{x_0^2}{4}\,\sin 2t\Big).\end{aligned}$$

In this formula we may substitute $t \to -t$ and apply the result to a state $\phi(x,t) = \exp(-\mathrm{i}Ht)\phi(x)$. Then, using (7.77) with $t \to -t$, we finally obtain the following formula:

$$\mathrm{e}^{-\mathrm{i}Ht}\,\mathrm{e}^{-\mathrm{i}x_0 p}\,\phi(x) = \exp\Big(\mathrm{i}\,\frac{x_0^2}{4}\,\sin 2t\Big)\,\mathrm{e}^{-\mathrm{i}(x_0\,\sin t)\,x}\,\mathrm{e}^{-\mathrm{i}(x_0\,\cos t)\,p}\,\mathrm{e}^{-\mathrm{i}Ht}\phi(x). \tag{7.80}$$

This formula has a very nice interpretation: It describes the time evolution of an initial state obtained from $\phi$ by a translation in terms of the time evolution of the original $\phi$. The action of the various operators on the right-hand side can be described as follows.

1. $\exp(-\mathrm{i}(x_0\,\cos t)\,p)$: The state $\phi$ at time $t$ is shifted by $x_0 \cos t$ in position space.
2. $\exp(-\mathrm{i}(x_0\,\sin t)\,x)$: Shift in momentum space by $-x_0 \sin t$.
3. Finally, the result is multiplied by a phase factor.

EXERCISE 7.11. *Show that the first two steps described above also have to be performed on a state $(x(t), p(t))$ in classical mechanics, when we calculate the time evolution of a shifted initial state in terms of the original state. The time-dependent phase factor (step 3) apparently has a purely quantum-theoretical origin.*

**Time evolution of a shifted initial state:**
Let $\phi(x,t)$ be the time evolution of an initial function $\phi(x)$ in the field of the harmonic oscillator. For $x_0 \in \mathbb{R}$ define the shifted function by $\phi_{x_0}(x) = \phi(x - x_0)$. Its time evolution is given by

$$\phi_{x_0}(x,t) = \exp\Bigl(\mathrm{i}\,\frac{x_0^2}{4}\,\sin 2t\Bigr)\,\mathrm{e}^{-\mathrm{i}(x_0\,\sin t)\,x}\,\phi(x - x_0\,\cos t, t) \tag{7.81}$$

We can draw some interesting conclusions from this result: Assume that $\phi = \phi_n$ is an eigenstate of the harmonic oscillator. The eigenstates have a trivial time evolution, they always remain centered at the origin, the expectation values of the position $x$ and momentum $p$ are zero for all times. Now, let's shift the eigenfunction to a new position $x_0$, that is, the new initial state is given by

$$\psi(x,0) = \phi_n(x - x_0). \tag{7.82}$$

Equation (7.81) gives the following result for the time evolution of the translated eigenstate:

$$\psi(x,t) = \exp\Bigl(\mathrm{i}\,\frac{x_0^2}{4}\,\sin 2t - \mathrm{i}E_n t - \mathrm{i}(x_0\,\sin t)\,x\Bigr)\,\phi_n(x - x_0\,\cos t). \tag{7.83}$$

In particular, the position probability density is just given by

$$|\psi(x,t)|^2 = |\phi_n(x - x_0\,\cos t)|^2. \tag{7.84}$$

Time evolution does not change the shape of $|\phi_n|^2$, it just translates the function to its classical position $x_0 \cos t$.

CD 5.19 shows the motion of the initially translated eigenstates $\phi_1$, $\phi_2$, and their superposition. The evolution of the shifted initial state shows the motion of $\phi_1 + \phi_2$ as in CD 5.5, while the center of the wave packet performs the oscillation $x_0\,\cos t$.

## 7.5. Motion of Gaussian Wave Packets

### 7.5.1. Coherent states

Here we apply the results of the previous section to the ground state $\phi_0(x) = \pi^{(-1/4)}\exp(-\frac{1}{2}x^2)$. Among the eigenstates it is distinguished also by the property that it is optimal with respect to the uncertainty relation

$$\Delta x \Delta p \geq \tfrac{1}{2}. \tag{7.85}$$

The ground state satisfies

$$\Delta x \Delta p = \tfrac{1}{2}. \tag{7.86}$$

For the time evolution of the initially translated ground state we can apply the results obtained in the previous section. We obtain

$$\psi(x,t) = \Big(\frac{1}{\pi}\Big)^{1/4} \exp\Big(\mathrm{i}\,\frac{x_0^2}{4}\,\sin 2t - \mathrm{i}\,\frac{t}{2}\Big)\,\exp\Big(\mathrm{i}\,p_t\,x - \frac{(x-x_t)^2}{2}\Big), \tag{7.87}$$

with $x_t = x_0\,\cos t$ and $p_t = -x_0\,\sin t$. This is a normalized Gaussian function centered at the average position $x_t$ and with average momentum $p_t$. Here $(x_t, p_t)$ describes the classical oscillation of a particle with initial position $x_0$ (in dimensionless units).

CD 5.10 contains movies of a coherent state in position space, momentum space, and in the energy representation. See also Color Plate 22.

We know already that the function $\psi(x,t)$ is again optimal with respect to the uncertainty relation (see Section 2.8.1). The state $\psi(x,t)$ hence satisfies Eq. (7.86) for all times. The states with minimal uncertainty are called *coherent states*. Their motion is most similar to the oscillation of a particle in classical mechanics. We collect these observations in the following box.

**Coherent states of a harmonic oscillator**:

The coherent states (states with minimal uncertainty) of a harmonic oscillator are shifted Gaussian functions, initially localized at $x_0$, which have the shape of the ground state. The maximum of a coherent state always follows the trajectory of a classical-mechanical particle that starts at $x_0$ with zero initial momentum. The wavelength of the phase always corresponds to the momentum of the classical particle. During its time evolution a coherent state retains its shape (the shape of the ground state) without spreading.

It is a consequence of Exercise 7.10 that the coherent states minimize the difference between the mean energy and the energy of the classical motion of $\langle x(t)\rangle_\psi$ and $\langle p(t)\rangle_\psi$ because the coherent states have minimal uncertainty. One even has the following result: $\psi$ is a coherent state if and only if

$$\tfrac{1}{2}\Big(\langle p\rangle^2_{\psi(t)} + \langle x\rangle^2_{\psi(t)}\Big) = \tfrac{1}{2}\langle p^2 + x^2\rangle_{\psi(t)}. \tag{7.88}$$

### 7.5.2. Arbitrary Gaussian function

For the harmonic oscillator potential, the Schrödinger equation with the Gaussian initial function

$$\phi(x) = \Big(\frac{a}{\pi}\Big)^{1/4} \exp\Big(-a\,\frac{x^2}{2}\Big) \tag{7.89}$$

has the solution

$$\psi(x,t) = \Bigl(\frac{a}{\pi}\Bigr)^{1/4} \bigl(\cos t + \mathrm{i}a \sin t\bigr)^{-1/2} \exp\Bigl(-a(t)\,\frac{x^2}{2}\Bigr), \tag{7.90}$$

where

$$a(t) = \frac{a\,\cos t + \mathrm{i}\,\sin t}{\cos t + \mathrm{i}a \sin t}. \tag{7.91}$$

This result can be obtained using the explicitly known integral kernel of the time evolution. This integral kernel (Mehler's kernel) will be obtained in a later section.

One has to be careful with the definition of the square root

$$\bigl(\cos t + \mathrm{i}a \sin t\bigr)^{-1/2} \tag{7.92}$$

in Eq. (7.90). It is necessary to take that branch of the square root that gives a continuous dependence on $t$. Hence for $a = 1$ this expression will simply become $\exp(-\mathrm{i}t/2)$.

By combining Eq. (7.90) with Eq. (7.81) one can easily find an expression for the time evolution of a translated Gaussian function.

## 7.6. Harmonic Oscillator in Two and More Dimensions

The wave function of a particle in two dimensions depends on a two-dimensional position variable $\mathbf{x} = (x_1, x_2)$. The Hamiltonian operator for the harmonic oscillator in two dimensions can be written as a sum of one-dimensional Hamiltonians

$$\begin{aligned} H &= -\frac{1}{2}\,\Delta + \frac{\mathbf{x}\cdot\mathbf{x}}{2} \\ &= -\frac{1}{2}\,\frac{d^2}{dx_1^2} + \frac{x_1^2}{2} - \frac{1}{2}\,\frac{d^2}{dx_2^2} + \frac{x_2^2}{2} \\ &= H_{x_1} + H_{x_2}, \end{aligned}$$

where $H_{x_i}$ $(i = 1, 2)$ acts only on the variable $x_i$. Hence we can make the same observations as in the case of free particles: The Schrödinger equation in two space dimensions can be solved by a product ansatz

$$\psi(\mathbf{x}, t) = \psi_1(x_1, t)\,\psi_2(x_2, t), \tag{7.93}$$

where each $\psi_i$ is a solution of the one-dimensional harmonic oscillator equation.

$$\begin{aligned} \mathrm{i}\frac{d}{dt}\psi &= \mathrm{i}\frac{d\psi_1}{dt}\psi_2 + \mathrm{i}\psi_1\frac{d\psi_2}{dt} \\ &= (H_{x_1}\psi_1)\psi_2 + \psi_1(H_{x_2}\psi_2) \\ &= (H_{x_1} + H_{x_2})\psi_1\psi_2 = H\psi. \end{aligned}$$

If $\psi_{n,m}$ is a product of two eigenfunctions $\phi_n$ and $\phi_m$ of the one-dimensional Hamiltonian,

$$\psi_{n,m}(\mathbf{x}) = \phi_n(x_1)\,\phi_m(x_2), \tag{7.94}$$

then $\psi_{n,m}(\mathbf{x})$ is an eigenfunction of $H$ belonging to the eigenvalue $E_{n,m}$

$$E_{n,m} = E_m + E_n = m + n + 1, \qquad H\,\psi_{n,m} = E_{n,m}\,\psi_{n,m}. \tag{7.95}$$

Hence it is clear that

$$\psi_{n,m}(\mathbf{x},t) = \psi_{n,m}(\mathbf{x})\,\mathrm{e}^{-\mathrm{i}E_{n,m}t} = \phi_n(x_1,t)\,\phi_m(x_2,t) \tag{7.96}$$

is a solution of the time-dependent Schrödinger equation. This solution is a product of solutions of one-dimensional equations for $x_1$- and $x_2$-coordinates, respectively. Of course, this does not mean that *every* solution of the two-dimensional oscillator is a product of one-dimensional solutions. Nevertheless, any solution can be written as an (infinite) linear combination of products as follows.

As a consequence of the fact that the functions $\phi_n$ form an orthonormal basis in the Hilbert space $L^2(\mathbb{R})$ it can be shown that the product functions $\psi_{n,m}$ form an orthonormal basis in $L^2(\mathbb{R}^2)$ (this is a property of the tensor product of Hilbert spaces). Hence every initial function $\phi \in L^2(\mathbb{R}^2)$ can be expanded as

$$\phi(\mathbf{x}) = \sum_{n=0}^{\infty}\sum_{m=0}^{\infty} c_{n,m}\,\psi_{n,m}(\mathbf{x}), \tag{7.97}$$

and the unique solution of the Schrödinger equation with initial condition $\psi(\mathbf{x},t) = \phi(\mathbf{x})$ is given by

$$\psi(\mathbf{x},t) = \sum_{n=0}^{\infty}\sum_{m=0}^{\infty} c_{n,m}\,\mathrm{e}^{-\mathrm{i}(n+m+1)\,t}\,\psi_{n,m}(\mathbf{x}). \tag{7.98}$$

The ground-state energy is $E_0 + E_0 = 1$ and every solution is periodic with period $2\pi$.

EXERCISE 7.12. *Generalize the above considerations to n-dimensions.*

If $\psi(x,t)$ is the time evolution of a one-dimensional harmonic oscillator, then its Fourier transform, the function $\hat{\psi}(k,t)$, is also a solution of the

Schrödinger equation of the harmonic oscillator in one dimension. Hence the wave function in phase space,

$$\Psi(x,k,t) = \psi(x,t)\,\hat{\psi}(k,t) \tag{7.99}$$

is a solution of the two-dimensional oscillator equation.

CD 5.14–5.18 show solutions of the two-dimensional harmonic oscillator. Among the various Gaussian wave packets, the coherent and squeezed states are of particular interest.

# 7.7. Theory of the Harmonic Oscillator

## 7.7.1. Supersymmetry

Define the operators

$$A = \frac{1}{\sqrt{2}}\,(x + \mathrm{i}p) = \frac{1}{\sqrt{2}}\Bigl(\frac{d}{dx} + x\Bigr), \tag{7.100}$$

$$A^\dagger = \frac{1}{\sqrt{2}}\,(x - \mathrm{i}p) = \frac{1}{\sqrt{2}}\Bigl(-\frac{d}{dx} + x\Bigr). \tag{7.101}$$

Here $p = -\mathrm{i}d/dx$ is the momentum operator, and $x$ denotes the position operator (the operator of multiplication of $\psi(x)$ by $x$). For suitable (differentiable) wave functions you can see by a partial integration that

$$\langle \phi, A\psi \rangle = \langle A^\dagger \phi, \psi \rangle. \tag{7.102}$$

Therefore, the operators $A$ and $A^\dagger$ are formally adjoint to each other.

$\boxed{\Psi}$ **Schwartz space.** For the mathematical investigation it is necessary to have a dense domain in the Hilbert space $L^2(\mathbb{R})$, which is invariant under the action of the operators $A$ and $A^\dagger$. Such a domain is the Schwartz space $\mathcal{S} = \mathcal{S}(\mathbb{R})$. The set $\mathcal{S}$ consists of infinitely differentiable functions. The functions (and all their derivatives) are required to go to zero faster than any inverse power of $|x|$, as $|x|$ tends to infinity. More precisely,

$$\mathcal{S} = \Bigl\{ f \in C^\infty(\mathbb{R}) \Bigm| \text{for all integers } k,\, l\text{:}\ \sup_{x\in\mathbb{R}} \Bigl| x^k \frac{d^l}{dx^l} f(x) \Bigr| < \infty \Bigr\}. \tag{7.103}$$

Typical examples of functions in $\mathcal{S}$ are Gauss functions $\exp(-ax^2)$, and all the oscillator eigenfunctions. Because all finite linear combinations of functions in $\mathcal{S}$ are again contained in $\mathcal{S}$, the set $\mathcal{S}$ is a linear subspace of the Hilbert space $L^2(\mathbb{R})$. Theorem 2.2 in Section 2.7 states that the linear subspace spanned by the functions

$$G_q(x) = \Bigl(\frac{1}{\pi}\Bigr)^{1/4} e^{\mathrm{i}qx} \exp\Bigl(-\frac{x^2}{2}\Bigr) \tag{7.104}$$

is dense in the Hilbert space. Because this subspace is contained in $\mathcal{S}$ you will be convinced that also $\mathcal{S}$ is a dense subspace of $L^2(\mathbb{R})$. As you probably know, this means that for every square-integrable function $\psi$ and for every $\epsilon > 0$ there is a function $f \in \mathcal{S}$ such that $\|\psi - f\| < \epsilon$.

Moreover, the operators $A$ and $A^\dagger$ leave $\mathcal{S}$ invariant, that is, with $f \in \mathcal{S}$ also $Af \in \mathcal{S}$, and similarly for $A^\dagger$. The invariance implies that one can define arbitrary powers of the operators $A$ and $A^\dagger$ on the domain $\mathcal{S}$. All the operators and commutators considered below are well defined on the domain $\mathcal{S}$.

By the way, another important property of the domain $\mathcal{S}$ is that the Fourier transform $\mathcal{F}$ maps $\mathcal{S}$ one-to-one onto itself. For any smooth, fast decaying function the Fourier transform is also smooth and fast decaying.

**The supersymmetric structure**: The canonical commutation relation

$$[x, p] = xp - px = \mathrm{i} \tag{7.105}$$

implies immediately

$$[A, A^\dagger] = 1 \quad \text{or} \quad AA^\dagger = A^\dagger A + 1. \tag{7.106}$$

The Hamiltonian of the harmonic oscillator has a very special structure. It can be written in terms of the operators $A$ and $A^\dagger$ as

$$H = \frac{1}{2}\,(p^2 + x^2) = A^\dagger A + \frac{1}{2}. \tag{7.107}$$

Hence, in order to find the eigenvalues of $H$, it is sufficient to determine the eigenvalues of $A^\dagger A$. ($E$ is an eigenvalue of $H$ if and only if $E - 1/2$ is an eigenvalue of $A^\dagger A$). In the following section, you will see that finding the eigenvalues of $A^\dagger A$ is made easy by the following observation.

**Spectral supersymmetry**:
The operators $A^\dagger A$ and $AA^\dagger$ have the same eigenvalues (except 0). Whenever $E$ is a nonzero eigenvalue of $A^\dagger A$ with eigenvector $\psi$, then $A\psi$ is an eigenvector of $AA^\dagger$ belonging to the same eigenvalue. Similarly, if $\phi$ is an eigenvector of $AA^\dagger$ belonging to the nonzero eigenvalue $E'$, then also $A^\dagger A$ has $E'$ as eigenvalue. The corresponding eigenvector of $A^\dagger A$ is given by $A^\dagger\phi$.

PROOF. The proof is very simple and holds quite generally for any two operators: Assume that $E \neq 0$ and

$$A^\dagger A\psi = E\psi. \tag{7.108}$$

Define $\phi = A\psi$. Then

$$AA^\dagger\phi = AA^\dagger(A\psi) = A(A^\dagger A\psi) = A(E\psi) = E(A\psi) = E\phi. \tag{7.109}$$

The proof of the second part is quite similar and is left to the reader. □

**Possible violation of spectral supersymmetry at $E = 0$:** We first note that $A^\dagger A\psi = 0$ implies $A\psi = 0$. In order to see this, let $\phi$ be any vector in the kernel of $A^\dagger$, that is, $A^\dagger\phi = 0$. Then

$$\langle \phi, A\psi\rangle = \langle A^\dagger\phi, \psi\rangle = 0 \tag{7.110}$$

for all $\psi$ in the (dense) domain of $A$. But this can only hold if $\phi$ is orthogonal to the range of $A$. (Thus, we have shown: For any densely defined linear operator $A$ the kernel $\operatorname{Ker} A^\dagger$ is orthogonal to the range $\operatorname{Ran} A$.) Hence $\phi = A\psi$ can only belong to the kernel of $A^\dagger$ if it is at the same time orthogonal to the range of $A$. This implies immediately that $A\psi = 0$. Now, if $E = 0$ in the proof of the spectral supersymmetry, then $\phi = A\psi = 0$ and hence one cannot know whether $AA^\dagger$ has the eigenvalue zero or not. (Remember that $AA^\dagger$ has the eigenvalue zero if and only if there is a *nonzero* vector $\phi$ such that $AA^\dagger\phi = 0$).

In the case that $A^*A$ has zero among its eigenvalues while $AA^*$ has not (or vice versa), we speak of a violation of the spectral supersymmetry at $E = 0$. You will learn in the next section that this indeed happens for the Hamiltonian of the harmonic oscillator.

EXERCISE 7.13. *Consider the classical motion of the harmonic oscillator*

$$H(x,p) = \tfrac{1}{2}\,(p^2 + x^2). \tag{7.111}$$

*Define the quantity*

$$a(t) = \frac{1}{\sqrt{2}}\,(x(t) + \mathrm{i}\,p(t)). \tag{7.112}$$

*Express the Hamiltonian function $H(x,p)$ in terms of $a(t)$ and $\overline{a(t)}$. Find the evolution equation for $a(t)$ and solve it.*

EXERCISE 7.14. *Using the canonical commutation relation $[A, A^\dagger] = 1$ and the rule (2.73), reduce the expression $[A, (A^\dagger)^n]$ to the simplest possible form. (Here $(A^\dagger)^n$ is the product of $n$ $A^\dagger$s.)*

### 7.7.2. The Spectrum of Eigenvalues

The spectral supersymmetry for $E \neq 0$ together with the relation

$$H = A^\dagger A + \tfrac{1}{2} = AA^\dagger - \tfrac{1}{2} \tag{7.113}$$

allows us to solve the eigenvalue problem of the harmonic oscillator almost without any calculation! We first note that all eigenvalues of $A^\dagger A$ must be non-negative. If $A^\dagger A\psi = E\psi$ then

$$\langle \psi, A^\dagger A\psi\rangle = E\|\psi\|^2. \tag{7.114}$$

The non-negativity of $E$ now follows from

$$\langle \psi, A^\dagger A \psi \rangle = \langle A\psi, A\psi \rangle = \|A\psi\|^2 \geq 0. \tag{7.115}$$

**Ground state and eigenvalues:**

If there is a state $\Omega \neq 0$ that is square-integrable and satisfies $A\Omega = 0$, then all the integers $n = 0, 1, 2, \dots$ are eigenvalues of $A^\dagger A$, and $E_n = n + \frac{1}{2}$ are eigenvalues of the harmonic oscillator Hamiltonian $H$. The corresponding eigenvectors are given by $(A^\dagger)^n \Omega$. The state $\Omega$ is called the ground state.

PROOF. The proof of this statement is obvious: If $A\Omega = 0$, and $\Omega \neq 0$, then $\Omega$ is an eigenvector of $A^\dagger A$ with eigenvalue 0. But because $AA^\dagger = A^\dagger A + 1$, the state $\Omega$ is at the same time an eigenstate of $AA^\dagger$ belonging to the eigenvalue 1. Because of the spectral supersymmetry, 1 is also an eigenvalue of $A^\dagger A$, the corresponding eigenvector is given by $A^\dagger \Omega$. But then $A^\dagger \Omega$ must be eigenvector of $AA^\dagger$ belonging to the eigenvalue 2. Repeating this argument, we find that $n$ is an eigenvalue of $A^\dagger A$ with associated eigenvector $(A^\dagger)^n \Omega$. □

Everything depends on whether there is a nonzero solution of $A\Omega = 0$. By the definition of $A$, the equation $A\Omega = 0$ is a simple first-order differential equation,

$$\frac{d}{dx}\Omega = -x\Omega, \tag{7.116}$$

which is obviously solved by

$$\Omega(x) = C \exp\Bigl(-\frac{x^2}{2}\Bigr). \tag{7.117}$$

The constant $C$ is arbitrary, we choose

$$C = \Bigl(\frac{1}{\pi}\Bigr)^{1/4} \tag{7.118}$$

in order to fulfill the normalization condition $\|\Omega\| = 1$.

We note that $A^\dagger$ and hence $AA^\dagger$ cannot have a zero eigenvalue, because then $A^\dagger A = AA^\dagger - 1$ would have the eigenvalue $-1$ in contradiction to the non-negativity of $A^\dagger A$ stated earlier. Indeed, the equation $A^\dagger f = 0$, that is, $f'(x) = x f(x)$, has the solution $\exp(x^2/2)$ which is not square-integrable.

So far, it has been shown that the numbers $E_n = n + 1/2$ are eigenvalues of $H$. It has not been shown that these are the only eigenvalues. This is related to the completeness of the eigenfunctions, which will be discussed briefly in Section 7.8.2 below.

### 7.7.3. The eigenvectors

Starting with the ground state $\Omega$ you can obtain the eigenvectors belonging to $E_n$, $n = 1, 2, \ldots$, by repeatedly applying the operator $A^\dagger$. If $\phi_n$ is a normalized eigenvector,

$$A^\dagger A \phi_n = n\phi_n, \qquad \|\phi_n\| = 1, \tag{7.119}$$

then

$$\begin{aligned} \|A^\dagger \phi_n\|^2 &= \langle A^\dagger \phi_n, A^\dagger \phi_n \rangle = \langle \phi_n, A A^\dagger \phi_n \rangle \\ &= \langle \phi_n, (A^\dagger A + 1)\phi_n \rangle = (n+1)\,\langle \phi_n, \phi_n \rangle = (n+1)\,\|\phi_n\|^2 \\ &= n+1. \end{aligned} \tag{7.120}$$

Hence the normalized eigenvector for the next eigenvalue is given by

$$\phi_{n+1} = \frac{1}{\sqrt{n+1}}\, A^\dagger \phi_n. \tag{7.121}$$

Starting with the normalized ground state, you will find by induction

$$\begin{aligned} \phi_0 &= \Omega, \\ \phi_n &= \frac{1}{\sqrt{n!}}\,(A^\dagger)^n\,\Omega. \end{aligned} \tag{7.122}$$

EXERCISE 7.15. *Calculate the first few eigenfunctions of the harmonic oscillator using the explicit expressions for $\Omega$ and $A^\dagger$. Compare the result with the Hermite functions in Eq. (7.34).*

EXERCISE 7.16. *Let $\phi_n$ be the (normalized) $n$th eigenstate of the harmonic oscillator. Prove that for $n \geq 1$,*

$$\phi_{n-1} = \frac{1}{\sqrt{n}}\, A\phi_n. \tag{7.123}$$

Because the operators $A^\dagger$ and $A$ serve to jump up and down the ladder of oscillator eigenstates, they are sometimes called *ladder operators.* The operator $A^\dagger$ generates the eigenstates out of the ground state. Each application of $A^\dagger$ creates a quantum of energy. Hence this operator is also called a *creation operator.* Correspondingly, $A$ is called an *annihilation operator.*

$\boxed{\Psi}$ Concluding this section we give a general proof for the fact that the $n$th eigenvector $\phi_n$ of the harmonic oscillator is given by the Hermite functions Eq. (7.34). This statement can be reformulated as

$$(A^\dagger)^n\,\Omega(x) = \frac{1}{\sqrt{2^n}}\, H_n(x)\,\Omega(x). \tag{7.124}$$

PROOF. Probably you tried Exercise 7.5, which asks you to verify the equation

$$H_{n+1}(x) = -\frac{d}{dx} H_n(x) + 2x\, H_n(x) \tag{7.125}$$

for Hermite polynomials. This relation is needed for the proof. The proof is best done with induction: Eq. (7.124) is true for $n = 0$ because $H_0(x) = 1$. Assuming that it is valid for $n = k$, let us prove the validity for $n = k + 1$.

$$\begin{aligned}
(A^\dagger)^{k+1}\,\Omega(x) &= A^\dagger\,(A^\dagger)^k\,\Omega(x) \\
&= A^\dagger \frac{1}{\sqrt{2^k}} H_k(x)\,\Omega(x) && \text{by the assumption for } k, \\
&= \frac{1}{\sqrt{2^k}}\frac{1}{\sqrt{2}}\Bigl(-\frac{d}{dx} + x\Bigr) H_n(x)\,\Omega(x) \\
&= \frac{1}{\sqrt{2^{k+1}}}\bigl(-H_k'\,\Omega - H_k\,\Omega' + x\,H_k\,\Omega\bigr) \\
&= \frac{1}{\sqrt{2^{k+1}}}\bigl(-H_k'\,\Omega + 2x\,H_k\,\Omega\bigr) && \text{by Eq. (7.116)}, \\
&= \frac{1}{\sqrt{2^{k+1}}} H_{k+1}\,\Omega(x) && \text{by Eq. (7.125)}.
\end{aligned}$$

This completes the proof of the assertion. □

## 7.8. Special Topic: More About Coherent States

The aim of the sections in the rest of this chapter is to prove some of the results mentioned above. These sections contain mathematically advanced material and may be omitted at first reading. The completeness of the eigenfunctions as well as some results about coherent states will be proved. In the next section, Mehler's formula, which gives the explicit form of the propagator (integral kernel of the time evolution), will be derived.

### 7.8.1. Coherent states

A useful dense subspace of the Hilbert space is the set $\mathfrak{D}$ consisting of all *finite* linear combinations of oscillator eigenstates.

$$\phi \in \mathfrak{D} \quad \text{if and only if} \quad \phi = \sum_{n=0}^{N} c_n \phi_n, \tag{7.126}$$

for some integer $N$ and suitable constants $c_n \in \mathbb{C}$, $n = 0, 1, \ldots, N$.

The operators $(A^\dagger)^n$ and $A^n$ are obviously well defined on $\mathfrak{D}$. Also the exponential operators

$$e^{zA} = \sum_{n=0}^{\infty} \frac{z^n}{n!} A^n, \qquad e^{zA^\dagger} = \sum_{n=0}^{\infty} \frac{z^n}{n!} (A^\dagger)^n, \qquad z \in \mathbb{C}, \tag{7.127}$$

are well defined on $\mathfrak{D}$ even for arbitrary complex parameters $z$. Because the operators $A$ and $A^\dagger$ satisfy the canonical commutation relations $[A, A^\dagger] = 1$, the exponential operators satisfy the Weyl relations

$$\mathrm{e}^{aA^\dagger + bA} = \mathrm{e}^{aA^\dagger}\,\mathrm{e}^{bA}\,\mathrm{e}^{ab/2} = \mathrm{e}^{bA}\,\mathrm{e}^{aA^\dagger}\,\mathrm{e}^{-ab/2}. \tag{7.128}$$

Here, $a$ and $b$ may be arbitrary complex numbers. Let us calculate the action of the exponential operators on the ground state $\Omega$. We have

$$\mathrm{e}^{zA}\,\Omega = \Omega \qquad (\text{because } A^n\Omega = 0 \text{ for } n = 1, 2, \dots), \tag{7.129}$$

$$\mathrm{e}^{zA^\dagger}\,\Omega = \sum_{n=0}^{\infty} \frac{z^n}{n!}\,(A^\dagger)^n\,\Omega = \sum_{n=0}^{\infty} \frac{z^n}{\sqrt{n!}}\,\phi_n. \tag{7.130}$$

EXERCISE 7.17. *Using the result of Exercise 7.14, find the simplest possible form for the commutator* $[A, \mathrm{e}^{zA^\dagger}]$.

EXERCISE 7.18. *Use the result of the previous exercise to prove that*

$$A\,(\mathrm{e}^{zA^\dagger}\Omega) = z\,(\mathrm{e}^{zA^\dagger}\Omega). \tag{7.131}$$

Equation (7.131) states that

$$\Psi_z = \mathrm{e}^{zA^\dagger}\,\Omega = \sum_{n=0}^{\infty} \frac{z^n}{\sqrt{n!}}\,\phi_n \tag{7.132}$$

is an eigenvector of the operator $A$ belonging to the eigenvalue $z$. So the vector $\Psi_z$ is interesting enough to deserve further investigation. Let us first calculate its norm. This is done easily because it has an expansion in the orthonormal set of the oscillator eigenfunctions:

$$\|\Psi_z\|^2 = \sum_{n=0}^{\infty} \left|\frac{z^n}{\sqrt{n!}}\right|^2 = \sum_{n=0}^{\infty} \frac{(|z|^2)^n}{n!} = \mathrm{e}^{|z|^2}, \tag{7.133}$$

and hence

$$\|\Psi_z\| = \exp\Bigl(\frac{|z|^2}{2}\Bigr). \tag{7.134}$$

Next, we calculate $\Psi_z(x)$ for real indices $z \in \mathbb{R}$. This can be done with a little trick. Remember that the momentum operator is given by

$$p = \frac{\mathrm{i}}{\sqrt{2}}\,(A^\dagger - A) \tag{7.135}$$

and hence, using the Weyl relation,

$$\begin{aligned} e^{-ip(z\sqrt{2})}\,\Omega(x) &= e^{zA^\dagger - zA}\,\Omega(x) \\ &= e^{-z^2/2}\,e^{zA^\dagger}\,e^{-zA}\,\Omega(x) \\ &= e^{-z^2/2}\,e^{zA^\dagger}\,\Omega(x) \\ &= e^{z^2}\,\Psi_z(x). \end{aligned} \tag{7.136}$$

On the other hand, for $z \in \mathbb{R}$, we have

$$e^{-ip(z\sqrt{2})}\,\Omega(x) = \Omega(x - z\sqrt{2}), \tag{7.137}$$

because $p$ is the generator of translations. Put this together to obtain

$$\Psi_z(x) = e^{z^2/2}\,\Omega(x - z\sqrt{2}), \quad z \in \mathbb{R}. \tag{7.138}$$

We see that $\Psi_z$ is (up to normalization) just a shifted ground state. The time evolution of translated ground states has been investigated earlier in Section 7.5.1. These states have been called coherent states.

**Coherent states:**

The functions

$$\begin{aligned} \phi^{(a)}(x) &= e^{-a^2/4}\,\Psi_{a/\sqrt{2}}(x) \\ &= \Big(\frac{1}{\pi}\Big)^{1/4} \exp\Big(-\frac{(x-a)^2}{2}\Big), \quad a \in \mathbb{R}, \end{aligned} \tag{7.139}$$

are just the coherent states of the harmonic oscillator. In the oscillator basis they have the expansion (*Glauber form*)

$$\phi^{(a)}(x) = \sum_{n=0}^{\infty} c_n\,\phi_n(x), \quad \text{with} \quad c_n = e^{-a^2/4}\,\frac{a^n}{\sqrt{2^n\,n!}}. \tag{7.140}$$

Moreover, the coherent states are eigenstates of the operator $A$ belonging to the eigenvalue $a/\sqrt{2}$.

### 7.8.2. Completeness of oscillator eigenfunctions

We can use the result above to show that the orthonormal set of oscillator eigenfunctions is a basis of the Hilbert space $L^2(\mathbb{R})$. Remember that an orthonormal set is a basis if every vector in the Hilbert space can be approximated by a (finite) linear combination of vectors from that set.

The set of all coherent states $\phi^{(a)}$ is not a linear subspace of the Hilbert space, because the sum $\phi^{(a)} + \phi^{(b)}$ (with $a \neq b$) is not a coherent state. But we can form the set $\mathcal{G}$ of all finite linear combinations of coherent states.

$$\psi \in \mathcal{G} \quad \text{if and only if} \quad \psi = \sum_{n=0}^{N} c_n \phi^{(a_n)}, \tag{7.141}$$

for some integer $N$ and suitable numbers $c_n \in \mathbb{C}$, $a_n \in \mathbb{R}$, $n = 0, 1, \dots, N$. Clearly, the set $\mathcal{G}$ is a linear subspace of the Hilbert space. It has been shown earlier that $\mathcal{G}$ is even a dense linear subspace of $L^2(\mathbb{R})$ (see the remark after Theorem 2.2 and note that the $\phi^{(a)}$ are just Gaussian functions of the type (2.106)).

The result (7.140) means that all shifted Gaussian functions $\phi^{(a)}$ can be approximated by finite linear combinations of the $\phi_n$. In mathematical terms this means that $\mathcal{G}$ is a subset of the closure $\overline{\mathfrak{D}}$ of the span of oscillator eigenfunctions. Because the closed set $\overline{\mathfrak{D}}$ contains a subset which is dense in the Hilbert space, it must be identical to the Hilbert space. (The closure of any dense set is the whole space). Hence every vector in the Hilbert space belongs to the closure of the span of the oscillator eigenfunctions, that is, every vector $\phi$ can be approximated by a linear combination of the $\phi_n$. This proves that the set of oscillator eigenfunctions is a basis.

The harmonic oscillator thus has a complete orthonormal basis of eigenfunctions. The basis vectors can all be obtained from the ground state $\Omega$ by repeated application of the creation operator $A^n$.

The completeness property also proves that the set of eigenvalues

$$\{E_n = n + 1/2 \mid n = 0, 1, 2, \dots\} \tag{7.142}$$

found by the simple algebraic argument in Section 7.7.2 indeed contains all eigenvalues of the harmonic oscillator, for if there were an eigenvalue $E'$ that is different from all the $E_n$, then the symmetry of $H$ would imply that the corresponding (nonzero) eigenfunction is orthogonal to all $\phi_n$. But this is impossible because the $\phi_n$ form a basis!

## 7.9. Special Topic: Mehler Kernel

We want to determine the action of the unitary time evolution

$$\exp(-\mathrm{i}Ht) = \exp\Bigl(-\frac{\mathrm{i}}{2}\,(p^2 + x^2)\,t\Bigr) = \exp\Bigl(-\mathrm{i}\,(A^\dagger A + \frac{1}{2}\,)\,t\Bigr) \tag{7.143}$$

on wave functions. It is sufficient to do this calculation on a dense set of wave functions because the action of a unitary operator can always be extended by continuity to the whole Hilbert space (as described in Section 2.5.3). We choose the dense set spanned by the (finite) linear combinations of the functions (see Theorem 2.2)

$$G_q(x) = \mathrm{e}^{\mathrm{i}qx}\,\Omega(x), \qquad q \in \mathbb{R}. \tag{7.144}$$

For an arbitrary function in this set we obtain, using the known temporal behavior of the position observable,

$$\begin{aligned}
\mathrm{e}^{-\mathrm{i}Ht}\, G_q &= \mathrm{e}^{-\mathrm{i}Ht}\, \mathrm{e}^{\mathrm{i}qx}\, \mathrm{e}^{\mathrm{i}Ht}\, \mathrm{e}^{-\mathrm{i}Ht}\Omega \\
&= \mathrm{e}^{\mathrm{i}qx(-t)}\, \mathrm{e}^{-\mathrm{i}t/2}\Omega \\
&= \mathrm{e}^{\mathrm{i}q(x\,\cos t - p\,\sin t)}\, \mathrm{e}^{-\mathrm{i}t/2}\Omega.
\end{aligned}$$

With $x\,\cos t - p\,\sin t = (\mathrm{e}^{-\mathrm{i}t}\, A^\dagger + \mathrm{e}^{\mathrm{i}t}\, A)/\sqrt{2}$ and the Weyl relations we obtain

$$\begin{aligned}
\mathrm{e}^{-\mathrm{i}Ht}\, G_q &= \mathrm{e}^{-\mathrm{i}t/2}\, \exp\Bigl(\frac{\mathrm{i}q}{\sqrt{2}}\,\mathrm{e}^{-\mathrm{i}t}\, A^\dagger\Bigr)\, \exp\Bigl(\frac{\mathrm{i}q}{\sqrt{2}}\,\mathrm{e}^{\mathrm{i}t}\, A\Bigr)\, \exp\Bigl(-\frac{q^2}{4}\Bigr)\,\Omega \\
&= \mathrm{e}^{-\mathrm{i}t/2}\, \exp\Bigl(-\frac{q^2}{4}\Bigr)\, \exp\Bigl(\frac{\mathrm{i}q}{\sqrt{2}}\,\mathrm{e}^{-\mathrm{i}t}\, A^\dagger\Bigr)\, \exp\Bigl(\frac{\mathrm{i}q}{\sqrt{2}}\,\mathrm{e}^{-\mathrm{i}t}\, A\Bigr)\,\Omega
\end{aligned}$$

Here we changed the factor in the exponential of function $A$, which does not matter because it is only applied to $\Omega$; see (7.129). Now we can apply the Weyl relation again to conclude

$$\begin{aligned}
\mathrm{e}^{-\mathrm{i}Ht}\, G_q &= \mathrm{e}^{-\mathrm{i}t/2}\, \exp\Bigl(-\frac{q^2}{4}\Bigr)\, \exp\Bigl(\frac{q^2}{4}\,\mathrm{e}^{-2\mathrm{i}t}\Bigr)\, \exp\Bigl(\frac{\mathrm{i}q}{\sqrt{2}}\,\mathrm{e}^{-\mathrm{i}t}\,(A^\dagger + A)\Bigr)\,\Omega \\
&= \mathrm{e}^{-\mathrm{i}t/2}\, \exp\Bigl(-\frac{q^2}{4}(1-\mathrm{e}^{-2\mathrm{i}t})\Bigr)\, \exp\Bigl(\mathrm{i}q\mathrm{e}^{-\mathrm{i}t}x\Bigr)\,\Omega.
\end{aligned}$$

On the other hand, comparison with the ansatz

$$\mathrm{e}^{-\mathrm{i}Ht}\, G_q(x) = \int_{-\infty}^{\infty} K_{\mathrm{osc}}(x,y,t)\, G_q(y)\, dy \tag{7.145}$$

leads to the relation

$$\begin{aligned}
&\int_{-\infty}^{\infty} K_{\mathrm{osc}}(x,y,t)\, \mathrm{e}^{-y^2/2}\, \mathrm{e}^{\mathrm{i}qy}\, dy \\
&\qquad = \exp\Bigl(-\mathrm{i}\frac{t}{2} - \frac{q^2}{4}(1-\mathrm{e}^{-2\mathrm{i}t}) + \mathrm{i}q\mathrm{e}^{-\mathrm{i}t}x - \frac{x^2}{2}\Bigr).
\end{aligned} \tag{7.146}$$

From this we can determine $K_{osc}(x,y,t)$ by an inverse Fourier transformation with respect to $y$:

$$\begin{aligned}
K_{\mathrm{osc}}(x,y,t)\, \mathrm{e}^{-y^2/2} &= \frac{1}{2\pi}\, \exp\Bigl(-\mathrm{i}\frac{t}{2} - \frac{x^2}{2}\Bigr) \\
&\quad \times \int_{-\infty}^{\infty} \mathrm{e}^{-\mathrm{i}qy}\, \exp\Bigl(-\frac{q^2}{4}\,(1-\mathrm{e}^{-2\mathrm{i}t}) + \mathrm{i}q\mathrm{e}^{-\mathrm{i}t}x\Bigr)\, dq.
\end{aligned} \tag{7.147}$$

Because this is an integral over a Gaussian function, it can be calculated explicitly. One obtains

$$K_{\mathrm{osc}}(x,y,t) = \frac{1}{\sqrt{\pi}}\, \frac{\mathrm{e}^{-\mathrm{i}t/2}}{(1-\mathrm{e}^{-2\mathrm{i}t})^{1/2}}\, \exp\Bigl(-\frac{(\mathrm{e}^{-\mathrm{i}t}x - y)^2}{1-\mathrm{e}^{-2\mathrm{i}t}} - \frac{x^2}{2} + \frac{y^2}{2}\Bigr). \tag{7.148}$$

A little trigonometric exercise converts this expression into

$$K_{\rm osc}(x,y,t) = \frac{1}{\sqrt{2\pi\mathrm{i}\,\sin t}}\,\exp\Bigl(\mathrm{i}\frac{x^2+y^2}{2}\,\cot t\;-\mathrm{i}\frac{xy}{\sin t}\Bigr). \tag{7.149}$$

A similar expression has been derived by Mehler in a different context in the 19th century. Therefore, the time evolution kernel for the harmonic oscillator is called the *Mehler kernel.*

As stated in Section 7.3.2, it is sufficient to know the time evolution for the time interval $(0,\pi)$ from which the rest follows by temporal symmetry. The Mehler kernel is well defined for $t \in (0,\pi)$, but singular at the borders of this interval. As you probably remember, any solution $\psi(x,t)$ of the harmonic oscillator is proportional to $\psi(\pm x, 0)$ at the times $t = 0, \pm\pi, \pm 2\pi, \ldots$. For these times the integral kernel must behave like a delta distribution. For example, at $t = 0$,

$$\psi(x) = \int_{-\infty}^{\infty} K_{\rm osc}(x,y,0)\,\psi(y)\,dy \tag{7.150}$$

and hence $K_{\rm osc}(x,y,0)$ is the Dirac delta function

$$K_{\rm osc}(x,y,0) = \delta(x-y). \tag{7.151}$$

*Chapter 8*

# Special Systems

**Chapter summary**: In some rare cases we can find a complete and explicit solution of the Schrödinger equation without having to use numerical methods. Our collection of analytically solvable problems so far includes only the free particle, the particle in a box, and the harmonic oscillator. In this chapter we add the free fall in a linear potential and the motion in a constant magnetic field.

The linear potential describes a constant force field, like the homogeneous gravitational field near the earth's surface. Whenever the free time evolution of the initial wave packet is known, the free fall can also be calculated exactly. The solution describes a uniformly accelerated motion and the expectation values of position and momentum behave in a classical way. Next we add a reflecting boundary condition in order to describe the quantum analog of a steel ball dancing on a horizontal pane of glass. In quantum mechanics this system has only discrete energies. The motion of wave packets in this situation is not periodic, because the frequencies of the eigenfunctions are incommensurable.

A charged particle in a constant magnetic field feels a force that is always perpendicular to its velocity. It is sufficient to treat this problem in the two-dimensional plane orthogonal to the direction of the magnetic field. The trajectories of classical particles are circles in that plane and the solutions of the two-dimensional Schrödinger equation all describe bound states. We determine the eigenvalues of the Hamiltonian operator by exploiting an analogy between the components of the velocity operator (which do not commute with each other) and the position and momentum observables for a harmonic oscillator. By this analogy we obtain results about the motion of arbitrary Gaussian wave packets and derive the integral kernel of the time evolution.

In quantum mechanics, the magnetic field has to be described by a (gauge-dependent) vector potential, but we expect that physical predictions should be independent of the chosen gauge. Thus, the interpretation of wave packets in magnetic fields becomes a tricky business, and one has to take into account the fact that the canonical momentum has no simple (gauge-invariant) relation with the velocity. For example, the Schrödinger equation in a constant magnetic field is not invariant under translations. Nevertheless, any translation can be compensated for by a gauge transformation, and therefore the system has a translational symmetry. As a consequence, the energy eigenvalues have an infinite multiplicity.

The constant magnetic field has a rotational symmetry, because the Hamiltonian commutes with the angular momentum operator. This leads us to an investigation of systems with sperical symmetry in two dimensions, which—after a transition to polar coordinates—can be solved in a basis of angular momentum eigenstates. The Schrödinger equation for the eigenvalues is thus reduced to an ordinary differential equation in the radial variable. This treatment prepares the ground for the more complicated reduction to angular momentum eigenspaces for three-dimensional systems in Book Two.

## 8.1. The Free Fall in a Constant Force Field

> *It is remarkable how great are the mathematical difficulties of this problem which, in classical mechanics, is one of the most basic and simple ones.*
>
> —S. Flügge

### 8.1.1. Classical mechanics

The homogeneous gravitational field near the earth's surface is usually described by a linear potential

$$V(\mathbf{x}) = \lambda z, \quad \mathbf{x} = (x, y, z) \in \mathbb{R}^3. \tag{8.1}$$

The force

$$\mathbf{F} = -\nabla V(\mathbf{x}) = -\lambda(0, 0, 1) \tag{8.2}$$

is independent of the position $\mathbf{x}$. Here $\lambda = mg$, where $m$ is the mass of the particle and $g$ is the gravitational acceleration. Of course, the linear potential also describes other situations with constant force fields—for example, the accelerated motion of an electron in a homogeneous electric field. In this case $\lambda = eE$, where $e$ is the charge of the electron, and $E$ describes the electric field strength.

The solution of the equations of motion in classical mechanics is indeed elementary. The resulting motion is the free fall. With the initial position $x(0) = x_0$ and initial momentum $p(0) = p_0$, the solution is given by

$$\begin{aligned} \mathbf{x}(t) &= \mathbf{x}_0 + \frac{\mathbf{p}_0}{m}\, t + \frac{\mathbf{F}}{2m}\, t^2, \\ \mathbf{p}(t) &= \mathbf{p}_0 + \mathbf{F}\, t. \end{aligned} \tag{8.3}$$

It describes a free motion $\mathbf{x}_0 + \frac{\mathbf{p}_0}{m}\, t$ where the position is translated in the direction of $\mathbf{F}$ by an amount proportional to $t^2$ and the momentum is translated by $\mathbf{F}\, t$. In the next section you will see that essentially the same observation can be made in the quantum-mechanical case.

### 8.1.2. The quantum time evolution

A particle in a linear potential belongs to the rare special systems for which the equations of motion can be solved both classically and quantum-mechanically.

CD 6.1 shows the quantum-mechanical free fall of a Gaussian wave packet in one dimension.

The main result of this section is the following:

**The Avron–Herbst formula**:

The Schrödinger equation for a particle under the influence of a constant force $\mathbf{F}$,

$$\mathrm{i}\frac{d}{dt}\psi = H\,\psi, \quad \text{with} \quad H = -\frac{1}{2m}\Delta - \mathbf{F}\cdot\mathbf{x}, \tag{8.4}$$

is solved by

$$\psi(\mathbf{x},t) = \exp\Bigl(-\mathrm{i}\frac{\mathbf{F}^2}{6m}t^3\Bigr)\,\exp(\mathrm{i}t\mathbf{F}\cdot\mathbf{x})\,\exp\Bigl(\frac{\mathrm{i}}{2m}\Delta\,t\Bigr)\,\psi_0\Bigl(\mathbf{x}-\frac{\mathbf{F}}{2m}t^2\Bigr), \tag{8.5}$$

where $\psi(\mathbf{x},0) = \psi_0(\mathbf{x})$ is an arbitrary initial state.

This formula allows us to determine the time evolution in a linear potential whenever we know the free time evolution. Before proving this result, let us have a closer look at it. According to Eq. (8.5), the solution at time $t$ can be obtained by the following procedure.

1. Take the initial function $\psi_0$ and perform a translation of its argument in the direction of $\mathbf{F}$ and by an amount proportional to $t^2$.
2. Let the resulting function evolve according to the free time evolution. This is described by the operator $\exp(\mathrm{i}\Delta t/2m)$, which in momentum space is just multiplication by $\exp(-\mathrm{i}\mathbf{p}^2t/2m)$.
3. Next, multiply the wave function with the phase factor $\exp(\mathrm{i}t\mathbf{F}\cdot\mathbf{x})$, which in momentum space describes a translation by the vector $\mathbf{F}t$.
4. Finally, multiply the result with a time-dependent phase factor (which does not depend on $x$).

We see that the quantum motion of a particle in a linear potential can be described in close analogy to the classical motion. The origin of the phase factor $\exp(-\mathrm{i}\mathbf{F}^2t^3/6m)$ is, of course, purely quantum mechanical.

CD 6.2 visualizes the Avron–Herbst formula. Here we can see the action of the four operations described above.

EXERCISE 8.1. *Why is the term* $\mathbf{F}t^2/2m$ *subtracted from* $\mathbf{x}$ *in the argument of* $\psi_0$ *in Eq.* (8.5) *while it is added to* $\mathbf{x}$ *in Eq.* (8.3)*?*

EXERCISE 8.2. *Use the results in Section 3.3.2 together with the Avron–Herbst formula in order to determine the time evolution of a Gaussian wave packet under the influence of a constant force.*

CD 6.3 and 6.4 show the free fall of Gaussian wave packets in two dimensions. The movies allow the comparison with the classical mechanical motion.

$\boxed{\Psi}$ For the proof of the Avron–Herbst formula, it is assumed that the constant field points in the $z$-direction, so that $V(\mathbf{x}) = -\mathbf{F}\cdot\mathbf{x} = \lambda z$. Then the Hamiltonian splits into a part $H_\perp$, which describes the free motion in the $xy$-plane, and a part

$$H_z := -\frac{1}{2m}\frac{d^2}{dz^2} + \lambda z, \tag{8.6}$$

which describes the free fall in the $z$-direction. It is easy to see that the operators $H_\perp$ and $H_z$ commute. Hence the time evolution can be separated as

$$e^{-iHt} = e^{-iH_\perp t}\, e^{-iH_z t}. \tag{8.7}$$

It is sufficient to consider the nontrivial part describing the free fall. For simplicity, we write the Hamiltonian $H_z$ in the form

$$H_z = \frac{p^2}{2m} + V, \quad p = -i\frac{d}{dz}, \quad V = \lambda z. \tag{8.8}$$

Here the operator $p$ is just the component of the momentum in the $z$-direction. If we perform a Fourier transform with respect to $z$, then $p$ is just the multiplication by $k$ in Fourier space and the potential $V$ corresponds to the differential operator $i\lambda d/dk$, according to Eq. (2.89). Thus, you can certainly verify the formula

$$H_z = \exp\Bigl(i\,\frac{p^3}{6\lambda m}\Bigr)\, V\, \exp\Bigl(-i\,\frac{p^3}{6\lambda m}\Bigr), \tag{8.9}$$

because in momentum space this is just identity

$$\frac{k^2}{2m} + i\lambda\frac{d}{dk} = \exp\Bigl(i\,\frac{k^3}{6\lambda m}\Bigr)\, i\lambda\frac{d}{dk}\, \exp\Bigl(-i\,\frac{k^3}{6\lambda m}\Bigr), \tag{8.10}$$

which can be obtained by an elementary differentiation. Hence you can see that the operators $V$ and $H_z$ are related by a unitary transformation. The corresponding time evolutions $\exp(-\mathrm{i}Vt)$ and $\exp(-\mathrm{i}H_z t)$ must therefore have the same relation,

$$\exp(-\mathrm{i}H_z t) = \exp\Bigl(\mathrm{i}\,\frac{p^3}{6\lambda m}\Bigr)\,\exp(-\mathrm{i}Vt)\,\exp\Bigl(-\mathrm{i}\,\frac{p^3}{6\lambda m}\Bigr). \tag{8.11}$$

Because the multiplication by the phase factor $\exp(-\mathrm{i}Vt) = \exp(-\mathrm{i}\lambda tz)$ amounts to a translation by $\lambda t$ in momentum space (see Section 2.6.1), we obtain

$$\exp(-\mathrm{i}H_z t) = \exp(-\mathrm{i}\lambda zt)\,\exp\Bigl(\mathrm{i}\,\frac{(p+\lambda t)^3}{6\lambda m}\Bigr)\,\exp\Bigl(-\mathrm{i}\,\frac{p^3}{6\lambda m}\Bigr). \tag{8.12}$$

Expanding the term $(p+\lambda t)^3$ and rearranging the expressions finally yields

$$\exp(-\mathrm{i}H_z t) = \exp\Bigl(-\mathrm{i}\,\frac{\lambda^2}{6m}\,t^3\Bigr)\,\exp(-\mathrm{i}\lambda zt)\,\exp\Bigl(-\mathrm{i}\,\frac{p^2}{2m}\,t\Bigr)\,\exp\Bigl(\mathrm{i}\,\frac{\lambda t^2}{2m}\,p\Bigr). \tag{8.13}$$

Notice that the last factor describes a translation $z \to z + \lambda t^2/2m$ in position space. Multiplying Eq. (8.13) by $\exp(-\mathrm{i}H_\perp t)$ finally gives the Avron–Herbst formula. □

### 8.1.3. Position and momentum operators

Let us determine the time evolution of the position and momentum observables,

$$\mathbf{x}(t) = \mathrm{e}^{\mathrm{i}Ht}\,\mathbf{x}\,\mathrm{e}^{-\mathrm{i}Ht}, \qquad \mathbf{p}(t) = \mathrm{e}^{\mathrm{i}Ht}\,\mathbf{p}\,\mathrm{e}^{-\mathrm{i}Ht}. \tag{8.14}$$

As mentioned earlier (Section 7.4.1), the knowledge of the time evolution of an observable allows us to calculate its expectation value at any time $t$ from the initial wave function. We determine the time evolution of $\mathbf{x}$ and $\mathbf{p}$ by solving the Heisenberg equations. First we need the commutators with the Hamiltonian

$$H = \frac{1}{2m}\,\mathbf{p}^2 - \mathbf{F}\cdot\mathbf{x}. \tag{8.15}$$

A little calculation gives the following commutation relations

$$\begin{aligned} \mathrm{i}\,[H, x_j] &= \frac{\mathrm{i}}{2m}\sum_k [p_k^2, x_j] \\ &= \frac{\mathrm{i}}{2m}\sum_k (-2\mathrm{i})\,\delta_{kj}\,p_k = \frac{1}{m}\,p_j, \end{aligned} \tag{8.16}$$

and

$$\begin{aligned} \mathrm{i}\,[H, p_j] &= \mathrm{i} \sum_k (-F_k)\,[x_k, p_j] \\ &= \mathrm{i} \sum_k (-F_k)\,\mathrm{i}\,\delta_{kj} = F_j. \end{aligned} \tag{8.17}$$

Hence the evolution equations for the time-dependent operators are

$$\frac{d}{dt}\,\mathbf{x}(t) = \frac{1}{m}\,\mathbf{p}(t), \qquad \frac{d}{dt}\,\mathbf{p}(t) = \mathbf{F}(t). \tag{8.18}$$

These equations look exactly like the corresponding equations in classical mechanics. The expressions corresponding to the classical solutions are

$$\mathbf{x}(t) = \mathbf{x} + \frac{1}{m}\,\mathbf{p}\,t + \frac{\mathbf{F}}{2m}\,t^2, \qquad \mathbf{p}(t) = \mathbf{p} + \mathbf{F}\,t. \tag{8.19}$$

As we are now dealing with operators you may wish to verify by differentiation that these expressions are indeed solutions of the Heisenberg equations (8.18). As a consequence of these observations we find that for any initial state the expectation values of position and momentum obey the laws of classical physics.

EXERCISE 8.3. *Compare the energy of the classical motion of the expectation values with the quantum-mechanical expectation value of the energy* $\langle H \rangle_\psi$.

## 8.2. Free Fall with Elastic Reflection at the Ground

In this section we describe a freely falling quantum particle that hits the earth's surface and is reflected elastically. This corresponds to the classical situation of a bouncing ball. The earth's surface is modeled by an infinite potential barrier, giving rise to a boundary condition at $z = 0$. In one-dimension we obtain (after a suitable scaling transformation) the stationary Schrödinger equation

$$-\frac{d^2}{dz^2}\,\psi(z) + z\,\psi(z) = E\,\psi(z) \tag{8.20}$$

in the region $z \geq 0$. We look for the eigenvalues and square-integrable eigenfunctions of the Hamiltonian

$$H = -\frac{d^2}{dz^2} + z \tag{8.21}$$

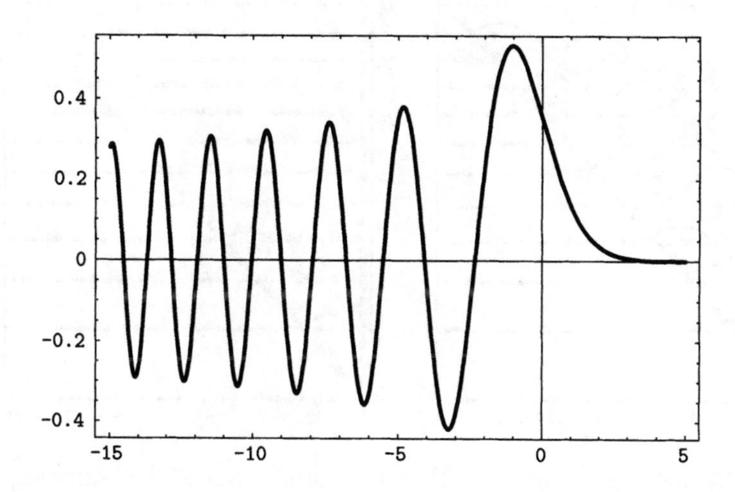


FIGURE 8.1. The Airy function $Ai(x)$ gives all the eigenfunctions of the bouncing ball.

on a suitable domain in the Hilbert space $L^2([0,\infty))$. The domain of definition of $H$ consists of functions that are twice differentiable (in a generalized sense) and satisfy Dirichlet boundary conditions at $z = 0$, that is,

$$\psi(0) = 0, \quad \text{for all } \psi \in \mathfrak{D}(H). \tag{8.22}$$

The eigenvalue equation $\psi'' - (z - E)\psi = 0$ is solved by the Airy functions $\mathrm{Ai}(z - E)$ and $\mathrm{Bi}(z - E)$. The Airy functions are well known and, for example, built into *Mathematica*. They can be expressed in terms of linear combinations of Bessel functions of order $\pm 1/3$. Only the function Ai is square-integrable on $[0, \infty)$, while $\mathrm{Bi}(z)$ diverges as $z \to \infty$. Figure 8.1 shows a plot of the function $z \to \mathrm{Ai}(z)$ for real $z$.

The eigenvalues of the problem (8.20) with boundary conditions can thus be determined by finding all those values of $E$ for which $\mathrm{Ai}(z - E)$ has a zero at $z = 0$. We have

$$\mathrm{Ai}(-E_n) = 0 \tag{8.23}$$

for an infinite ordered set of numbers $E_n$, $n = 0, 1, 2, \ldots$, all satisfying $E_n > 0$. There is no simple formula for the zeros of Airy functions, but the zeros can be found numerically (e.g., with the help of *Mathematica*). The results are given in the box below.

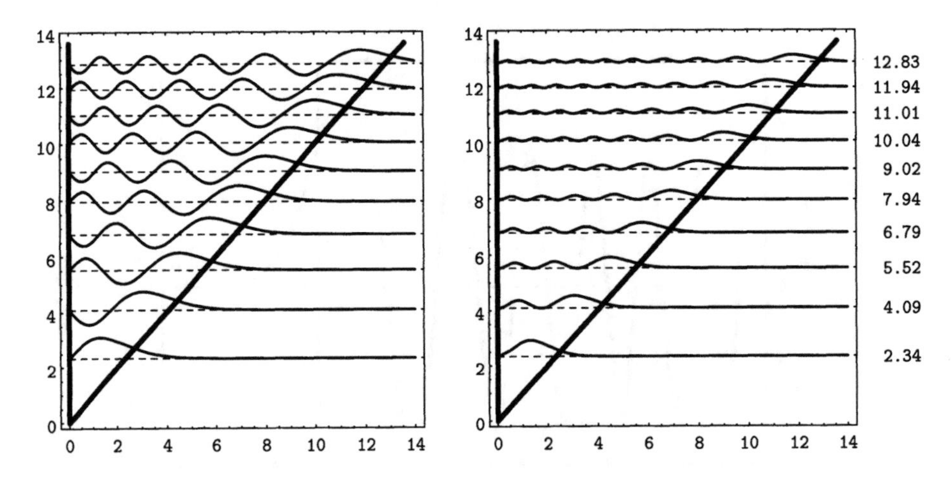


FIGURE 8.2. The normalized eigenfunctions of the bouncing ball problem and the corresponding position probability densities.

**Bouncing ball:**

The Schrödinger operator $H$ given by Eq. (8.21) with Dirichlet boundary condition at $z = 0$ has a basis of eigenfunctions $\psi_n$ in the Hilbert space $L^2([0,\infty))$ which are shifted Airy functions. We have

$$\psi_n(z) = c_n \mathrm{Ai}(z - E_n), \tag{8.24}$$

where $c_n$ are suitable normalization constants (see Fig. 8.2). The positive real numbers $E_n$ have to be determined from the condition $\mathrm{Ai}(-E_n) = 0$. The first zeros of the Airy function Ai thus determine the energy eigenvalues as

$$\begin{aligned} E_0 &= 2.338108, & E_5 &= 9.022651, \\ E_1 &= 4.087949, & E_6 &= 10.04017, \\ E_2 &= 5.520560, & E_7 &= 11.00852, \\ E_3 &= 6.786708, & E_8 &= 11.93602, \\ E_4 &= 7.944134, & E_9 &= 12.82878. \end{aligned}$$

It is clear that the solutions of the time-dependent Schrödinger equation can be obtained as linear combinations of the eigenfunctions,

$$\psi(z,t) = \sum a_n \, \psi_n(z) \, \mathrm{e}^{-\mathrm{i}E_n t}, \tag{8.25}$$

where the coefficients $a_n$ have to be determined from the expansion of the initial function $\psi(z,0) = \psi_0(z)$ into the basis of eigenfunctions, that is,

$$a_n = \langle \psi_n, \psi_0 \rangle = c_n \int_0^\infty \mathrm{Ai}(z - E_n)\, \psi_0(z)\, dz. \tag{8.26}$$

Here the constants $c_n$ have to be determined such that $\int_0^\infty |\psi_n(z)|^2\, dz = 1$. Because there is no simple relation between different energies, superpositions of two or more eigenfunctions will not describe a periodic motion. (The time dependence will only be quasi-periodic).

EXERCISE 8.4. *(See Exercise 1.10.) Let $\psi_1$ and $\psi_2$ be two eigenstates of a Hamiltonian $H$, belonging to eigenvalues $E_1$ and $E_2$, respectively. The superposition $a\psi_1 + b\psi_2$ has a periodic time dependence if and only if the eigenvalues are commensurable, that is,*

$$\frac{E_1}{E_2} = \frac{n}{m} \quad \text{for some integers } n \text{ and } m. \tag{8.27}$$

EXERCISE 8.5. *The position probability density for a superposition of two eigenstates is always periodic in time with period $T = \pi/|E_2 - E_1|$.*

EXERCISE 8.6. *Consider again the superposition $a\psi_1 + b\psi_2$ of two eigenstates with respective energies, $E_1$ and $E_2$. By adding a suitable constant to the potential (gauge transformation) it is always possible to make this superposition periodic in time.*

EXERCISE 8.7. *Find the eigenvalues and eigenfunctions of the operator*

$$H = -\frac{1}{2}\frac{d^2}{dz^2} + g\, z, \qquad (\text{for some } g \in \mathbb{R}). \tag{8.28}$$

*Use a suitable scale transformation of the Schrödinger equation* (8.20).

CD 6.5 and 6.6 show oscillating states of the bouncing ball problem, which consist of superpositions of two eigenstates. Hence there is a suitable gauge transformation that makes the time evolution periodic. Only CD 6.6.2 shows a nonperiodic superposition of three eigenstates.

CD 6.7 shows the motion of wave packets that are initially Gaussian. Figure 8.3 shows the expectation value of the position as a function of time for the wave packet in CD 6.7.1 and 6.7.2. This illustrates well that the motion is nonperiodic.

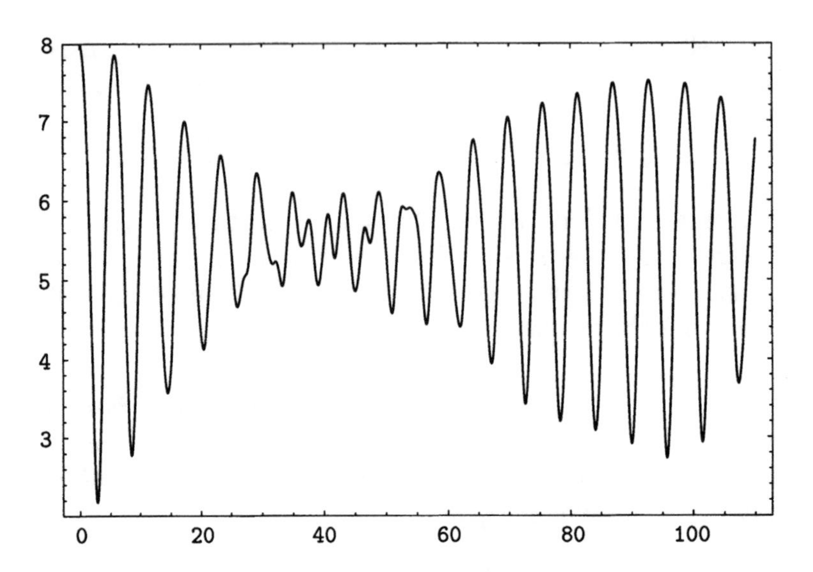


FIGURE 8.3. The mean position for an initially Gaussian wave packet in a linear potential with a Dirichlet boundary condition at $z = 0$. When the oscillation is large, the wave packet gets well localized near the classical turning points. When the average position oscillates with a small amplitude, the wave packet is distributed over the classically allowed region.

## 8.3. Magnetic Fields in Two Dimensions

Here we consider a particle in the presence of a magnetic field with constant direction. This allows the reduction of the Schrödinger equation to an equation in two spatial dimensions.

If the magnetic field strength $\vec{B}(\mathbf{x})$ has a direction independent of $\mathbf{x} \in \mathbb{R}^3$, we choose the coordinate system in such a way that the $x_3$-axis points in the direction of the magnetic field, that is,

$$\vec{B}(\mathbf{x}) = (0, 0, B(\mathbf{x})). \tag{8.29}$$

The condition (see Eq. (4.40))

$$0 = \operatorname{div} \vec{B} = \frac{\partial B_1}{\partial x_1} + \frac{\partial B_2}{\partial x_2} + \frac{\partial B_3}{\partial x_3} = \frac{\partial B}{\partial x_3} \tag{8.30}$$

implies immediately that $B(\mathbf{x})$ does not depend on $x_3$. Thus, the situation is in fact two-dimensional, that is, invariant with respect to translations along the $x_3$-axis.

The quantum-mechanical formalism requires the introduction of a non-unique vector potential $\vec{A}$. We write

$$\vec{A}(\mathbf{x}) = (A_1(x_1, x_2), A_2(x_1, x_2), 0) \tag{8.31}$$

and find for the function $B$

$$B(x_1, x_2) = \frac{\partial}{\partial x_1} A_2(x_1, x_2) - \frac{\partial}{\partial x_2} A_1(x_1, x_2). \tag{8.32}$$

For this situation the Schrödinger equation reads as follows

$$\mathrm{i}\frac{d}{dt}\psi(\mathbf{x}, t) = \left(\frac{1}{2}p_3^2 + \frac{1}{2}\left(p_1 - \frac{q}{c}A_1\right)^2 + \frac{1}{2}\left(p_2 - \frac{q}{c}A_2\right)^2\right)\psi(\mathbf{x}, t). \tag{8.33}$$

It can be simplified by writing $\psi$ as a product

$$\psi(\mathbf{x}, t) = \phi(x_1, x_2, t)\,\xi(x_3, t). \tag{8.34}$$

Inserting this product into the Schrödinger equation, we find that $\psi$ is a solution if $\xi$ solves the free equation with respect to the variable $x_3$,

$$\mathrm{i}\frac{d}{dt}\xi(x_3, t) = -\frac{1}{2}\frac{d^2}{dx_3^2}\,\xi(x_3, t), \tag{8.35}$$

and if $\phi$ solves the two-dimensional Schrödinger equation in the $x_1x_2$-plane:

$$\mathrm{i}\frac{d}{dt}\phi(\mathbf{x}, t) = \frac{1}{2}\left(-\mathrm{i}\nabla - \frac{q}{c}\vec{A}(\mathbf{x})\right)^2\phi(\mathbf{x}, t), \quad \mathbf{x} = (x_1, x_2) \in \mathbb{R}^2. \tag{8.36}$$

(Here we switched to a two-dimensional notation: $\nabla = (\partial_1, \partial_2)$ is the two-dimensional gradient and $\vec{A} = (A_1, A_2)$).

**Magnetic field in two dimensions**:

In two dimensions, the magnetic field strength $B$ is a scalar function that can be interpreted as the third component of a vector $(0, 0, B)$. It can be obtained from a vector potential $\vec{A} = (A_1, A_2)$,

$$B(\mathbf{x}) = \frac{\partial}{\partial x_1} A_2(\mathbf{x}) - \frac{\partial}{\partial x_2} A_1(\mathbf{x}), \quad \mathbf{x} = (x_1, x_2) \in \mathbb{R}^2. \tag{8.37}$$

The Schrödinger equation has the usual form,

$$\mathrm{i}\frac{d}{dt}\phi = \frac{1}{2}\left(\mathbf{p} - \frac{q}{c}\vec{A}\right)^2\phi, \quad \mathbf{p} = -\mathrm{i}\nabla = -\mathrm{i}\left(\frac{\partial}{\partial x_1}, \frac{\partial}{\partial x_2}\right). \tag{8.38}$$

For the vector potential describing a magnetic field in two dimensions, we may choose the expression

$$\vec{A}(x_1, x_2) = (-x_2, x_1)\int_0^1 s\,B(\mathbf{x}s)\,ds. \tag{8.39}$$

This is the Poincaré gauge in two dimensions; see Eq. (4.51). It has the property

$$\vec{A}(\mathbf{x}) \cdot \mathbf{x} = 0. \tag{8.40}$$

Usually, the magnetic vector potential is nonzero even in regions where the magnetic field strength $B(\mathbf{x})$ vanishes. Assume that the two-dimensional magnetic field $B(\mathbf{x})$ is nonzero only in some finite (bounded) region and that it has a nonvanishing flux,

$$\int B(\mathbf{x})\, d^2x \neq 0. \tag{8.41}$$

From Stoke's law we obtain

$$\oint \vec{A}(\mathbf{x}) \cdot d\vec{s} = \int \operatorname{curl} \vec{A}(\mathbf{x})\, d^2x = \int B(\mathbf{x})\, d^2x, \tag{8.42}$$

where the circulation is taken along a large circle outside the support of $B$. Hence we see that the vector potential $\vec{A}(\mathbf{x})$ cannot vanish everywhere on the circle, no matter which gauge we choose. The vector potential keeps influencing the wave function even in regions which are far away from the support of $B$.

The preceding remark is best illustrated by the Aharonov–Bohm effect. CD 6.20 shows the scattering of a wave packet at an infinitely long solenoid. The magnetic field is confined to the interior of the solenoid, which is modeled by a circular Dirichlet barrier. The vector potential is nonzero outside the solenoid and produces interference effects that depend on the flux of the magnetic field through the solenoid. See also Color Plate 17. A more detailed explanation can be found on the CD-ROM.

## 8.4. Constant Magnetic Field

### 8.4.1. The Schrödinger equation

We assume that the magnetic field is constant. The problem may be reduced to two dimensions as described in the previous section. The field strength of a constant field in two dimensions is given by a scalar constant $B \in \mathbb{R}$,

$$B(\mathbf{x}) = B, \qquad \text{for all } \mathbf{x} = (x_1, x_2) \in \mathbb{R}^2. \tag{8.43}$$

For the vector potential we make the following choice (Poincaré gauge):

$$\vec{A}(\mathbf{x}) = \frac{B}{2}\,(-x_2, x_1). \tag{8.44}$$

In this case the Poincaré gauge coincides with the Coulomb gauge (Section 4.7.3).

CD 6.11 shows several views of the magnetic vector potential for the constant field. The vectors $\vec{A}(\mathbf{x})$ are always orthogonal to $\mathbf{x}$, that is, $\mathbf{x} \cdot \vec{A}(\mathbf{x}) = 0$. Usually, we visualize a vector field in two dimensions by arrows, although one could use the color map too.

The Schrödinger equation, which we want to discuss now, is

$$\mathrm{i}\frac{d}{dt}\psi(\mathbf{x},t) = \left(\frac{1}{2}\left(p_1 + \frac{B}{2}x_2\right)^2 + \frac{1}{2}\left(p_2 - \frac{B}{2}x_1\right)^2\right)\psi(\mathbf{x},t). \tag{8.45}$$

In order to simplify the notation, we have redefined $\frac{q}{c}B \to B$.

EXERCISE 8.8. *Show that the vector potential* $\vec{A}_2(\mathbf{x}) = (-Bx_2, 0)$ *also describes the constant magnetic field* (8.43). *Find the gauge transformation linking* $\vec{A}$ *and* $\vec{A}_2$, *that is, find a scalar function* $g$ *such that*

$$\vec{A}_2(\mathbf{x}) = \vec{A}(\mathbf{x}) + \nabla g(\mathbf{x}). \tag{8.46}$$

### 8.4.2. The velocity operators

Looking at the Schrödinger equation above it seems reasonable to introduce the abbreviations

$$v_1 := p_1 + \frac{B}{2}x_2, \qquad v_2 := p_2 - \frac{B}{2}x_1. \tag{8.47}$$

With this notation, the Hamiltonian of a particle in a constant magnetic field in two dimensions becomes

$$H = \tfrac{1}{2}(v_1^2 + v_2^2). \tag{8.48}$$

The operators $v_i$ are components of a *velocity operator*. This can be seen as follows. If we define, as usual, the time-dependent position operator

$$\mathbf{x}(t) = \mathrm{e}^{\mathrm{i}Ht}\,\mathbf{x}\,\mathrm{e}^{-\mathrm{i}Ht}, \tag{8.49}$$

we find

$$\frac{d}{dt}\,x_i(t) = \mathrm{e}^{\mathrm{i}Ht}\,\mathrm{i}\,[H, x_i]\,\mathrm{e}^{-\mathrm{i}Ht} = v_i(t). \tag{8.50}$$

Here the commutators of the position operators $x_i$ with $H$ can be calculated, for example, as follows

$$\begin{aligned} \mathrm{i}\,[H, x_1] &= \frac{\mathrm{i}}{2}\,[v_1^2, x_1] \qquad (\text{because } [v_2^2, x_1] = 0) \\ &= \frac{\mathrm{i}}{2}\,(v_1\,[v_1, x_1] + [v_1, x_1]\,v_1) \\ &= \frac{\mathrm{i}}{2}\,(v_1\,[p_1, x_1] + [p_1, x_1]\,v_1) \\ &= \frac{\mathrm{i}}{2}\,(v_1\,(-\mathrm{i}) + (-\mathrm{i})\,v_1) \\ &= v_1. \end{aligned}$$

A very peculiar feature of the velocity operator in the presence of a magnetic field is the noncommutativity of its components. We find

$$\begin{aligned} [v_1, v_2] &= \Big[p_1 + \frac{B}{2}\,x_2\,,\, p_2 - \frac{B}{2}\,x_1\Big] \\ &= -\frac{B}{2}\,[p_1, x_1] + \frac{B}{2}\,[x_2, p_2] = \mathrm{i}B. \end{aligned} \tag{8.51}$$

That means that the operators $v_1$ and $v_2$ are canonically conjugate variables. We are going to explore the consequences of this observation in the following sections.

EXERCISE 8.9. *For a particle in an arbitrary magnetic field, the velocity operator is given by*

$$\frac{d}{dt}\,\mathbf{x}(t) = \mathbf{v}(t) = \mathbf{p}(t) - \frac{q}{c}\,\vec{A}(\mathbf{x}). \tag{8.52}$$

In a magnetic field we cannot rely on our intuition about the time evolution of wave packets (Section 3.5). The local wavelength in a wave packet gives information about the canonical momentum which—in a magnetic field—is usually different from the velocity. Therefore, looking at the phase tells little about the dynamic properties of the state. In particular, it is not very instructive to prepare a wave packet with a certain average *momentum* (as, e.g., in Section 3.3.2) because the resulting motion would depend on the chosen gauge. For example, if we put the Gaussian initial function

$$\psi(\mathbf{x}) = N\,\exp\Big(-\frac{(\mathbf{x}-\mathbf{x}_0)^2}{2}\Big) \tag{8.53}$$

(which has average momentum 0) in a region where the magnetic vector potential is $\vec{A}(\mathbf{x}) \approx \vec{A}_0$, then the wave packet will start moving with the approximate initial velocity $-\frac{q}{c}\vec{A}_0$. For an inhomogeneous vector potential the task of preparing a wave packet with a certain average initial *velocity* is rather nontrivial.

## 8.5. Energy Spectrum in a Constant Magnetic Field

There is a complete formal analogy between the Hamiltonian operator

$$H = \frac{1}{2}\,(v_1^2 + v_2^2), \quad \text{where } [v_1, v_2] = \mathrm{i}B, \tag{8.54}$$

and the Hamiltonian of the one-dimensional harmonic oscillator

$$H_{\text{osc}} = \frac{1}{2}\,(p^2 + x^2), \quad \text{where } [x, p] = \mathrm{i}. \tag{8.55}$$

Therefore, it is no surprise that the operators $H$ and $H_{osc}$ have essentially the same spectrum of eigenvalues.

**Energy eigenvalues in a constant magnetic field:**

The two-dimensional Hamiltonian with a constant magnetic field

$$H = \frac{1}{2}\Bigl(p_1 + \frac{B}{2}x_2\Bigr)^2 + \frac{1}{2}\Bigl(p_2 - \frac{B}{2}x_1\Bigr)^2 \tag{8.56}$$

has the eigenvalues

$$E_n = |B|\,\Bigl(n + \frac{1}{2}\Bigr), \quad n = 0, 1, 2, 3, \ldots . \tag{8.57}$$

A normalized energy eigenstate with the lowest possible energy $E_0 = |B|/2$ is given by

$$\phi_0(\mathbf{x}) = \Bigl(\frac{|B|}{2\pi}\Bigr)^{1/2} \exp\Bigl(-\frac{|B|}{4}\,x^2\Bigr). \tag{8.58}$$

PROOF. We define the operator

$$A = \sqrt{\frac{1}{2\,|B|}}\,(v_1 + \mathrm{i}\,v_2), \tag{8.59}$$

and calculate the product

$$\begin{aligned} A^\dagger A &= \frac{1}{2\,|B|}\,(v_1 - \mathrm{i}\,v_2)\,(v_1 + \mathrm{i}\,v_2) \\ &= \frac{1}{2\,|B|}\,(v_1^2 + v_2^2 + \mathrm{i}\,(v_1\,v_2 - v_2\,v_1)) \\ &= \frac{1}{2\,|B|}\,(v_1^2 + v_2^2 - B). \end{aligned}$$

and similarly,

$$AA^\dagger = \frac{1}{2\,|B|}\,(v_1^2 + v_2^2 + B).$$

Assuming $B > 0$, the Hamiltonian can be written as

$$H = B\,(A^\dagger A + \tfrac{1}{2}). \tag{8.60}$$

Now the same argument that has been carried through for the harmonic oscillator in Section 7.7.1 allows us to conclude that the operator $H$ has the eigenvalues

$$E_n = B\,(n + \tfrac{1}{2}), \quad n = 0, 1, 2, 3, \dots, \tag{8.61}$$

provided that the equation

$$A\,\phi_0 = 0 \tag{8.62}$$

has a nontrivial solution $\phi_0$. This equation is a first-order differential equation,

$$\Bigl(-\mathrm{i}\,\frac{\partial}{\partial x_1} + \frac{B}{2}\,x_2 + \frac{\partial}{\partial x_2} - \mathrm{i}\,\frac{B}{2}\,x_1\Bigr)\,\phi_0(\mathbf{x}) = 0. \tag{8.63}$$

Because this equation consists of a sum of terms involving only $x_1$ and of terms involving only $x_2$, we may use a product ansatz. Inserting

$$\phi_0(\mathbf{x}) = \psi_1(x_1)\,\psi_2(x_2)$$

we obtain for the factors the equations

$$-\mathrm{i}\,\Bigl(\frac{\partial}{\partial x_1} + \frac{B}{2}\,x_1\Bigr)\,\psi_1(x_1) = 0, \qquad \Bigl(\frac{\partial}{\partial x_2} + \frac{B}{2}\,x_2\Bigr)\,\psi_2(x_2) = 0.$$

These equations are identical and may be compared to the equation (7.116) for the ground state of the harmonic oscillator. Hence we find, with $\psi_1(x) = \psi_2(x) = \exp(-Bx^2/4)$, that

$$\phi_0(\mathbf{x}) = \exp\Bigl(-\frac{B}{4}\,(x_1^2 + x_2^2)\Bigr) \tag{8.64}$$

is a solution of $A\phi_0 = 0$ and hence of $H\phi_0 = (B/2)\phi_0$. After normalization this gives Eq. (8.58) for $B > 0$. For $B < 0$ an analogous reasoning can be carried through with $A$ and $A^\dagger$ interchanged. □

The harmonic oscillator and the constant magnetic field look very similar on a formal level. The components of the velocity in a magnetic field correspond to the position and momentum of the harmonic oscillator.

| magnetic field in two dimensions | | harmonic oscillator in one dimension |
|---|---|---|
| $v_1$ | $\longleftrightarrow$ | $x$ |
| $v_2$ | $\longleftrightarrow$ | $p$ |

Nevertheless, the two systems are quite different from a physical point of view. The particle in a constant magnetic field has a symmetry under translations, which the harmonic oscillator in phase space does not. The harmonic oscillator force distinguishes the coordinate origin as an equilibrium point, but in the homogeneous magnetic field all points are the same. Although the origin $\mathbf{x} = (0,0)$ appears to be distinguished by the property $\vec{A}(\mathbf{x}) = (0,0)$, this is only due to our particular choice of $\vec{A}$ and does not correspond to a physical property of the system. A gauge transformation can shift the zero of $\vec{A}$ to any other place in $\mathbb{R}^2$. Because a gauge transformation leads to a physically equivalent description, the origin is not different from any other point.

EXERCISE 8.10. *Find a gauge transformation that transforms our vector potential* $\vec{A} = (B/2)(-x_2, x_1)$ *into*

$$\vec{A}'(x_1, x_2) = \frac{B}{2}(-x_2 + a_2, x_1 - a_1). \tag{8.65}$$

The translational symmetry will bring us to the conclusion that all eigenvalues $E_n$ in the constant magnetic field are infinitely degenerate.

## 8.6. Translational Symmetry in a Magnetic Field

### 8.6.1. Classical motion

A short look at the classical motion in the presence of a constant magnetic field will be useful for comparison with the quantum-mechanical time evolution. The classical Hamiltonian equations

$$\begin{aligned}
\dot{x}_1(t) &= \frac{\partial}{\partial p_1} H(\mathbf{x}, \mathbf{p}) = p_1(t) + \frac{B}{2} x_2(t) \\
\dot{x}_2(t) &= \frac{\partial}{\partial p_2} H(\mathbf{x}, \mathbf{p}) = p_2(t) - \frac{B}{2} x_1(t) \\
\dot{p}_1(t) &= -\frac{\partial}{\partial x_1} H(\mathbf{x}, \mathbf{p}) = \frac{B}{2}\left(p_2(t) - \frac{B}{2} x_1\right) \\
\dot{p}_2(t) &= -\frac{\partial}{\partial x_2} H(\mathbf{x}, \mathbf{p}) = -\frac{B}{2}\left(p_1(t) + \frac{B}{2} x_2\right)
\end{aligned} \tag{8.66}$$

can be rewritten as

$$\ddot{x}_1(t) = B\,\dot{x}_2(t),$$
$$\ddot{x}_2(t) = -B\,\dot{x}_1(t), \tag{8.67}$$

which is the two-dimensional form of the Lorentz force equation $\ddot{\mathbf{x}} = \dot{\mathbf{x}} \times \vec{B}$. In this form the equations of motion are manifestly invariant with respect to translations.

The absolute value of the velocity is constant. This can be seen from

$$H(\mathbf{x}, \mathbf{p}) = \tfrac{1}{2}(\dot{x}_1^2 + \dot{x}_2^2) \tag{8.68}$$

and the fact that the Hamiltonian function is a constant of motion (conservation of the energy). The Lorentz force and hence the acceleration is always orthogonal to the velocity and for a constant field it is independent of the position. From this we infer that the classical particle performs a circular motion with constant angular velocity. Indeed, you can verify that the expressions

$$\dot{x}_1(t) = v_1 \cos Bt + v_2 \sin Bt$$
$$\dot{x}_2(t) = v_2 \cos Bt - v_1 \sin Bt \tag{8.69}$$

are solutions of Eq. (8.67) with initial velocity

$$(\dot{x}_1(0), \dot{x}_2(0)) = (v_1, v_2). \tag{8.70}$$

For the position we obtain

$$x_1(t) = \overline{x}_1 - \frac{1}{B}\,\dot{x}_2(t),$$
$$x_2(t) = \overline{x}_2 + \frac{1}{B}\,\dot{x}_1(t). \tag{8.71}$$

The classical orbit $t \to (x_1(t), x_2(t))$ of the particle is thus a circle with center $(\overline{x}_1, \overline{x}_2)$ and radius

$$R = \frac{|\mathbf{v}|}{|B|} = \frac{\sqrt{v_1^2 + v_2^2}}{|B|}. \tag{8.72}$$

If we look at this circle from above, the motion of the particle will be counterclockwise for $B < 0$ (i.e., if we look in the direction of $\vec{B} = (0, 0, B)$), and clockwise for $B > 0$.

Given the position and the velocity at any time, the center of the circle (the classical center of motion) can be obtained from

$$(\overline{x}_1, \overline{x}_2) = \Big(x_1(t) + \frac{1}{B}\,\dot{x}_2(t), x_2(t) - \frac{1}{B}\,\dot{x}_1(t)\Big). \tag{8.73}$$

These quantities are therefore constants of motion.

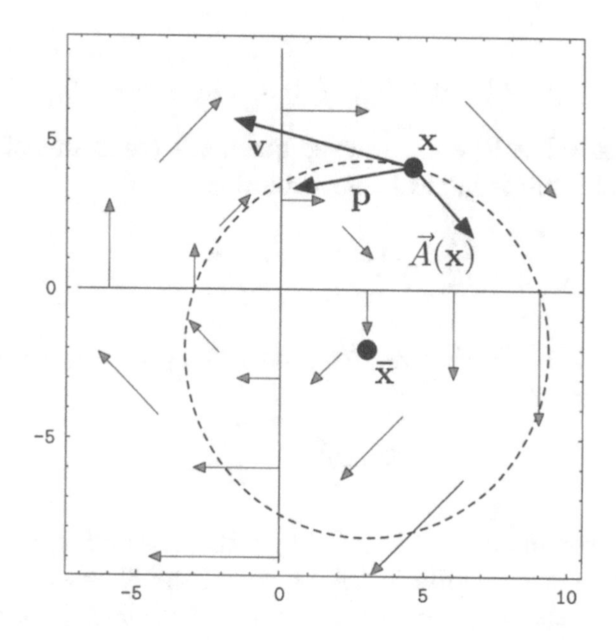


FIGURE 8.4. Classical motion of a particle in a constant magnetic field $B = -1$. The gray arrows indicate the vector potential. The dashed circle is the classical path. The relation between canonical momentum, velocity and vector potential is $\mathbf{p} = \mathbf{v} + \vec{A}(\mathbf{x})$.

CD 6.13 shows the classical motion in the field $B = -1$ (see also Fig. 8.4). The particle moves counterclockwise on a circle. The radius of the circle depends on the absolute value of the initial velocity. The animations also visualize the momentum, the velocity, and the vector potential at the position of the particle.

### 8.6.2. Symmetry under translations

The quantum-mechanical operators that correspond to the center of the classical orbit are

$$\overline{x}_1 = x_1 + \frac{1}{B} v_2 = \frac{1}{2} x_1 + \frac{1}{B} p_2, \tag{8.74}$$

$$\overline{x}_2 = x_2 - \frac{1}{B} v_1 = \frac{1}{2} x_2 - \frac{1}{B} p_1. \tag{8.75}$$

(As usual, we denote the quantum-mechanical operators by the same symbols as the classical quantities). The operators $\overline{x}_i$ commute with the operators $v_j$ and hence with $H$. For example,

$$[v_1, \overline{x}_1] = [p_1 + \tfrac{B}{2} x_2, \tfrac{1}{2} x_1 + \tfrac{1}{B} p_2] = \tfrac{1}{2} [p_1, x_1] + \tfrac{1}{2} [x_2, p_2] = 0, \tag{8.76}$$

and similarly,

$$[v_i, \overline{x}_j] = 0, \quad [H, \overline{x}_j] = 0, \quad \text{for } i, j = 1, 2. \tag{8.77}$$

Hence the observables $\overline{x}_j$ are conserved quantities (constants of motion) also under the quantum-mechanical time evolution

$$\frac{d}{dt}\overline{x}_j(t) = 0, \quad \text{for } j = 1, 2. \tag{8.78}$$

EXERCISE 8.11. *Verify that the operators $\overline{x}_1$ and $\overline{x}_2$ satisfy canonical commutation relations*

$$[\overline{x}_1, \overline{x}_2] = -\mathrm{i}\frac{1}{B}. \tag{8.79}$$

The canonical commutation relations imply that the unitary transformations generated by $\overline{x}_1$ cause shifts in the values of the observable $\overline{x}_2$ (and vice versa). What is the meaning of these transformations? In order to answer this question, let us calculate the action of these transformations on any state $\psi(x_1, x_2)$. Because

$$\exp(-\mathrm{i}a\overline{x}_1) = \exp(-\mathrm{i}\,\tfrac{a}{2}\,x_1 - \mathrm{i}\,\tfrac{a}{B}\,p_2) = \exp(-\mathrm{i}\,\tfrac{a}{2}\,x_1)\,\exp(-\mathrm{i}\,\tfrac{a}{B}\,p_2) \tag{8.80}$$

we find

$$\exp(-\mathrm{i}a\overline{x}_1)\,\psi(x_1, x_2) = \exp(-\mathrm{i}\,\tfrac{a}{2}\,x_1)\,\exp(-\mathrm{i}\,\tfrac{a}{B}\,p_2)\,\psi(x_1, x_2) \tag{8.81}$$

$$= \exp(-\mathrm{i}\,\tfrac{a}{2}\,x_1)\,\psi(x_1, x_2 - \tfrac{a}{B}). \tag{8.82}$$

Hence the transformation generated by $\overline{x}_1$ is a shift in the variable $x_2$ together with a multiplication by a phase factor. This phase factor changes the momentum $p_1$ of the particle, but not the state of motion. In particular, because $\overline{x}_1$ commutes with $v_1$, the value of the velocity in $x_1$-direction is not changed by this transformation.

Similarly, the transformation $\exp(-\mathrm{i}b\overline{x}_2)$ is a translation by $-b/2$ in the direction of $x_1$:

$$\exp(-\mathrm{i}b\overline{x}_2)\,\psi(x_1, x_2) = \exp(-\mathrm{i}\,\tfrac{b}{2}\,x_2)\,\psi(x_1 + \tfrac{b}{B}, x_2). \tag{8.83}$$

CD 6.14 presents interactive pictures and movies showing the translations generated by the operators $\overline{x}_1$ and $\overline{x}_2$. We show that vertical and horizontal translations do not commute (because of the Weyl relations). The animation of a translated ground state (8.58) shows that this type of translations changes the momentum of the particle (while it does not change the energy).

EXERCISE 8.12. *Show that the gauge transformation which shifts the vector potential to the point* $(0, \frac{a}{B})$ *cancels the phase factor* $\exp(-\mathrm{i}\,\frac{a}{2}\,x_1)$ *caused by the translation* $\exp(-\mathrm{i}a\overline{x}_1)$ *(see Exercise 8.10).*

EXERCISE 8.13. *The unitary transformation*

$$\psi \longrightarrow \phi = \exp(-\mathrm{i}a\overline{x}_1)\,\psi \tag{8.84}$$

*changes the expectation value of* $\overline{x}_2$. *Assuming* $\|\psi\| = 1$, *show that*

$$\langle \overline{x}_2 \rangle_\phi = \langle \overline{x}_2 \rangle_\psi + \frac{a}{B}. \tag{8.85}$$

**Translations in a constant magnetic field:**
The operator $\overline{x} = (\overline{x}_1, \overline{x}_2)$ defined in Eq. (8.74) generates translations in the following sense. For arbitrary $\mathbf{w} = (w_1, w_2) \in \mathbb{R}^2$ define the unitary operator

$$\exp(-\mathrm{i}B\,\overline{\mathbf{x}} \times \mathbf{w}) = \exp(-\mathrm{i}B\,(\overline{x}_1 w_2 - \overline{x}_2 w_1)). \tag{8.86}$$

This operator shifts a wave packet by $\mathbf{w}$ in position space and at the same time adds the vector $\vec{A}(\mathbf{w}) = \frac{B}{2}(-w_2, w_1)$ to the momentum:

$$\begin{aligned}\exp(-\mathrm{i}B\,\overline{\mathbf{x}} \times \mathbf{w})\,\psi(\mathbf{x}) &= \exp(-\mathrm{i}\,\tfrac{B}{2}\,\mathbf{x} \times \mathbf{w})\,\psi(\mathbf{x} - \mathbf{w}) \\ &= \exp(-\mathrm{i}\mathbf{x} \cdot \vec{A}(\mathbf{w}))\,\psi(\mathbf{x} - \mathbf{w}).\end{aligned} \tag{8.87}$$

PROOF. With the definition of $\overline{\mathbf{x}}$, the operator $\overline{\mathbf{x}} \times \mathbf{w}$ can be written as

$$\overline{\mathbf{x}} \times \mathbf{w} = \frac{1}{2}\mathbf{x} \times \mathbf{w} + \frac{1}{B}\mathbf{p} \cdot \mathbf{w}, \tag{8.88}$$

and the two summands commute,

$$[\mathbf{x} \times \mathbf{w}, \mathbf{p} \cdot \mathbf{w}] = 0. \tag{8.89}$$

Hence we find

$$\exp(-\mathrm{i}B\,\overline{\mathbf{x}} \times \mathbf{w})\,\psi(\mathbf{x}) = \exp(-\mathrm{i}\,\tfrac{B}{2}\,\mathbf{x} \times \mathbf{w})\,\exp(-\mathrm{i}\mathbf{p} \cdot \mathbf{w}) \tag{8.90}$$

from which the result Eq. (8.87) follows immediately. □

The translation described above changes the wave packet, but it does not change the velocity (because the generator $\overline{x}$ commutes with the velocity operators $v_i$). A pure shift in position space (generated by the operator $\mathbf{p} = -\mathrm{i}\nabla$), whenever it is performed in the presence of a magnetic vector potential, does change the dynamic state of the particle. Shifting the particle from $\mathbf{x}_0$ to a new position $\mathbf{x}_0 + \mathbf{w}$ changes the velocity from $\mathbf{p} - \vec{A}(\mathbf{x}_0)$ to $\mathbf{p} - \vec{A}(\mathbf{x}_0 + \mathbf{w})$. For the constant magnetic field we have $\vec{A}(\mathbf{x}_0 + \mathbf{w}) =$

$\vec{A}(\mathbf{x}_0) + \vec{A}(\mathbf{w})$. On the other hand, the translation generated by $\overline{\mathbf{x}}$ performs a simultaneous shift by $\vec{A}(\mathbf{w})$ in momentum space. This cancels precisely the change of the velocity resulting from the pure shift in position space, so that the velocity is not changed at all.

### 8.6.3. Infinite degeneracy of eigenvalues

The particle in a constant magnetic field is a system that is invariant under translations. In the quantum-mechanical formalism, which involves a vector potential that is not translationally invariant, this is only an invariance up to a gauge transformation. Nevertheless, all physically measurable quantities do not depend on the choice of the gauge. Hence, in particular, the energy of a state is the same as that of a shifted state. Let $\psi$ be an eigenstate of $H$,

$$H\,\psi = E_n\,\psi. \tag{8.91}$$

Because the generators of translations $\overline{x}_1$ and $\overline{x}_2$ commute with $H$, we find that

$$H\,\mathrm{e}^{-\mathrm{i}a\overline{x}_j}\,\psi = \mathrm{e}^{-\mathrm{i}a\overline{x}_j}\,H\,\psi = E_n\,\mathrm{e}^{-\mathrm{i}a\overline{x}_j}\,\psi. \tag{8.92}$$

Hence the shifted eigenstate is again an eigenstate with the same energy. A Gaussian function (the ground state) and all its translations span an infinite-dimensional subspace of $H$. Therefore, the ground state and hence every eigenvalue has infinite multiplicity.

### 8.6.4. Translation preserving the center of motion

The velocity operators $v_1$ and $v_2$ also generate unitary transformations:

$$\begin{aligned} \mathrm{e}^{-\mathrm{i}av_1}\,\psi(x_1,x_2) &= \mathrm{e}^{-\mathrm{i}ap_1}\,\mathrm{e}^{-\mathrm{i}aBx_2/2}\,\psi(x_1,x_2) \\ &= \mathrm{e}^{-\mathrm{i}aBx_2/2}\,\psi(x_1-a,x_2). \end{aligned} \tag{8.93}$$

We have used that the summands in $v_1 = p_1 + Bx_2/2$ commute, and hence the exponential function $\mathrm{e}^{-\mathrm{i}av_1}$ can be factorized. A similar calculation gives

$$\mathrm{e}^{-\mathrm{i}av_2}\,\psi(x_1,x_2) = \mathrm{e}^{\mathrm{i}aBx_1/2}\,\psi(x_1,x_2-a). \tag{8.94}$$

Hence the operators $v_i$ generate translations in the direction of the coordinate $x_i$. But because the operators $v_i$ do not commute with the Hamiltonian $H$, this type of translation changes the dynamic state of the particle. The operator $v_1$ is canonically conjugate to $v_2$ and therefore a transformation generated by $v_1$ changes the values of $v_2$. This is done in such a way that the classical centers $\overline{x}_i$ remain unchanged (because $v_i$ commutes with $\overline{x}_j$).

EXERCISE 8.14. *Combine the two types of translations (with generators* $\overline{x}_j$ *and* $v_k$*, respectively) to obtain a Gaussian state with initial position* $\mathbf{x}_0$ *and velocity* $\mathbf{v}_0$ *from the centered Gaussian function* $\exp(-a\,x^2/2)$.

CD 6.15 lets you experiment with the unitary translations generated by the velocity operators $v_1$ and $v_2$. The noncommutativity of the components of the velocity, Eq. (8.51), implies Weyl relations for the unitary groups.

# 8.7. Time Evolution in a Constant Magnetic Field

## 8.7.1. Time-dependence of the quantum-mechanical operators

The solutions Eq. (8.71),

$$x_1(t) = \overline{x}_1 + \frac{1}{B}\, v_1 \sin Bt - \frac{1}{B}\, v_2 \cos Bt, \tag{8.95}$$

$$x_2(t) = \overline{x}_2 + \frac{1}{B}\, v_1 \cos Bt + \frac{1}{B}\, v_2 \sin Bt, \tag{8.96}$$

with $\overline{x}_j$ and $v_j$ interpreted as quantum-mechanical observables are also solutions of the quantum-mechanical evolution equation

$$\frac{d}{dt}\, x_j(t) = \mathrm{i}\,[H, x_j(t)] \tag{8.97}$$

for the observables

$$x_j(t) = \mathrm{e}^{\mathrm{i}Ht}\, x_j\, \mathrm{e}^{-\mathrm{i}Ht}. \tag{8.98}$$

This can easily be verified by calculating the time-derivative and comparing it with the commutator. For example,

$$\frac{d}{dt}\, x_1(t) = -v_1 \cos Bt - v_2 \sin Bt \tag{8.99}$$

is the time-derivative of Eq (8.95). For the commutator we have

$$\mathrm{i}\,[H, x_1(t)] = -\frac{1}{B}\,\mathrm{i}\,[H, v_1]\,\sin Bt + \frac{1}{B}\,\mathrm{i}\,[H, v_2]\,\cos Bt. \tag{8.100}$$

The equality of (8.99) and (8.100) follows immediately from the relations

$$\mathrm{i}\,[H, v_1] = B\, v_2, \qquad \mathrm{i}\,[H, v_2] = -B\, v_1. \tag{8.101}$$

The quantum-mechanical time evolution of the velocity observables $v_1$ and $v_2$ is given by

$$\begin{aligned} v_1(t) &= v_1 \cos Bt + v_2 \sin Bt, \\ v_2(t) &= v_2 \cos Bt - v_1 \sin Bt, \end{aligned} \tag{8.102}$$

in agreement with the classical solution Eq. (8.69) and in complete analogy with the corresponding result Eq. (7.72) for $x(t)$ and $p(t)$ in the field of a harmonic oscillator.

### 8.7.2. Motion on circles

As an example, let us calculate the time evolution of a state that is shifted by $\exp(-\mathrm{i}av_1)$. This group of operators generates translations (in $x_2$-direction) leaving the center of motion $(\overline{x}_1, \overline{x}_2)$ invariant. We consider an arbitrary initial state $\psi(x_1, x_2)$ and write, as usual,

$$\psi(x_1, x_2, t) = \mathrm{e}^{-\mathrm{i}Ht}\,\psi(x_1, x_2).$$

For the time evolution of the shifted initial state we obtain

$$\begin{aligned} \mathrm{e}^{-\mathrm{i}Ht}\,\mathrm{e}^{-\mathrm{i}av_1}\,\psi(x_1, x_2) &= \mathrm{e}^{-\mathrm{i}Ht}\,\mathrm{e}^{-\mathrm{i}av_1}\,\mathrm{e}^{\mathrm{i}Ht}\,\mathrm{e}^{-\mathrm{i}Ht}\,\psi(x_1, x_2) \\ &= \mathrm{e}^{-\mathrm{i}av_1(-t)}\,\mathrm{e}^{-\mathrm{i}Ht}\,\psi(x_1, x_2) \\ &= \mathrm{e}^{-\mathrm{i}a(\cos Bt)v_1 + \mathrm{i}a(\sin Bt)v_2}\,\psi(x_1, x_2, t). \end{aligned} \tag{8.103}$$

Now we use the Weyl relation for the canonically conjugate operators $v_1$ and $v_2$:

$$\mathrm{e}^{\mathrm{i}a_1v_1 + \mathrm{i}a_2v_2} = \mathrm{e}^{\mathrm{i}a_1a_2B/2}\,\mathrm{e}^{\mathrm{i}a_1v_1}\,\mathrm{e}^{\mathrm{i}a_2v_2}. \tag{8.104}$$

Using this formula and the temporary abbreviation $c = a\cos Bt$ and $s = a\sin Bt$ in Eq. (8.103) gives

$$\begin{aligned} \mathrm{e}^{-\mathrm{i}cv_1 + \mathrm{i}sv_2}\,\psi(x_1, x_2, t) &= \mathrm{e}^{-\mathrm{i}stB/2}\,\mathrm{e}^{-\mathrm{i}cv_1}\,\mathrm{e}^{\mathrm{i}sv_2}\,\psi(x_1, x_2, t) \\ &= \mathrm{e}^{-\mathrm{i}stB/2}\,\mathrm{e}^{-\mathrm{i}cBx_2/2}\,\mathrm{e}^{-\mathrm{i}cp_1}\,\mathrm{e}^{-\mathrm{i}sBx_1/2}\,\mathrm{e}^{\mathrm{i}sp_2}\,\psi(x_1, x_2, t) \\ &= \mathrm{e}^{-\mathrm{i}cBx_2/2}\,\mathrm{e}^{-\mathrm{i}sBx_1/2}\,\mathrm{e}^{-\mathrm{i}cp_1}\,\mathrm{e}^{\mathrm{i}sp_2}\,\psi(x_1, x_2, t) \\ &= \mathrm{e}^{-\mathrm{i}(sBx_1 + cBx_2)/2}\,\psi(x_1 - c, x_2 + s, t) \end{aligned}$$

EXAMPLE 8.7.1. Let the initial state $\psi(\mathbf{x})$ be the centered ground state

$$\psi(\mathbf{x}, t) = \mathrm{e}^{-\mathrm{i}Bt/2}\exp\Bigl(-\frac{B}{4}\,\mathbf{x}^2\Bigr). \tag{8.105}$$

Then $\phi(\mathbf{x})$ is a Gaussian function centered at $\mathbf{x}_0 = (a, 0)$. Let us consider the corresponding classical situation: A particle that is initially at rest at the origin gets shifted toward $\mathbf{x}_0$ in such a way that the center of the orbit remains the same. Hence the shifted particle must have the initial velocity $\mathbf{v} = (0, -Ba)$ and hence it performs the circular motion $\mathbf{x}(t) = (a\cos Bt, -a\sin Bt) = \mathbf{x}_t$ with velocity $\dot{\mathbf{x}}(t) = -Ba\,(\sin Bt, \cos Bt)$. According to the classical equations of motion, the canonical momentum of the particle is

$$\begin{aligned}\mathbf{p}(t) = \dot{x}(t) + \vec{A}(\mathbf{x}(t)) &= \frac{B}{2}\,(-\overline{x}_2, \overline{x}_1) + \frac{1}{2}\,\dot{\mathbf{x}}(t) \\ &= \frac{1}{2}\,\dot{\mathbf{x}}(t) = -a\,\frac{B}{2}\,(\sin Bt, \cos Bt) \\ &= \mathbf{p}_t.\end{aligned}$$

The quantum-mechanical solution describes a very similar behavior,

$$\phi(\mathbf{x}, t) = \mathrm{e}^{-\mathrm{i}Bt/2} \exp\Bigl(\mathrm{i}\mathbf{p}_t \cdot \mathbf{x} - \frac{B}{4}\,(\mathbf{x} - \mathbf{x}_t)^2\Bigr), \tag{8.106}$$

that is, a Gaussian function centered at $\mathbf{x}_t$ with average momentum $\mathbf{p}_t$.

Let us collect our results in the following box:

**Time evolution of a shifted initial state:**

Let

$$\psi(\mathbf{x}, t) = \exp(-\mathrm{i}Ht)\psi(\mathbf{x}) \tag{8.107}$$

be the time evolution of an arbitrary initial state $\psi$ in the constant magnetic field in two dimensions. Then the time evolution of the shifted initial state

$$\phi(\mathbf{x}) = \mathrm{e}^{-\mathrm{i}av_1}\,\psi(\mathbf{x}) = \mathrm{e}^{-\mathrm{i}aBx_2/2}\,\psi(x_1 - a, x_2) \tag{8.108}$$

is given by

$$\phi(\mathbf{x}, t) = \exp(\mathrm{i}\mathbf{p}_t \cdot \mathbf{x})\,\psi(\mathbf{x} - \mathbf{x}_t, t) \tag{8.109}$$

where

$$\mathbf{x}_t = a(\cos Bt, -\sin Bt) \tag{8.110}$$

$$\mathbf{p}_t = -a\,\frac{B}{2}\,(\sin Bt, \cos Bt). \tag{8.111}$$

With the translation operators $\exp(-\mathrm{i}a\overline{x}_i)$ and $\exp(-\mathrm{i}av_i)$ we can prepare an initial state with an arbitrary velocity at an arbitrary position if we start with a centered initial state with average velocity 0; see Exercise 8.14.

CD 6.16 shows the motion of various Gaussian initial states. The states are shifted ground states and move (like the coherent states of the harmonic oscillator) without changing the shape of their position probability density. The centers of the wave packets perform a classical motion as indicated in the movies.

### 8.7.3. Rotational symmetry

Consider again the Hamiltonian for a constant magnetic field in the Poincaré gauge:

$$H = \frac{1}{2}\Bigl(p_1 + \frac{B}{2}x_2\Bigr)^2 + \frac{1}{2}\Bigl(p_2 - \frac{B}{2}x_1\Bigr)^2. \tag{8.112}$$

By expanding the squares we arrive at the following expression

$$H = \frac{1}{2}(p_1^2 + p_2^2) + \frac{1}{2}\Bigl(\frac{B}{2}\Bigr)^2 (x_1^2 + x_2^2) - \frac{B}{2}(x_1 p_2 - x_2 p_1). \tag{8.113}$$

In the last expression we recognize the angular momentum operator (Eq. (4.24)),

$$L = x_1 p_2 - x_2 p_1. \tag{8.114}$$

The other summands represent the Hamiltonian of a two-dimensional harmonic oscillator with oscillator frequency $\omega = B/2$,

$$H_{\text{osc}}(\omega) = \tfrac{1}{2}(\mathbf{p}^2 + \omega^2 \mathbf{x}^2). \tag{8.115}$$

Hence the Hamiltonian operator in a constant magnetic field in two-dimensions can be written as

$$H = H_{\text{osc}}\Bigl(\frac{B}{2}\Bigr) - \frac{B}{2}L. \tag{8.116}$$

We also note that the Hamiltonian of the two-dimensional harmonic oscillator commutes with $L$,

$$[H_{\text{osc}}, L] = 0, \qquad \text{and hence} \quad [H, L] = 0. \tag{8.117}$$

This is related to the fact that the Hamiltonian Eq. (8.113) is spherically symmetric, as will be discussed in the next chapter. As a consequence, the canonical angular momentum $L$ is a constant of motion.

For the vector potential $\vec{A}(\mathbf{x}) = \frac{B}{2}(-x_2, x_1)$ the relation

$$\mathbf{p} \cdot \vec{A}(\mathbf{x}) = \frac{B}{2}L \tag{8.118}$$

shows that $\mathbf{p} \cdot \vec{A}$ is a conserved quantity. For example, for a particle which at $t = 0$ is located at the origin $\mathbf{x} = 0$, the canonical angular momentum $L$ is zero and hence the canonical momentum is always orthogonal to the vector potential.

### 8.7.4. Unitary time evolution

The Hamiltonian for a constant magnetic field in the Poincaré gauge is the sum of two commuting operators $H_{\text{osc}}$ and $L$. Hence the time evolution operator $\exp(-\mathrm{i}Ht)$ can be written as a product of a rotation with a harmonic oscillator evolution,

$$e^{-iHt} = e^{i(Bt/2)L}\, e^{-iH_{\rm osc}(B/2)t}. \tag{8.119}$$

This allows us to determine the time evolution in a magnetic field simply by calculating the time evolution in a harmonic oscillator potential and then performing a rotation.

EXERCISE 8.15. *Can you hit the origin with a wave packet? Prepare a wave packet at some given point* $\mathbf{x}_0 \in \mathbb{R}^2$ *in such a way that the center of the wave packet (the mean position) at some later time hits the origin* $\mathbf{x} = (0,0)$. *(Hint: Consider a Gaussian wave packet centered at* $\mathbf{x}_0$ *with average momentum* $p = 0$ *in the Poincaré gauge.)*

Ψ Because the motion in a constant magnetic field is an oscillation times a rotation, it is easy to determine the integral kernel of the time evolution from the Mehler kernel of the harmonic oscillator.

**Propagator for the constant magnetic field:**

Choosing the Poincaré gauge $\vec{A}(\mathbf{x}) = (B/2)\,(-x_2, x_1)$, the time evolution in the constant magnetic field $B$ in two dimensions can be written as an integral operator

$$e^{-iHt}\,\psi(\mathbf{x}) = \int_{\mathbb{R}^2} K(\mathbf{x},\mathbf{y},t)\,\psi(\mathbf{y})\,d^2y, \tag{8.120}$$

where the integral kernel is given by

$$K(\mathbf{x},\mathbf{y},t) = \frac{B}{4\pi i\,\sin(Bt/2)}\,\exp\Bigl(i(\mathbf{x}-\mathbf{y})^2\frac{B}{4}\,\cot\frac{Bt}{2} - i\frac{B}{2}\,\mathbf{x}\times\mathbf{y}\Bigr). \tag{8.121}$$

PROOF. We start with the evolution kernel (propagator) of the one-dimensional harmonic oscillator with oscillator frequency $\omega = B/2$. This is Mehler's kernel

$$K_{\rm osc}(x,y,t) = \sqrt{\frac{B}{2}}\,\frac{1}{\sqrt{2\pi i\sin(Bt/2)}} \times \exp\Bigl(i(x^2+y^2)\,\frac{B}{4}\,\cot\frac{Bt}{2} - i\frac{B}{2}\,\frac{xy}{\sin(Bt/2)}\Bigr).$$

(We applied a scaling transformation to Eq. (7.149) in order to take account of the oscillator frequency $B/2$). We obtain the kernel of the two-dimensional harmonic oscillator by forming a product of two Mehler kernels because the oscillator Hamiltonian in two dimensions is just the sum of one-dimensional operators:

$$\begin{aligned} K_{\mathrm{osc}}^{2\mathrm{D}}(\mathbf{x},\mathbf{y},t) &= K_{\mathrm{osc}}(x_1,y_1,t)\,K_{\mathrm{osc}}(x_2,y_2,t) \\ &= \frac{B}{4\pi\mathrm{i}\sin(Bt/2)}\exp\Bigl(\mathrm{i}(\mathbf{x}^2+\mathbf{y}^2)\,\frac{B}{4}\cot\frac{Bt}{2}-\mathrm{i}\frac{B}{2}\,\frac{\mathbf{x}\cdot\mathbf{y}}{\sin(Bt/2)}\Bigr). \end{aligned} \tag{8.122}$$

The propagator in a magnetic field can be obtained from this by applying the rotation corresponding to the operator $\exp(\mathrm{i}BtL/2)$. This is a rotation through an angle $Bt/2$ around the origin in $\mathbb{R}^2$ which leads to replacement

$$x_1 \longrightarrow x_1 \cos\frac{Bt}{2} - x_2 \sin\frac{Bt}{2},$$
$$x_2 \longrightarrow x_1 \sin\frac{Bt}{2} + x_2 \cos\frac{Bt}{2},$$

in the argument of wave functions. In the argument of the exponential function above the rotation leaves $\mathbf{x}^2$ invariant, while

$$\frac{\mathbf{x}\cdot\mathbf{y}}{\sin(Bt/2)} \longrightarrow \mathbf{x}\cdot\mathbf{y}\cot\frac{Bt}{2} - \mathbf{x}\times\mathbf{y}.$$

Substituting this into Eq. (8.122) finally gives the result Eq. (8.121). □

With the results of this section we are able to calculate the motion of arbitrary Gaussian wave packets in the constant magnetic field. CD 6.17 shows the motion of several "squeezed" states. While the centers of the wave packets perform the well-known circular motion, the shape of the position probability density oscillates between a sharp and a flat distribution. Like every motion in the constant magnetic field, the time evolution is periodic in time.

CD 6.18 shows the time evolution of eigenstates of the harmonic oscillator in a magnetic field. If the oscillator eigenstate is not rotationally symmetric, then the additional rotation generated by the term $-BL/2$ in the Hamiltonian becomes clearly visible. CD 6.19 does a similar visualization for oscillating states. In all cases the time evolution is an oscillator motion (with oscillator frequency $B/2$) and a simultaneous rotation through an angle $Bt/2$.

## 8.8. Systems with Rotational Symmetry in Two Dimensions

### 8.8.1. Rotations

We consider a quantum-mechanical system that is described by wave functions in the Hilbert space $\mathfrak{H} = L^2(\mathbb{R}^2)$. The rotations are generated by the angular momentum operator

$$L = \mathbf{x} \times \mathbf{p} = x_1 p_2 - x_2 p_1. \tag{8.123}$$

A rotation through an angle $\varphi$ around the coordinate origin in two dimensions is described by the rotation matrix

$$\mathbf{R}(\varphi) = \begin{pmatrix} \cos\varphi & -\sin\varphi \\ \sin\varphi & \cos\varphi \end{pmatrix}. \tag{8.124}$$

This matrix rotates vectors in the positive (counterclockwise) direction. The unitary group $\exp(-\mathrm{i}L\varphi)$ describes a rotation of a function $\psi \in \mathfrak{H}$ as follows

$$\mathrm{e}^{-\mathrm{i}L\varphi}\psi(\mathbf{x}) = \psi(\mathbf{R}(-\varphi)\mathbf{x}), \qquad \text{with} \quad \mathbf{x} = \begin{pmatrix} x_1 \\ x_2 \end{pmatrix}. \tag{8.125}$$

Notice that in order to rotate a function counter-clockwise one has to perform a clockwise rotation of its argument.

$\boxed{\Psi}$ The formula (8.125) can be easily verified. First one notes that

$$U(\varphi) : \psi(\mathbf{x}) \longrightarrow \psi(\mathbf{R}(-\varphi)\mathbf{x}) \tag{8.126}$$

defines a unitary group. Then one calculates its generator by differentiating with respect to $\varphi$. Using the chain rule one obtains

$$\begin{aligned} \mathrm{i}\,\frac{d}{d\varphi}\,\psi(\mathbf{R}(-\varphi)\mathbf{x})\Big|_{\varphi=0} &= \mathrm{i}\,\nabla\psi(\mathbf{R}(-\varphi)\mathbf{x}) \cdot \frac{d}{d\varphi}\mathbf{R}(-\varphi)\mathbf{x}\Big|_{\varphi=0} \\ &= \mathrm{i}\,\nabla\psi(\mathbf{x}) \cdot \begin{pmatrix} x_2 \\ -x_1 \end{pmatrix} \\ &= -\,\mathrm{i}\left(x_1 \frac{\partial}{\partial x_2} - x_2 \frac{\partial}{\partial x_1}\right)\psi(\mathbf{x}) \\ &= L\psi(\mathbf{x}). \end{aligned} \tag{8.127}$$

This shows that $L$ is indeed the generator of the unitary group $U(\varphi)$, that is, $U(\varphi) = \exp(-\mathrm{i}L\varphi)$. □

A physical system with Hamiltonian $H$ is said to be *rotationally invariant* or *spherically symmetric* if

$$\mathrm{e}^{\mathrm{i}L\varphi}\, H\, \mathrm{e}^{-\mathrm{i}L\varphi} = H. \tag{8.128}$$

This implies in particular that $H$ and $L$ commute,

$$[H\,,\, L] = H\,L - L\,H = 0, \tag{8.129}$$

and that the angular momentum is a conserved quantity (see Section 6.11),

$$L(t) = \mathrm{e}^{\mathrm{i}Ht}\, L\, \mathrm{e}^{-\mathrm{i}Ht} = L. \tag{8.130}$$

EXERCISE 8.16. *Verify that the free Hamiltonian*

$$H_0 = -\frac{\partial^2}{\partial x_1^2} - \frac{\partial^2}{\partial x_2^2}$$

*is rotationally invariant.*

EXERCISE 8.17. *The Hamiltonian $H = H_0 + V$ with a spherically symmetric potential $V(\mathbf{x}) = V(r)$, $r = \sqrt{x_1^2 + x_2^2}$, is rotationally invariant. The same is true for the Hamiltonian with a constant magnetic field in the Poincaré gauge.*

### 8.8.2. Polar coordinates

In order to treat spherically symmetric situations it is useful to introduce polar coordinates. Polar coordinates $(r, \varphi)$ specify a distance $r$ from the coordinate origin and an angle $\phi$ with a fixed direction (say, the $x_1$-axis of a Cartesian coordinate system). Hence the relation between Cartesian coordinates $(x_1, x_2)$ and polar coordinates $(r, \varphi)$ is given by the following set of equations:

$$\begin{aligned} x_1(r,\varphi) &= r\,\cos\varphi, \qquad & r(x_1,x_2) &= |\mathbf{x}| = \sqrt{x_1^2 + x_2^2}, \\ x_2(r,\varphi) &= r\,\sin\varphi, \qquad & \varphi(x_1,x_2) &= \arctan(x_1,x_2), \end{aligned} \tag{8.131}$$

Here the arctan as a function of two arguments is defined by:

$$\arctan(x_1,x_2) = \arctan\Bigl(\frac{x_2}{x_1}\Bigr) + \begin{cases} \pi, & \text{for } x_1 < 0,\ x_2 \ge 0, \\ -\pi, & \text{for } x_1 < 0,\ x_2 < 0, \\ 0, & \text{for } x_1 > 0. \end{cases} \tag{8.132}$$

Its values are always in the interval $(-\pi, \pi]$. This definition can be extended by continuity to $x_1 = 0$, as long as $x_2 \neq 0$. There is a discontinuity along the negative $x_1$-axis (the half-line with $x_1 < 0$, $x_2 = 0$): The value jumps from $\pi$ to $-\pi$ when the point $(x_1, x_2)$—coming from the upper half-plane ($x_2 > 0$)—crosses the negative $x_1$-axis. Thus, the angular coordinate on a circle cannot be defined globally in a continuous way. At the coordinate origin, the function $\arctan(x_1, x_2)$ (and hence the angular coordinate) remains totally undefined. There is no consistent way to extend the above definition in order to include that point.

Any function $\psi(x_1, x_2)$ can be transformed to polar coordinates by setting

$$\phi(r,\varphi) = \psi(x_1(r,\varphi), x_2(r,\varphi)). \tag{8.133}$$

Next we define the unit vectors in the direction of the polar coordinate lines

$$\mathbf{e}_r = (\cos\varphi, \sin\varphi) = \frac{\mathbf{x}}{r},$$
$$\mathbf{e}_\varphi = (-\sin\varphi, \cos\varphi) = \frac{d\mathbf{e}_r}{d\varphi}. \tag{8.134}$$

With the help of these expressions we can write the action of the gradient operator on functions $\phi(r, \varphi)$ in polar coordinates as

$$\nabla = \mathbf{e}_r \frac{\partial}{\partial r} + \frac{1}{r} \mathbf{e}_\varphi \frac{\partial}{\partial \varphi}. \tag{8.135}$$

From this we may compute the expression in polar coordinates for the angular momentum operator and the Laplacian:

$$L = -\mathrm{i} r \mathbf{e}_r \times \nabla = -\mathrm{i} \frac{\partial}{\partial \phi}, \tag{8.136}$$

$$\Delta = \nabla \cdot \nabla = \frac{\partial^2}{\partial r^2} + \frac{1}{r} \frac{\partial}{\partial r} - \frac{1}{r^2} L^2. \tag{8.137}$$

Hence we obtain the Schrödinger operator with a rotationally symmetric potential

$$H = -\frac{1}{2} \Delta + V(r)$$
$$= \frac{1}{2} \left( -\frac{\partial^2}{\partial r^2} - \frac{1}{r} \frac{\partial}{\partial r} \right) + V(r) - \frac{1}{2} \frac{1}{r^2} \frac{\partial^2}{\partial \phi^2}. \tag{8.138}$$

### 8.8.3. Eigenvalue problem in polar coordinates

We want to solve the eigenvalue equation

$$H\, \phi(r, \varphi) = E\, \phi(r, \varphi). \tag{8.139}$$

Because $H$ is a sum of a term that contains only radial derivatives and a term that contains the angular derivative $L = -\mathrm{i}\partial/\partial\varphi$, we may try a product ansatz

$$\phi(r, \varphi) = f(r)\, \chi(\varphi), \tag{8.140}$$

where $\chi$ is an eigenfunction of the angular momentum $L$, that is,

$$L\chi(\varphi) = \ell\chi(\varphi). \tag{8.141}$$

This is a differential equation in the variable $\varphi$ with the solution

$$\chi(\varphi) = \exp(\mathrm{i}\ell\varphi) \tag{8.142}$$

for any $\ell \in \mathbb{C}$. But the wave function $\phi = f\chi$ should of course be in the domain of the angular momentum operator $L$, which is originally defined as $-\mathrm{i}x \times \nabla$ on $\mathbb{R}^2$. In order to be differentiable, the wave function in particular has to be continuous. For the representation in polar coordinates this means that the wave function should depend on $\varphi$ in such a way that

$$\phi(r,\varphi) = \phi(r,\varphi + 2\pi). \tag{8.143}$$

Hence $\chi$ must be periodic with period $2\pi$, which is only the case if $\ell$ is an integer:

$$\ell = 0, \pm 1, \pm 2, \pm 3, \dots . \tag{8.144}$$

Inserting (8.140) into the eigenvalue equation we obtain for the function $f$ the equation

$$\frac{1}{2}\left(-f'' - \frac{1}{r} f' + \frac{\ell^2}{r^2} f\right) + V(r)\, f = E\, f \tag{8.145}$$

(where the prime denotes the derivative with respect to $r$). This equation is called the stationary *radial Schrödinger equation.*

## 8.9. Spherical Harmonic Oscillator

We treat the spherical harmonic oscillator in polar coordinates in two dimensions. This will also be an important step in the treatment of the constant magnetic field.

CD 5.20 lets you play with the eigenstates of the harmonic oscillator, which are simultaneously eigenstates of the angular momentum operator. In this section we are going to derive an analytic expression for these states.

The potential of the spherical oscillator is

$$V(\mathbf{x}) = \frac{\omega^2}{2}\left(x_1^2 + x_2^2\right) = \frac{\omega^2}{2} r^2 \equiv V(r). \tag{8.146}$$

The radial Schrödinger equation for this system is

$$\frac{1}{2}\left(-f'' - \frac{1}{r} f' + \frac{\ell^2}{r^2} f + \omega^2 r^2 f\right) = E\, f. \tag{8.147}$$

The substitution

$$s = |\omega|\, r^2, \qquad g(s) = f(r), \qquad \lambda = \frac{2}{|\omega|} E \tag{8.148}$$

converts this equation into

$$-4s\, g'' - 4\, g' + \left(\frac{\ell^2}{s} + s\right) g = \lambda\, g. \tag{8.149}$$

Using the ansatz

$$g(s) = \mathrm{e}^{-s/2}\, s^{|\ell|/2}\, u(s) \tag{8.150}$$

we obtain (after a boring little calculation) a new equation for $u$:

$$-s\,u'' + (s - |\ell| - 1)\,u' + \Bigl(\frac{|\ell|+1}{2} - \frac{\lambda}{4}\Bigr)\,u = 0. \tag{8.151}$$

This is precisely *Kummer's equation*

$$-s\,u'' + (s - b)\,u' + a\,u = 0, \tag{8.152}$$

which has the two linearly independent solutions,

$$u(s) = {}_1F_1(a,b,s), \quad \text{and} \quad u(s) = U(a,b,s). \tag{8.153}$$

The *confluent hypergeometric function* ${}_1F_1$ satisfies

$${}_1F_1(a,b,0) = 1, \qquad \frac{d}{ds}{}_1F_1(a,b,s)\Big|_{s=0} = \frac{a}{b}, \tag{8.154}$$

while the second solution behaves in leading order for small $s$ like $s^{1-b}$ (for $b > 1$), and like $-\ln s$ (for $b = 1$). Hence we obtain the following solution of the radial Schrödinger equation,

$$f(r) = \mathrm{e}^{-|\omega|\,r^2/2}\,r^{|\ell|}\,{}_1F_1\Bigl(\frac{|\ell|+1}{2} - \frac{E}{2\,|\omega|}, |\ell|+1, |\omega|\,r^2\Bigr), \tag{8.155}$$

and a second solution with $U$ instead of ${}_1F_1$. However, the second solution behaves asymptotically for small $r$ like $r^{-|\ell|}$ and is therefore not square-integrable for $\ell \neq 0$. This solution has to be excluded because we can only accept solutions that can be normalized. For $\ell = 0$, the second solution is square-integrable because it has only a logarithmic divergence at $r = 0$. Nevertheless, it has to be excluded because it can be shown that the domain of the Hamiltonian operator contains only bounded functions.

The hypergeometric function ${}_1F_1$ diverges exponentially, as $r \to \infty$, unless $a = -n$ with $n = 0, 1, 2\ldots$, where it is a polynomial of degree $n$ in $s$. In this case the exponential factor in (8.155) makes $f(r)$ square-integrable at infinity. Hence we can only obtain square-integrable solutions, if we assume

$$\frac{|\ell|+1}{2} - \frac{E}{2\,|\omega|} = -n, \quad \text{where} \quad n = 1, 2, 3, \ldots\,, \tag{8.156}$$

or

$$E \equiv E_{n,\ell} = |\omega|(2n + |\ell| + 1). \tag{8.157}$$

We collect our results in the following box:

**Spherical harmonic oscillator:**

The spherical harmonic oscillator in two dimensions,

$$H_{\mathrm{osc}}(\omega) = \tfrac{1}{2}\,(-\Delta + \omega^2\, r^2), \qquad \omega \in \mathbb{R}, \tag{8.158}$$

has the eigenvalues

$$E_{n,\ell} = |\omega|(2n + |\ell| + 1), \tag{8.159}$$

with $n = 0, 1, 2, \dots$ and $\ell = 0, \pm 1, \pm 2, \dots$.
The corresponding eigenfunctions are

$$\begin{aligned} \psi_{n,\ell}(r,\varphi) &= f_{n,\ell}(r)\, \mathrm{e}^{\mathrm{i}\ell\varphi}, \\ f_{n,\ell}(r) &= N_{n,\ell}\, \mathrm{e}^{-|\omega|\, r^2/2}\, r^{|\ell|}\, {}_1F_1(-n, |\ell| + 1, |\omega|\, r^2). \end{aligned} \tag{8.160}$$

($N_{n,\ell}$ is an appropriate normalization constant.) The eigenfunctions $\psi_{n,\ell}$ of the Hamiltonian are simultaneously eigenfunctions of the angular momentum operator $L$ belonging to the eigenvalue $\ell$.

EXERCISE 8.18. *Compare the eigenvalues* (8.159) *with the result obtained in Section 7.6.*

## 8.10. Angular Momentum Eigenstates in a Magnetic Field

The solution of the harmonic oscillator problem enables us to write down the solution in case of the constant magnetic field. The eigenvectors of the spherical harmonic oscillator are also eigenvectors of the angular momentum operator $L$ and hence eigenvectors of the operator

$$H = H_{\mathrm{osc}}\Bigl(\frac{B}{2}\Bigr) - \frac{B}{2}\, L. \tag{8.161}$$

With $\psi_{n,\ell}(r,\varphi)$ as in Eq. (8.160) we have

$$H\, \psi_{n,\ell} = \Bigl(\frac{|B|}{2}\, (2n + |\ell| + 1) - \frac{B}{2}\, \ell\Bigr)\, \psi_{n,\ell}. \tag{8.162}$$

The set of all possible energies (the *energy spectrum*) is thus given by the set of all numbers

$$\frac{|B|}{2}\, (2n + 1 + |\ell| - (\operatorname{sgn} B)\, \ell) \quad \text{for which} \quad n = 0, 1, 2, \dots, \ell \in \mathbb{Z}. \tag{8.163}$$

This is just the set

$$\sigma(H) = \Bigl\{ |B|\, (k + \tfrac{1}{2}) \;\Big|\; k = 0, 1, 2, \dots \Bigr\}. \tag{8.164}$$

Each of the eigenvalues in this set can be obtained by an infinite number of combinations of $n$ and $\ell$. In this way we recover our result about the infinite degeneracy of eigenvalues in a constant magnetic field (Section 8.6.3). For example, the ground-state energy $E = |B|/2$ is obtained for $n = 0$ and all $\ell \geq 0$ (if $B < 0$), respectively all $\ell \leq 0$ (if $B > 0$). Because ${}_1F_1(0, b, s) = 1$, the eigenfunctions belonging to the ground-state energy are given by

$$N_{0,\ell}\, \mathrm{e}^{-|B|\, r^2/2}\, r^{\ell}\, \mathrm{e}^{\mathrm{i}\ell\varphi}, \quad \text{with} \quad \ell = \begin{cases} 0, 1, 2, \dots\,, & \text{for } B < 0, \\ 0, -1, -2, \dots\,, & \text{for } B > 0. \end{cases} \tag{8.165}$$

*Chapter 9*

# One-Dimensional Scattering Theory

**Chapter summary**: In classical as well as in quantum mechanics the motion of particles in external fields falls into one of two categories: Either the motion stays within an approximately finite region for all times or the particle escapes toward infinity and behaves asymptotically like a free particle. Here we treat the latter case.

We only consider the simplest case of a scattering process in one dimension. A particle is shot toward a target which is represented by a force field with a finite range. As usual, we are interested in the temporal behavior of wave packets (and their Fourier transforms). As in the case of free particles, the wave packets are formed as superpositions of solutions with a well-defined energy. These energy eigenfunctions are not square-integrable. Asymptotically they are made of pieces that represent incoming and outgoing plane waves to the left and to the right of the target.

A wave packet hitting a target usually gets dispersed into all directions. In one dimension, the particle can be either reflected or transmitted. The probabilities for these events are determined by the reflection and transmission coefficients. We use the energy representation derived in Section 3.8 to explain how these scattering coefficients determine the asymptotic behavior of the wave packets. In a few cases, explicit expressions for the scattering coefficients can be derived. The most common examples are rectangular steps, barriers, and wells.

Again we have numerous opportunities to point out differences between classical and quantum mechanics. A particularly striking phenomenon is the tunnel effect, the ability of a quantum particle to pass through a repulsive barrier, even if the energy is too low to allow a transition in classical mechanics. The tunnel effect is used in the scanning tunneling microscope to obtain images of solid surfaces with a resolution showing single atoms (see the gallery on the CD-ROM).

## 9.1. Asymptotic Behavior

In a typical scattering experiment, particles emerging from an accelerator are shot toward a target, for example, a metal foil. Here you will learn about the simplest case where the scattering takes place in one space dimension.

It is assumed that the interaction of the particles with the target can be described by an electrostatic potential $V(x)$ that becomes asymptotically constant,

$$\lim_{x\to+\infty} V(x) = V_+, \qquad \lim_{x\to-\infty} V(x) = V_-, \tag{9.1}$$

so that the particles feel no force in regions far away from the target. Because we can always perform a space reflection $x \to -x$, we assume without loss of generality that

$$V_- \le V_+. \tag{9.2}$$

It is much easier to treat the case where the limits $V_\pm$ are approached sufficiently fast. Usually, one requires something like

$$\int_0^{\pm\infty} |V(x) - V_\pm|\, dx < \infty, \tag{9.3}$$

but most of our examples will even satisfy

$$V(x) = \begin{cases} V_+, & x \ge R, \\ V_-, & x \le -R, \end{cases} \tag{9.4}$$

with some $R \ge 0$.

As a first step in the solution of the time-dependent Schrödinger equation, let us investigate the stationary equation

$$\Bigl(-\frac{1}{2}\frac{\partial^2}{\partial x^2} + V(x)\Bigr)\, \psi(E,x) = E\,\psi(E,x). \tag{9.5}$$

Depending on the numerical value of $E$, this equation may or may not have square-integrable solutions. It will turn out that for certain ranges of the energy the solutions $\psi(E,x)$ behave similar to plane waves (they are bounded and oscillate, but have no statistical interpretation). In order to describe more localized phenomena you may then proceed by forming wave packets as in the case of free particles: With a suitable amplitude function $g(E)$ describing the contribution of the energy $E$, a wave packet is defined by

$$\psi(x,t) = \int \psi(E,x)\, \mathrm{e}^{-\mathrm{i}Et}\, g(E)\, dE. \tag{9.6}$$

Under certain conditions we expect the function $\psi$ to be a square-integrable solution of the time-dependent Schrödinger equation.

For the following it is useful to define the function

$$k(E) = \begin{cases} \sqrt{2E}, & \text{for } E \ge 0, \\ \mathrm{i}\sqrt{-2E}, & \text{for } E \le 0. \end{cases} \tag{9.7}$$

For $E \geq 0$ this is just the momentum of a free particle with energy $E$. The function $k$ is either real and non-negative ($E \geq 0$) or purely imaginary with a positive imaginary part ($E < 0$).

Moreover, for $E \neq 0$, we define the functions

$$\underrightarrow{\omega}(E,x) = \frac{1}{\sqrt{2\pi k(E)}} \exp(ik(E)x), \tag{9.8}$$

$$\underleftarrow{\omega}(E,x) = \frac{1}{\sqrt{2\pi k(E)}} \exp(-ik(E)x), \tag{9.9}$$

which are solutions of the stationary free Schrödinger equation.

For $E > 0$ these functions are normalized plane waves with a normalization factor chosen to simplify the formulas of the energy representation (see, e.g., Eq. (3.79)). The arrow indicates the direction of the phase velocity which in one dimension is either to the left or to the right.

For $E < 0$ the behavior of these functions is radically different: The function $\underrightarrow{\omega}$ *decreases* exponentially for $x \to \infty$ (increases for $x \to -\infty$), while $\underleftarrow{\omega}$ *increases* exponentially for $x \to \infty$ (decreases for $x \to -\infty$). Even for negative energies the functions $\underrightarrow{\omega}$ and $\underleftarrow{\omega}$ are solutions of the stationary free Schrödinger equation. But due to their exponential increase toward one direction they cannot be used to form square-integrable wave packets and have to be rejected for the description of free particles. The Schrödinger equation has no useful solutions for negative energies.

For a constant potential, $V(x) = V_0$, the general solution of the stationary Schrödinger equation is a linear combination of the plane waves with energies shifted by $V_0$,

$$\psi(E,x) = a\, \underrightarrow{\omega}(E-V_0,x) + b\, \underleftarrow{\omega}(E-V_0,x). \tag{9.10}$$

For potentials that are only asymptotically constant the functions $\underrightarrow{\omega}$ and $\underleftarrow{\omega}$ will be useful for describing the asymptotic behavior of the solutions at large distances from the interaction region: If the potential approaches the constant values $V_\pm$, as $x \to \pm\infty$, we expect the solutions of the stationary Schrödinger equation to behave as follows:

$$\psi(E,x) \to \begin{cases} a\, \underrightarrow{\omega}(E-V_-,x) + b\, \underleftarrow{\omega}(E-V_-,x), & \text{as } x \to -\infty, \\ c\, \underrightarrow{\omega}(E-V_+,x) + d\, \underleftarrow{\omega}(E-V_+,x), & \text{as } x \to \infty. \end{cases} \tag{9.11}$$

One needs the assumption (9.3) in order to prove that all solutions of the stationary Schrödinger equation have this asymptotic behavior.

The behavior of a classical particle with energy $E$ suggests to distinguish between the following cases:

1. $E > V_+ \geq V_-$: Asymptotically for $x \to \pm\infty$, the particle behaves like a free particle (constant potential)

2. $V_+ > E > V_-$: The particle behaves like a free particle for $x \to -\infty$, but no propagation is possible in the asymptotic region $x \to \infty$.

3. $V_- > E$: The particle cannot escape to infinity, but it may exist in regions where $V(x) < E$.

These cases also serve to classify the asymptotic behavior of the solutions of the Schrödinger equation.

**Case 1**: In the first case, $E > V_+ \geq V_-$, we have oscillating plane waves on both sides. Far to the left, the plane wave with positive momentum, $\underset{\rightarrow}{\omega}(E-V_-, x)$, corresponds to particles moving in the right direction, that is, toward the interaction region, and is therefore called the *incoming wave*. The term $\underset{\leftarrow}{\omega}(E-V_-, x)$ describes a plane wave with momenta pointing to the left, away from the target. This is an *outgoing wave*. Far to the right of the target these roles are reversed and it is the wave $\underset{\rightarrow}{\omega}(E-V_+, x)$ which is outgoing, while the wave $\underset{\leftarrow}{\omega}(E-V_+, x)$, which moves in the left direction, is incoming. It is the goal of scattering theory to determine the coefficients $b$ and $c$ of the outgoing waves in terms of the coefficients $a$ and $d$ of the incoming waves.

**Case 2**: In the case $V_+ > E > V_-$ the particle will be asymptotically free only for $x \to -\infty$. Hence we expect the asymptotic behavior

$$\psi(E, x) \to a\, \underset{\rightarrow}{\omega}(E-V_-, x) + b\, \underset{\leftarrow}{\omega}(E-V_-, x), \quad \text{as } x \to -\infty. \tag{9.12}$$

For $x \to +\infty$, the asymptotic form (9.11) is a linear combination of an exponentially increasing part $\underset{\leftarrow}{\omega}(E-V_+, x)$ and an exponentially decreasing part $\underset{\rightarrow}{\omega}(E-V_+, x)$. The exponentially increasing function cannot be used to form square-integrable wave packets. For a physically acceptable solution we must require $c = 0$. The only useful asymptotic form for $x \to \infty$ is

$$\psi(E, x) \to d\, \underset{\rightarrow}{\omega}(E-V_+, x), \quad \text{as } x \to \infty. \tag{9.13}$$

**Case 3**: When the energy is less than $V_-$, then all solutions of the stationary Schrödinger equation are either exponentially decaying or exponentially increasing. A solution $\psi(E, x)$ has a useful interpretation only if it has the asymptotic form

$$\psi(E, x) \to \begin{cases} a\, \underset{\leftarrow}{\omega}(E-V_-, x), & (x \to \infty), \\ d\, \underset{\rightarrow}{\omega}(E-V_+, x), & (x \to \infty). \end{cases} \tag{9.14}$$

This condition on the asymptotic behavior is very special. It will not be possible to find such a solution for all values of $E$. If we can find, for some value of $E < V_-$, a solution $\psi(E, x)$ with this behavior, then $\psi(E, x)$ is square-integrable and hence has a probability interpretation. $E$ is an eigenvalue of the Hamiltonian, and $\psi$ the corresponding eigenfunction.

Solutions that are asymptotically equal to a linear combination of oscillating plane waves cannot be square-integrable. Hence we expect no energy eigenvalues $E$ with $E > V_-$, as long as (9.12) describes the asymptotic form of the solutions for $x \to -\infty$.

# 9.2. Example: Potential Step

## 9.2.1. Continuity condition

Here we illustrate the above considerations by a simple, exactly solvable example — the potential step

$$V(x) = \begin{cases} V_-, & x < 0, \\ V_+, & x \geq 0, \end{cases} \qquad \text{with } V_- < V_+. \tag{9.15}$$

A solution of the stationary Schrödinger equation for any value of $E$ can be obtained from the ansatz

$$\psi(E,x) = \begin{cases} a\, \underset{\rightarrow}{\omega}(E-V_-,x) + b\, \underset{\leftarrow}{\omega}(E-V_-,x), & x < 0, \\ c\, \underset{\rightarrow}{\omega}(E-V_+,x) + d\, \underset{\leftarrow}{\omega}(E-V_+,x), & x \geq 0. \end{cases} \tag{9.16}$$

For arbitrary coefficients $a, b, c$, and $d$, the parts of this function are solutions of the stationary Schrödinger equation in a constant potential $V_+$ resp. $V_-$. In order to obtain a solution for all $x$, we have to choose the coefficients in such a way that the pieces can be glued together at $x = 0$. Because the Schrödinger equation is of second order, a solution $\psi$ must be twice differentiable (in a generalized sense). Hence the parts of the solution for $x > 0$ and $x < 0$ must be glued together at $x = 0$ in a differentiable way. Let us therefore require that $\psi$ and $\psi'$ be continuous at $x = 0$. This assumption leads to a system of linear equations for the coefficients $a$, $b$, $c$, and $d$. Using the abbreviation

$$k_\pm = k(E-V_\pm), \tag{9.17}$$

we obtain the linear system

$$\sqrt{k_+}\,(a+b) = \sqrt{k_-}\,(c+d) \qquad \text{(continuity of } \psi) \tag{9.18}$$

$$\sqrt{k_-}\,(a-b) = \sqrt{k_+}\,(c-d) \qquad \text{(continuity of } \psi') \tag{9.19}$$

which enables us to compute the coefficients $b$, $c$ of the outgoing waves in terms of the coefficients $a$ and $d$ of the incident waves.

### 9.2.2. Energies higher than the step size

For energies $E > V_+$ the solution $\psi(E,x)$ is a linear combination of plane waves on both sides of the step. Let us first consider the situation where a particle approaches the barrier from the left. Hence the solution $\psi(E,x)$ should have no incoming part in the region $x > 0$. We choose $a = 1$ and $d = 0$ in Eq. (9.16) and write the solution in the form:

$$\psi(E,x) = \begin{cases} \underrightarrow{\omega}(E-V_-,x) + \underleftarrow{R}(E)\,\underleftarrow{\omega}(E-V_-,x), & x < 0, \\ \underrightarrow{T}(E)\,\underrightarrow{\omega}(E-V_+,x), & x \geq 0. \end{cases} \tag{9.20}$$

The linear system (9.18) can easily be solved for the coefficients $b = \underleftarrow{R}(E)$ and $c = \underrightarrow{T}(E)$,

$$\underleftarrow{R}(E) = \frac{k_- - k_+}{k_- + k_+}, \qquad \underrightarrow{T}(E) = \frac{2\sqrt{k_+ k_-}}{k_+ + k_-}. \tag{9.21}$$

$\underleftarrow{R}(E)$ is called *reflection coefficient*, $\underrightarrow{T}(E)$ *transmission coefficient* at energy $E$ for scattering from the left. The quantity $|\underleftarrow{R}(E)|^2$ is interpreted as the *reflection probability* at energy $E$, and $|\underrightarrow{T}(E)|^2$ is the *transmission probability* at energy $E$. We find

$$|\underleftarrow{R}(E)|^2 + |\underrightarrow{T}(E)|^2 = 1, \quad \text{for } E \geq V_+ > 0. \tag{9.22}$$

The interactive pictures in CD 7.1 show the reflection and transmission coefficients as functions of the energy and of the step size.

### 9.2.3. Total reflection

The ansatz (9.20) can also be used for energies $V_- < E \leq V_+$. The solution for $x \geq 0$ consists only of an exponentially decaying part. Now the quantity $k_+ = k(E-V_+) = \mathrm{i}\sqrt{2(V_+ - E)}$ is purely imaginary and hence $\underleftarrow{R}(E)$ becomes a complex number with absolute value 1,

$$|\underleftarrow{R}(E)|^2 = \frac{k_-^2 + |k_+|^2}{k_-^2 + |k_+|^2} = 1. \tag{9.23}$$

Thus, the particle is reflected with probability 1.

Note that although the total energy is smaller than the potential energy in the region $x \geq 0$, there is a nonvanishing probability of finding the particle in that region. This probability decays exponentially with the distance from the step. There are no propagating particles in that region.

If we let $V_+$ tend to infinity, then $k_+ \to \infty$ and hence $\underrightarrow{T}(E)\, \underrightarrow{\omega}(E{-}V_+, x)$ tends to zero, that is, the part of the solution $\psi(E,x)$ in the region $x > 0$ tends to zero. We find

$$\underleftarrow{R}(E) \to -1, \quad \text{for } V_+ \to \infty, \tag{9.24}$$

and therefore the solution in the limit $V_+ \to \infty$ becomes

$$\psi(E,x) = \begin{cases} \underrightarrow{\omega}(E{-}V_-, x) - \underleftarrow{\omega}(E{-}V_-, x), & x < 0, \\ 0, & x \geq 0. \end{cases} \tag{9.25}$$

This is the solution of the Schrödinger equation with a Dirichlet boundary condition at $x = 0$. Hence the potential step in the limit of infinite step size becomes a Dirichlet wall.

Finally, we note that for $E < 0$ the Schrödinger equation has no bounded solution because the solution that decays exponentially in the region $x \geq 0$ has an exponentially increasing part in the region $x \leq 0$.

### 9.2.4. Scattering from the right

For $E > V_+$ another type of solution can be obtained by writing

$$\psi(E,x) = \begin{cases} \underleftarrow{T}(E)\, \underleftarrow{\omega}(E{-}V_-, x), & x < 0, \\ \underleftarrow{\omega}(E{-}V_+, x) + \underrightarrow{R}(E)\, \underrightarrow{\omega}(E{-}V_+, x), & x \geq 0. \end{cases} \tag{9.26}$$

In the right region $x \geq 0$ this solution describes a plane wave with momentum $-k_+$ (incoming wave), and an outgoing plane wave with momentum $+k_+$ (reflected wave). In the region $x < 0$ there is only a transmitted plane wave with momentum $-k_-$ (outgoing to the left). The step goes down and still something is reflected.

The continuity of $\psi$ and $\psi'$ at $x = 0$ implies (for $E > V_+$)

$$\underrightarrow{R}(E) = \frac{k_+ - k_-}{k_+ + k_-} = -\underleftarrow{R}(E), \tag{9.27}$$

$$\underleftarrow{T}(E) = \frac{2\sqrt{k_+ k_-}}{k_+ + k_-} = \underrightarrow{T}(E). \tag{9.28}$$

CD 7.2 presents several interactive pictures showing the plane wave solution $\psi(E,x)$ in the various situations described above.

## 9.3. Wave Packets and Eigenfunction Expansion

The solutions $\psi(E,x)$ for $E > V_-$ have a sharp energy, but they are not square-integrable. In order to obtain solutions with a proper quantum-mechanical interpretation we have to form wave packets. This is done as in the case of free particles by a continuous superposition of solutions with sharp energies.

### 9.3.1. Energy representation in a constant potential

For wave packets in a constant potential, the considerations of Section 3.8 can be carried through with only minor modifications. A wave packet $\psi(x)$ can be written as an integral over the plane waves

$$\psi(x) = \int_{V_-}^{\infty} \Big( \underset{\rightarrow}{\omega}(E{-}V_-,x)\, \underset{\rightarrow}{g}(E) + \underset{\leftarrow}{\omega}(E{-}V_-,x)\, \underset{\leftarrow}{g}(E) \Big)\, dE. \tag{9.29}$$

with suitable functions $\underset{\rightarrow}{g}$ and $\underset{\leftarrow}{g}$, which are square-integrable in energy space. Here the summand with "$\rightarrow$" describes the part of the wave packet moving in the right direction; the part with "$\leftarrow$" moves in the left direction. You should be aware of the fact that only the energies $E \geq V_-$ are allowed in the constant potential $V_-$, and hence the condition for the square-integrability reads

$$\|\psi\|^2 = \int_{V_-}^{\infty} (|\underset{\rightarrow}{g}(E)|^2 + |\underset{\leftarrow}{g}(E)|^2)\, dE. \tag{9.30}$$

The energy representation of the wave packet is the following two-component wave function in energy space:

$$g(E) = \begin{pmatrix} \underset{\rightarrow}{g}(E) \\ \underset{\leftarrow}{g}(E) \end{pmatrix}, \quad E > V_-. \tag{9.31}$$

For the following, it is instructive to express the Fourier transform of the wave packet $\psi(x)$ with the help of the energy representation.

At $t = 0$, the part of the wave packet that moves in the right direction is given by

$$\begin{aligned} \underset{\rightarrow}{\psi}(x) &= \int_{V_-}^{\infty} \underset{\rightarrow}{\omega}(E - V_-,x)\, \underset{\rightarrow}{g}(E)\, dE \\ &= \int_{0}^{\infty} \underset{\rightarrow}{\omega}(E,x)\, \underset{\rightarrow}{g}(E + V_-)\, dE. \end{aligned} \tag{9.32}$$

Now we can perform the variable substitution $E = k^2/2$ and assume $k > 0$ since the momenta of the right-moving part are positive. Inserting the definition of $\underrightarrow{\omega}$, we obtain

$$\underrightarrow{\psi}(x) = \frac{1}{\sqrt{2\pi}} \int_0^\infty \mathrm{e}^{ikx} \frac{1}{\sqrt{k}} \underrightarrow{g}\Big(\frac{k^2}{2} + V_-\Big)\, dk. \tag{9.33}$$

This can be extended to an integral from $-\infty$ to $+\infty$ if we assume that the integrand is zero for $k < 0$. Hence we see immediately that the wave packet $\underrightarrow{\psi}(x)$ has the Fourier transform

$$\hat{\underrightarrow{\psi}}(k) = \begin{cases} \frac{1}{\sqrt{k}} \underrightarrow{g}(\frac{k^2}{2} + V_-), & k \geq 0, \\ 0, & k < 0. \end{cases} \tag{9.34}$$

For the left-moving part we find similarly

$$\begin{aligned} \underleftarrow{\psi}(x) &= \frac{1}{\sqrt{2\pi}} \int_0^\infty \mathrm{e}^{-ikx} \frac{1}{\sqrt{k}} \underleftarrow{g}\Big(\frac{k^2}{2} + V_-\Big)\, dk \\ &= \frac{1}{\sqrt{2\pi}} \int_{-\infty}^0 \mathrm{e}^{ikx} \frac{1}{\sqrt{k}} \underleftarrow{g}\Big(\frac{k^2}{2} + V_-\Big)\, dk, \end{aligned} \tag{9.35}$$

and hence

$$\hat{\underleftarrow{\psi}}(k) = \begin{cases} 0, & k \geq 0, \\ \frac{1}{\sqrt{k}} \underleftarrow{g}(\frac{k^2}{2} + V_-), & k < 0. \end{cases} \tag{9.36}$$

### 9.3.2. Wave packets in a step potential

Let us now form a wave packet from the solutions $\psi(E, x)$ in a step potential. We take some energy distribution $g(E)$ with support in the region $E > V_-$ and define

$$\psi(x) = \int_{V_-}^\infty \psi(E, x)\, g(E)\, dE. \tag{9.37}$$

For $\psi(E, x)$ as in Eq. (9.20) this wave packet consists of several parts. Let us define the functions

$$\underrightarrow{\psi}^{\mathrm{in}}(x) = \int_{V_-}^\infty \underrightarrow{\omega}(E - V_-, x)\, g(E)\, dE, \qquad x < 0, \tag{9.38}$$

$$\underleftarrow{\psi}^{\mathrm{r}}(x) = \int_{V_-}^\infty \underleftarrow{\omega}(E - V_-, x)\, \underleftarrow{R}(E)\, g(E)\, dE, \qquad x < 0, \tag{9.39}$$

$$\underrightarrow{\psi}^{\mathrm{t}}(x) = \int_{V_-}^\infty \underrightarrow{\omega}(E - V_+, x)\, \underrightarrow{T}(E)\, g(E)\, dE, \qquad x \geq 0. \tag{9.40}$$

The wave packet $\psi$ is given in terms of these functions by

$$\psi(x) = \begin{cases} \underrightarrow{\psi}^{\mathrm{in}}(x) + \underleftarrow{\psi}^{\mathrm{r}}(x), & \text{for } x < 0, \\ \underrightarrow{\psi}^{\mathrm{t}}(x), & \text{for } x \geq 0. \end{cases} \tag{9.41}$$

A solution of the time-dependent Schrödinger equation with the step potential is given by

$$\psi(x,t) = \int_{V_+}^{\infty} \psi(E,x)\, \mathrm{e}^{-\mathrm{i}Et}\, g(E)\, dE. \tag{9.42}$$

As indicated by the arrows, the wave packets $\underset{\rightarrow}{\psi}^{\mathrm{in}}$ and $\underset{\rightarrow}{\psi}^{\mathrm{t}}$ move in the right direction, while $\underset{\leftarrow}{\psi}^{\mathrm{r}}$ moves in the left direction. It is now easy to describe a wave packet that moves toward the potential step and splits into a reflected and a transmitted part. You will read more about this in the next section.

CD 7.3 shows how the components $\underset{\rightarrow}{\psi}^{\mathrm{in}}$, $\underset{\leftarrow}{\psi}^{\mathrm{r}}$, and $\underset{\rightarrow}{\psi}^{\mathrm{t}}$ can be added to form a wave packet that describes the scattering at a potential step. The section CD 7.4 contains several movies of Gaussian wave packets hitting a potential step. Even if the energies are strictly higher than the size of the step, the particle gets reflected with a certain probability. The moment of reflection shows the interference of the incoming and the reflected waves in the region $x < 0$.

## 9.4. Potential Step: Asymptotic Momentum Distribution

Let us discuss the time evolution of the wave packet (9.42) at a potential step, assuming for simplicity $V_- = 0$. Figure 9.1 shows the typical behavior of a wave packet with energies higher than $V_+$. The illustration shows the position and momentum distributions before, during, and after the scattering at a potential step. The incoming wave packet moves toward the step and has a rather wide distribution around its average position. Therefore, the distribution of the momenta (and hence of the energies) is rather sharp, so that all energies contributing to the incoming wave packet are well above $V_+$. (The illustration shows the energy $E = k_0^2/2$ in relation to the size of the potential step). As long as the wave packet does not hit the step, the momentum distribution is stationary, but it starts to change as soon as the wave packet feels the potential jump. The peak at $k_0$ shifts to a lower average momentum accounting for the energy necessary to pass the potential jump. A small peak corresponding to the reflected part of the wave packet emerges at $-k_0$. Some ripples appear in the position probability density. They come from the interference of the incoming and the reflected wave packets. After the scattering, the wave packet has split into a large transmitted part and a small reflected part, both moving away from the step. The momentum distribution remains stationary from here on, because the propagation in a constant potential is physically indistinguishable from the free motion.

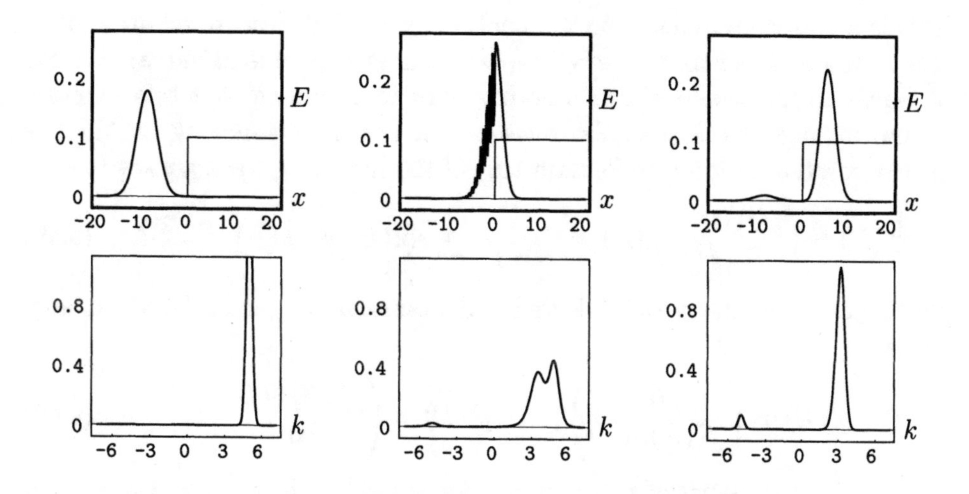


FIGURE 9.1. The position probability density $|\psi(x)|^2$ (first row) and the momentum probability density $|\hat{\psi}(k)|^2$ (second row) for the scattering at a potential step. You can see the wave packet before it hits the step, during the interaction with the step, and after the scattering process. The energies are strictly larger than the size of the step, hence the existence of a reflected part is a pure quantum phenomenon.

CD 7.5 explains how to calculate the momentum distribution of the scattered waves from the given initial wave packet and the scattering coefficients.

In the following, this scattering process is described in a slightly more quantitative way. The first task is to prepare an incoming wave packet. Take, for simplicity, a Gaussian function. At $t = 0$, it should be located sufficiently far to the left, so that it is negligibly small in the region of the step. "Incoming" here means that the average momentum of the Gaussian points to the right, that is, toward the step. Thus, the incoming wave packet is something like

$$\underset{\rightarrow}{\psi}^{\text{in}}(x) = \left(\frac{a}{\pi}\right)^{1/4} \mathrm{e}^{ik_0 x} \exp\left(-\frac{a\,(x-x_0)^2}{2}\right) \tag{9.43}$$

with $x_0 \ll -1/\sqrt{a} < 0$ ($a$ describes the width of the position distribution). The momentum representation of this wave packet is

$$\underset{\rightarrow}{\hat{\psi}}^{\text{in}}(k) = \left(\frac{1}{a\pi}\right)^{1/4} \mathrm{e}^{-ikx_0} \exp\left(-\frac{(k-k_0)^2}{2a}\right). \tag{9.44}$$

This is a Gaussian function in momentum space that is centered around $k_0$. It is certainly clever to choose $k_0 \gg \sqrt{a}$ because then—due to the rapid decay of Gaussian functions—there is no big error in assuming that this function is zero for negative $k$ (i.e., the wave packet has no left-moving part). The energy representation (see Section 3.8) of the incoming wave packet is

$$g^{\text{in}}(E) = \begin{pmatrix} g(E) \\ 0 \end{pmatrix}, \quad g(E) = \frac{1}{\sqrt{k(E)}} \, \hat{\underset{\rightarrow}{\psi}}^{\text{in}}(k(E)), \quad k(E) = \sqrt{2E}. \tag{9.45}$$

This gives immediately the reflected and transmitted wave packets in energy space

$$g^{\text{r}}(E) = \begin{pmatrix} 0 \\ \underset{\leftarrow}{R}(E)\, g(E) \end{pmatrix}, \qquad g^{\text{t}}(E) = \begin{pmatrix} \underset{\rightarrow}{T}(E)\, g(E) \\ 0 \end{pmatrix}. \tag{9.46}$$

How do these wave packets look in position space? If the reflection coefficient $R$ were a constant independent of $E$, then the reflected wave packet (9.39) would be obtained from the incoming wave packet (9.38) by replacing $\underset{\rightarrow}{\omega}$ with $\underset{\leftarrow}{\omega}$. This amounts to a reflection at the origin because it is just the substitution $x \to -x$, which changes $\underset{\rightarrow}{\omega}$ into $\underset{\leftarrow}{\omega}$. Hence the wave packet $\underset{\leftarrow}{\psi}^{\text{r}}$ would be a Gaussian centered at $-x_0$ (i.e., far to the right) with average momentum $-k_0$ pointing to the left. The norm of the Gaussian would be diminished by the value of $R^2$ (which is less than 1). Now, in fact $R$ depends on $E$, and hence the reflected wave packet becomes distorted. (In case of Fig. 9.1 the reflection coefficient varies only slightly with the energies under consideration, hence the reflected wave packet still looks pretty much like a Gaussian).

Similarly, the function $\underset{\rightarrow}{\psi}^{\text{t}}$ is a slightly distorted Gaussian-like wave packet. Its total energy is about the same as the energy of the incoming Gaussian, but because the transmitted wave packet moves in the potential $V_+$, the momentum distribution is shifted from $k_0$ to

$$k_0' = \sqrt{k_0^2 - 2V_+}. \tag{9.47}$$

The wave packet $\psi(x,t)$ in (9.42) initially consists only of the incoming part because the reflected part is located in the region $x > 0$, and the transmitted part in the region $x < 0$

$$\psi(x, t{=}0) \approx \begin{cases} \underset{\rightarrow}{\psi}^{\text{in}}(x), & x < 0, \\ 0, & x \geq 0. \end{cases} \tag{9.48}$$

The incoming and reflected parts are going to exchange their positions during the scattering process because they move into opposite directions. When the incoming and reflected waves overlap, we see an interference pattern in the region $x < 0$ (but there is no interference pattern in the region $x > 0$). Also, the transmitted part $\underset{\rightarrow}{\psi}^{\mathrm{in}}(x,t)$ will sooner or later enter the region $x > 0$. Much later, when the parts of the wave packet have left the vicinity of the step, we have the following situation:

$$\psi(x,t) \approx \begin{cases} \underset{\leftarrow}{\psi}^{\mathrm{r}}(x), & x < 0, \\ \underset{\rightarrow}{\psi}^{\mathrm{t}}(x), & x \geq 0, \end{cases} \approx \underset{\leftarrow}{\psi}^{\mathrm{r}}(x) + \underset{\rightarrow}{\psi}^{\mathrm{t}}(x). \tag{9.49}$$

This wave packet is outgoing. All parts move away from the potential step. In the energy representation, the outgoing wave packet is simply given by

$$g^{\mathrm{out}}(E) = \begin{pmatrix} \underset{\rightarrow}{T}(E)g(E) \\ \underset{\leftarrow}{R}(E)g(E) \end{pmatrix}. \tag{9.50}$$

The time evolution in the energy representation is just a multiplication by $\exp(-\mathrm{i}Et)$, corresponding to a stationary energy distribution.

With the energy distribution, the momentum distribution also becomes stationary as soon as the wave packet $\psi(x,t)$ is sufficiently far away from the interaction region and all its parts move freely. The momentum distribution of the outgoing wave packet can be calculated easily from the energy representation, as demonstrated in Section 9.3.1. Because the transmitted part moves in the constant potential $V_+$ and the reflected part moves freely, we obtain

$$|\hat{\psi}^{\mathrm{out}}(k)|^2 = |k| \begin{cases} \left|\underset{\rightarrow}{T}(\frac{k^2}{2}+V_+)\, g(\frac{k^2}{2}+V_+)\right|^2, & k \geq 0, \\ \left|\underset{\leftarrow}{R}(\frac{k^2}{2}) g(\frac{k^2}{2})\right|^2, & k < 0. \end{cases} \tag{9.51}$$

This distribution shows a peak at $-k_0$, which stems from the reflected wave packet and a peak at $k_0'$ from the transmitted wave packet. From the asymptotic momentum distribution we can infer the asymptotic position probability density of the scattered wave packet as in Section 3.6.

CD 7.6 contains several movies of wave packets at a potential step. Position space and momentum space are shown simultaneously. You can observe how the the wave packet in momentum space approaches its asymptotic form when the scattered waves in position space move away from the potential step.

## 9.5. Scattering Matrix

In scattering theory, we want to determine the relation between incoming and outgoing wave packets. A general incoming wave packet with energies greater than $V_+$ is described by

$$\psi^{\mathrm{in}}(x) = \int_{V_+}^{\infty} \Big( \underset{\rightarrow}{\omega}(E - V_-, x)\, g_1(E) + \underset{\leftarrow}{\omega}(E - V_+, x)\, g_2(E) \Big)\, dE. \tag{9.52}$$

It consists of a part describing a wave packet moving in the right direction in the constant potential $V_-$, and a part moving in the left direction in the potential $V_+$. The time evolution of $\psi^{\text{in}}$ describes two incoming wave packets which are initially (in the distant past) located far to the left, resp. far to the right. The interaction with the potential converts the incoming wave packet in an outgoing wave packet of the form

$$\psi^{\text{out}}(x) = \int_{V_+}^{\infty} \Big( \underleftarrow{\omega}(E-V_-,x)\, h_1(E) + \underrightarrow{\omega}(E-V_+,x)\, h_2(E) \Big)\, dE. \tag{9.53}$$

The relation between the coefficients of the outgoing parts and the coefficients of the incident parts is given by the *scattering matrix*

$$S(E) = \begin{pmatrix} \underleftarrow{T}(E) & \underrightarrow{R}(E) \\ \underleftarrow{R}(E) & \underrightarrow{T}(E) \end{pmatrix}. \tag{9.54}$$

We have

$$\begin{pmatrix} h_1(E) \\ h_2(E) \end{pmatrix} = S(E) \begin{pmatrix} g_1(E) \\ g_2(E) \end{pmatrix}. \tag{9.55}$$

The outgoing wave packet depends linearly on the incoming wave,

$$\psi^{\text{out}} = S\, \psi^{\text{in}}. \tag{9.56}$$

The linear operator $S$ is called the *scattering operator*. The scattering matrix is unitary, $S(E)S(E)^\dagger = \mathbf{1}_2$ (the $2\times 2$ unit matrix). In particular, this implies

$$|h_1(E)|^2 + |h_2(E)|^2 = |g_1(E)|^2 + |g_2(E)|^2,$$

and hence

$$\begin{aligned} \|\psi^{\text{out}}\|^2 &= \int_{V_+}^{\infty} (|h_1(E)|^2 + |h_2(E)|^2)\, dE \\ &= \int_{V_+}^{\infty} (|g_1(E)|^2 + |g_2(E)|^2)\, dE = \|\psi^{\text{in}}\|^2, \end{aligned}$$

that is, the scattering operator $S$ is unitary.

The considerations so far are not typical for the rectangular step. CD 7.7 shows that the scattering at a smooth potential without sudden jumps exhibits the same phenomena. The considerations above are valid in this case too.

# 9.6. Transition Matrix

## 9.6.1. Potential step

If the potential has a step at $x = s$, we can again use the ansatz

$$\psi(E,x) = \begin{cases} a\,\underset{\rightarrow}{\omega}(E-V_-,x) + b\,\underset{\leftarrow}{\omega}(E-V_-,x), & x < s, \\ c\,\underset{\rightarrow}{\omega}(E-V_+,x) + d\,\underset{\leftarrow}{\omega}(E-V_+,x), & x \geq s. \end{cases} \tag{9.57}$$

The continuity of $\psi$ and $\psi'$ at $x = s$ leads to the linear system of equations

$$\sqrt{k_+}\,(a\,\mathrm{e}^{\mathrm{i}k_-s} + b\,\mathrm{e}^{-\mathrm{i}k_-s}) = \sqrt{k_-}\,(c\,\mathrm{e}^{\mathrm{i}k_+s} + d\,\mathrm{e}^{-\mathrm{i}k_+s}), \tag{9.58}$$

$$\sqrt{k_-}\,(a\,\mathrm{e}^{\mathrm{i}k_-s} - b\,\mathrm{e}^{-\mathrm{i}k_-s}) = \sqrt{k_+}\,(c\,\mathrm{e}^{\mathrm{i}k_+s} - d\,\mathrm{e}^{-\mathrm{i}k_+s}). \tag{9.59}$$

We can solve this system for $c$ and $d$,

$$\begin{pmatrix} c \\ d \end{pmatrix} = T(V_-, V_+, s) \begin{pmatrix} a \\ b \end{pmatrix}, \tag{9.60}$$

where

$$T(V_-, V_+, s) = \frac{1}{2\sqrt{k_+k_-}} \begin{pmatrix} (k_- + k_+)\,\mathrm{e}^{\mathrm{i}(k_- - k_+)s} & -(k_- - k_+)\,\mathrm{e}^{-\mathrm{i}(k_- + k_+)s} \\ -(k_- - k_+)\,\mathrm{e}^{\mathrm{i}(k_- + k_+)s} & (k_- + k_+)\,\mathrm{e}^{-\mathrm{i}(k_- - k_+)s} \end{pmatrix} \tag{9.61}$$

is the *transition matrix* for scattering at a potential step from $V_-$ to $V_+$ at $x = s$. As usual, $k_\pm$ is the momentum at energy $E$ with respect to the potentials $V_\pm$, that is, for $E > V_\pm$,

$$k_\pm = k(E - V_\pm) = \sqrt{2(E - V_\pm)}. \tag{9.62}$$

The transition matrix connects the coefficients $c$ and $d$ of the solution on the right side of the step with the coefficients $a$ and $b$ on the left side. This is particularly useful if one wants to calculate this relation in the case of several steps.

## 9.6.2. Two potential steps

As an example, consider a potential $V$ of the form

$$V(x) = \begin{cases} V_-, & \text{for } x < -r, \\ V_0, & \text{for } -r \leq x < R, \\ V_+, & \text{for } R \leq x. \end{cases} \tag{9.63}$$

CD 7.9–7.14 show the scattering at a potential barrier that is defined by $V_+ = V_- = 0$ and $V_0 > 0$. The potential barrier can be regarded as a succession of two potential steps. The movies treat this process in close analogy to the scattering at a single step. It is shown how the reflection and transmission coefficients determine the behavior of the solutions in position space and in momentum space. Below you will learn how to calculate the scattering coefficients for this situation. Notice that the coefficients now depend on the energy, the height, and the thickness of the barrier. Again, the considerations are not typical for the rectangular barrier. CD 7.15 has several examples of numerically calculated solutions at smoothed potential barriers.

We easily find the connection between the solutions in the asymptotic regions $x < -r$ and $x \geq R$. The solution of the stationary Schrödinger equation at energy $E$ in the middle region $-r \leq x < R$ is given by a linear combination of $\underset{\leftarrow}{\omega}$ and $\underset{\rightarrow}{\omega}$ as

$$\psi(E,x) = c_1\, \underset{\rightarrow}{\omega}(E-V_0,x) + c_2\, \underset{\leftarrow}{\omega}(E-V_0,x), \qquad \text{in } -r \leq x < R.$$

The coefficients $c_1$ and $c_2$ can be determined from the coefficients $a$ and $b$ of the asymptotic solution to the left with the help of the corresponding transition matrix,

$$\begin{pmatrix} c_1 \\ c_2 \end{pmatrix} = T(V_-, V_0, -r) \begin{pmatrix} a \\ b \end{pmatrix}. \tag{9.64}$$

Similarly, the solution in $x \geq R$ is determined by $c_1$, $c_2$ as

$$\begin{pmatrix} c \\ d \end{pmatrix} = T(V_0, V_+, R) \begin{pmatrix} c_1 \\ c_2 \end{pmatrix}. \tag{9.65}$$

Hence we find

$$\begin{pmatrix} c \\ d \end{pmatrix} = T(V_0, V_+, R)\, T(V_-, V_0, -r) \begin{pmatrix} a \\ b \end{pmatrix}. \tag{9.66}$$

For scattering from the left we use as before the ansatz $a = 1$, $d = 0$, and write $b = \underset{\leftarrow}{R}(E)$ and $c = \underset{\rightarrow}{T}(E)$. The linear equation

$$\begin{pmatrix} \underset{\rightarrow}{T}(E) \\ 0 \end{pmatrix} = T(V_0, V_+, R)\, T(V_-, V_0, -r) \begin{pmatrix} 1 \\ \underset{\leftarrow}{R}(E) \end{pmatrix} \tag{9.67}$$

can be solved for the reflection and transmission coefficients, $\underleftarrow{R}(E)$ and $\underrightarrow{T}(E)$,

$$\underleftarrow{R}(E) = \frac{k_0\,(k_- - k_+)\,\cos k_0 L + \mathrm{i}\,(k_0^2 - k_- k_+)\,\sin k_0 L}{k_0\,(k_- + k_+)\,\cos k_0 L - \mathrm{i}\,(k_0^2 + k_- k_+)\,\sin k_0 L}\,\exp(-2\mathrm{i}k_- r), \quad (9.68)$$

$$\underrightarrow{T}(E) = \frac{2\,k_0\,\sqrt{k_- k_+}\,\exp(-\mathrm{i}(k_- r + k_+ R))}{k_0\,(k_- + k_+)\,\cos k_0 L - \mathrm{i}\,(k_0^2 + k_- k_+)\,\sin k_0 L}, \quad (9.69)$$

with the obvious abbreviations $k_\pm = k(E - V_\pm)$, $k_0 = k(E - V_0)$, $L = r + R$ (= distance between the two steps).

### 9.6.3. The potential barrier

Let us apply the results of the preceeding section to the scattering at a *potential barrier*, which is a two-step potential with $V_\pm = 0$ and $V_0$ positive,

$$V(x) = \begin{cases} V_0, & \text{for } |x| < R, \\ 0, & \text{for } |x| \geq R, \end{cases} \qquad V_0 > 0. \quad (9.70)$$

Here we consider a scattering process with energies higher than the size $V_0$ of the barrier. The scattering coefficients $\underleftarrow{R}$ and $\underrightarrow{T}$ for the scattering from the left can be obtained from Eqs. (9.68) and (9.69). They depend not only on the energy $E$, but also on the height $V_0$ and the width $R$ of the barrier. An example is shown in Color Plate 23, which shows the scattering coefficients as functions of the parameter $V_0$. The energy and the width have the values $E = 18$ and $R = 1$, respectively.

Color Plate 23 shows that the reflection coefficient is zero at certain values of $V_0$, for example, at $V_0 \approx 16.8$. The vanishing of the reflection coefficient means that in this case there is no reflected wave at all. How is this possible? It does not happen for a single-step potential. But in the case of a two-step potential we can interpret the reflected wave as a superposition of two parts. Each part comes from the reflection at one of the steps. The two reflected parts have the same energy (wavelength) and they interfere with each other in the region $x < -R$. A zero in the reflection coefficient obviously means that the two parts cancel each other (destructive interference). This can only happen if they have the same amplitude and the phase difference $\pi$ (so that one part is the negative of the other).

In the example of Color Plate 23 the reflection probability has a local maximum, for example, at $V_0 \approx 15$. In this case, the two reflected waves are approximately in phase so that they can interfere constructively. Color Plate 24 shows the solutions of the Schrödinger equation with energy $E = 18$ for the two situations. You can see that the solution with $V_0 = 16.8$ has no reflected part, because the incoming wave on the left-hand side has the same amplitude as the transmitted wave on the right-hand side. Inside the barrier, the wave function exhibits a remarkable *resonance phenomenon*, which does not occur for the smaller value of $V_0$. The solution belonging to $V_0 = 15$ has a large reflected part which interferes strongly with the incoming wave to the left of the barrier.

The knowledge of the reflection and transmission coefficients helps us to explain the behavior of wave packets. Color Plate 25 shows an incoming wave packet (a Gaussian function) in the energy representation. It is centered around $E = 18$ in energy space (hence the maximum of the momentum distribution is at $k = 6$). The energy representation has only one component because the left-moving part of incoming wave packet is approximately zero. In our example, the energy distribution is so narrow that essentially all energies in the wave packet are above $V_0$ in both situations. Hence, according to classical physics, we would expect no reflection at all. In quantum mechanics, however, a significant part of the wave packet is reflected at the barrier. Color Plate 25 also shows the scattering coefficients. They are plotted as functions of the energy in the relevant region of the energy space. You can see that the support of the wave packet in energy space is near a zero of the reflection coefficient for $V_0 = 16.8$, and near a maximum of reflection coefficient for $V_0 = 16.8$.

The energy representation of the reflected wave packet is given by the product $\underleftarrow{R}(E)\, g(E)$; the transmitted part is $\underrightarrow{T}(E)\, g(E)$. From the energy representation we can determine the Fourier transformation of the outgoing wave essentially by a variable transformation (as described in Section 3.8). The result is shown in Fig. 9.2.

The two situations ($V_0 = 16.8$ and $V_0 = 15$) produce quite different momentum distributions in the reflected wave packet. First of all, there is the seemingly paradoxical fact that a smaller barrier produces a larger reflected part. For the larger value of $V_0$, the energies of the wave packet are near a zero of the reflection coefficient. Hence the reflected part $\underleftarrow{R}(E)\, g(E)$ is rather small and has a zero. You can see this most clearly in Color Plate 25.

Color Plate 26 shows the scattering of our wave packet at the barrier. You can see that for the larger value $V_0 = 16.8$ the amplitude of the reflected part is indeed smaller. The resonance that appears already in the plane wave of Color Plate 24 implies that the probability of finding the particle inside the barrier is quite large during the whole scattering process. As a result, the transmitted wave is delayed in comparison with the nonresonant situation ($V_0 = 15$). The *time delay* is typical for resonance phenomena.

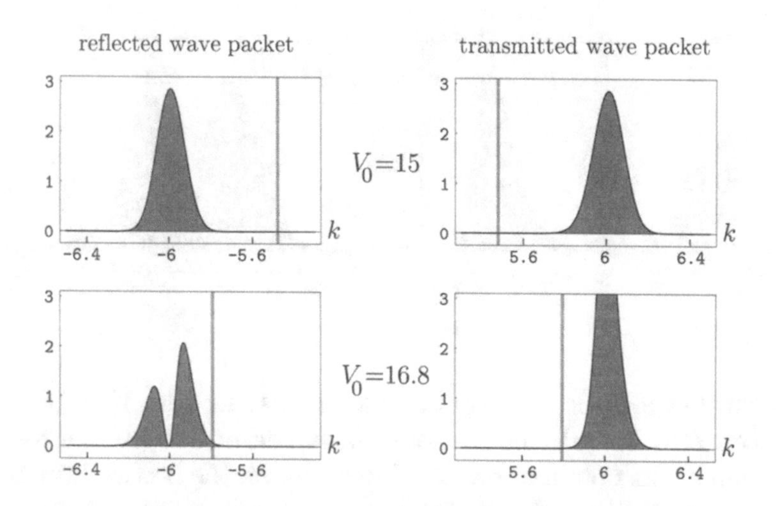


FIGURE 9.2. Scattering at a potential barrier. Absolute values of the Fourier transforms of the reflected and transmitted wave packets for two values of the parameter $V_0$. The vertical line at $|k| = \sqrt{2V_0}$ indicates the size of $V_0$.

CD 7.9 presents interactive versions of Color Plates 23 and 24. It shows the dependence of scattering coefficients and sharp-energy solutions on the parameters describing the potential barrier. In CD 7.10 you can see movies showing the scattering of a wave packet at barriers of varying height. Color Plate 26 shows snapshots from these movies. CD 7.12 shows movies of the wave packet in both position and momentum space. You can see how the momentum distribution predicted by the scattering coefficients is approached by the Fourier transform of the scattered wave packet. Finally, CD 7.13 allows you to investigate the asymptotic momentum distribution by interactively changing the parameters $R$, $V_0$, and the energy of the incoming wave packet.

From the momentum distribution Fig. 9.2 we can obtain the asymptotic behavior of the scattered wave packet in position space, because the asymptotic behavior is described by the free Schrödinger equation. You can verify this statement by looking at Fig. 9.3.

CD 7.14 investigates the influence of the thickness of the barrier on the scattering of a Gaussian wave packet. This is discussed along the same lines as in this section. The reflection coefficient depends periodically on the thickness $R$; see also Fig. 9.4.

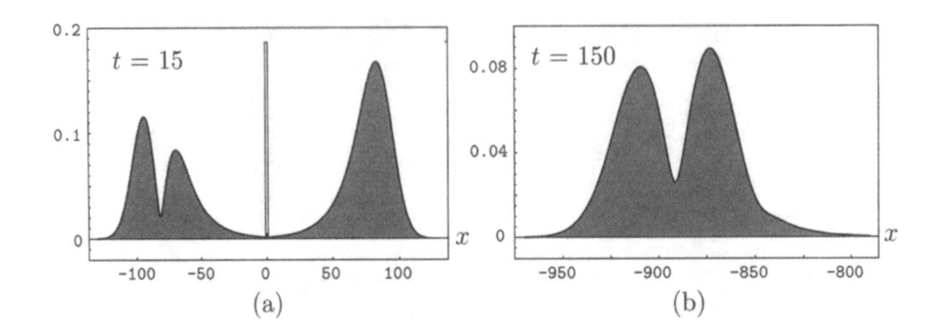


FIGURE 9.3. Scattering at a potential barrier with $V_0 = 16.8$ and $R = 1$. (a) The scattered wave packet has left the region of the potential barrier. From now on the free evolution governs the behavior, and the momentum distribution is constant as in Fig. 9.2. The shape of the position distribution is already similar to the asymptotic momentum distribution shown in Fig. 9.4. (b) Reflected part of the wave packet at a later time. The wave packet in position space is essentially a scaled version of the wave packet in momentum space.

## 9.7. The Tunnel Effect

The formulas (9.68) and (9.69) for the reflection and transmission coefficients with a potential of the form (9.70) can be used to illustrate some important quantum-mechanical effects. With positive energies $E$ smaller than the height $V_0$ of the barrier, a classical particle would be reflected. A quantum-mechanical particle has a certain chance to "tunnel" through the barrier.

CD 7.11 shows that a wave packet has a certain chance to penetrate a thin barrier, even if the barrier is much higher than all energies contributing to the wave packet. The movies CD 7.12.3 and 7.12.4 visualize the same phenomenon also in momentum space. CD 7.16.2 is a movie of a Gaussian wave packet penetrating a rectangular barrier in two dimensions; see also Color Plate 27.

Eqs. (9.68) and (9.69) remain true for $E < V_0$, we only have to take into account that $k_0$ is now purely imaginary,

$$k_0 = \mathrm{i}\sqrt{2(V_0 - E)} = \mathrm{i}\kappa_0, \qquad \text{for } 0 \le E \le V_0. \tag{9.71}$$

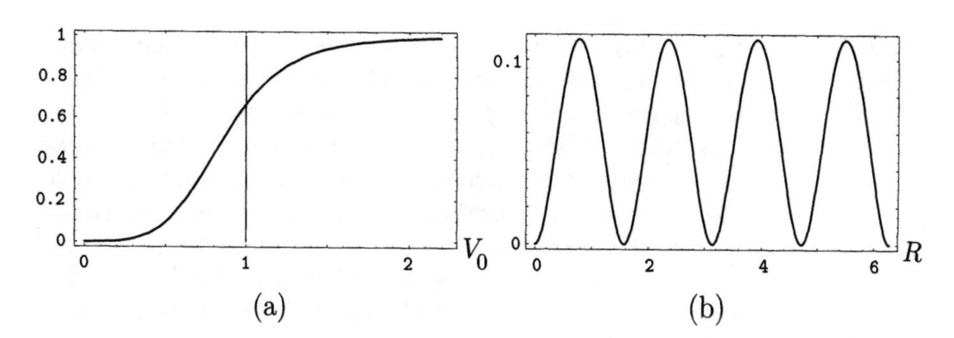


(a) (b)

FIGURE 9.4. (a) Reflection probability at a fixed energy ($E = 1$) as a function of $V_0$ with fixed $R = 1$. (b) Reflection probability as a function of the thickness $R$, keeping $V_0 = 1/2$ fixed.

The transmission coefficient (9.69) now becomes (with $k_- = k_+ = k$, $r = R$, $L = 2R$)

$$\underset{\rightarrow}{T}(E) = \frac{2\,\kappa_0\,k\,\exp(-\mathrm{i}kL)}{2\kappa_0\,k\,\cosh\kappa_0 L + \mathrm{i}\,(\kappa_0^2 - k^2)\,\sinh\kappa_0 L}, \tag{9.72}$$

and hence the transition probability $|\underset{\rightarrow}{T}(E)|^2$ does not vanish.

The exponentially decaying part of the wave function in the classically forbidden region $-R \leq x < R$ is nonzero at $x = R$, and thus there is a nonvanishing probability for the particle to penetrate the barrier (*tunnel effect*). This is a pure quantum effect because classical particles would be reflected at energies which are too low to traverse the obstacle.

Figure 9.4 shows the reflection probability $|\underset{\leftarrow}{R}(E)|^2$ at a fixed energy ($E = 1$). It still depends on the thickness $R$ and on the height $V_0$ of the potential barrier. The tunnel effect implies that the reflection probability is less than 1 even for barriers higher than the energy ($V_0 > E$). The vanishing of the reflection probability for certain values of $R$ is caused by the destructive interference between waves reflected at the first step and waves reflected at the second step.

The tunnel effect is used in the scanning tunneling microscope (STM). CD 7.18 shows the tunneling of an electron through a narrow vacuum gap between the tip of a sharp needle and a metal surface. The STM works by scanning the surface with the needle. During the scan the vertical position of the needle is adjusted in order to keep the tunneling current constant. In that way one can obtain images of single atoms on the surface. It is also possible to map the average position probability of surface electrons. CD 7.19 explains in somewhat more detail how the STM works and shows a gallery of spectacular images.

## 9.8. Example: Potential Well

### 9.8.1. Bound-state energies

The rectangular *potential well* is a potential barrier with a negative middle potential,

$$V(x) = \begin{cases} -V_0, & \text{for } |x| < R, \\ 0, & \text{for } |x| \geq R, \end{cases} \qquad V_0 > 0. \tag{9.73}$$

Now it makes sense to consider also the energy region $-V_0 < E < 0$. The relation (9.66) between the coefficients of the solution on the left-hand side of the well and the coefficients on the right-hand side still remains correct, that is,

$$\begin{pmatrix} c \\ d \end{pmatrix} = T(-V_0, 0, R)\, T(0, -V_0, -R) \begin{pmatrix} a \\ b \end{pmatrix} . \tag{9.74}$$

The coefficient $a$ belongs to the solution $\underset{\rightarrow}{\omega}$ in the region $x < -R$, which increases exponentially for $x \to -\infty$ if the energy is negative. A solution with nonzero coefficient $a$ thus cannot be used in quantum mechanics because it does not belong to the Hilbert space of square-integrable functions. A similar observation leads to the exclusion of the coefficient $d$ because it belongs to the exponentially increasing solution $\underset{\leftarrow}{\omega}$ in the region $x \geq R$. Therefore, only a function satisfying the constraint

$$\psi(E, x) = \begin{cases} b\, \underset{\leftarrow}{\omega}(E, x), & \text{for } x < 0, \\ c\, \underset{\rightarrow}{\omega}(E, x), & \text{for } x \geq R, \end{cases} \tag{9.75}$$

is acceptable as a solution of the stationary Schrödinger equation. We thus arrive at the condition

$$\begin{pmatrix} c \\ 0 \end{pmatrix} = T(-V_0, 0, R)\, T(0, -V_0, -R) \begin{pmatrix} 0 \\ b \end{pmatrix} . \tag{9.76}$$

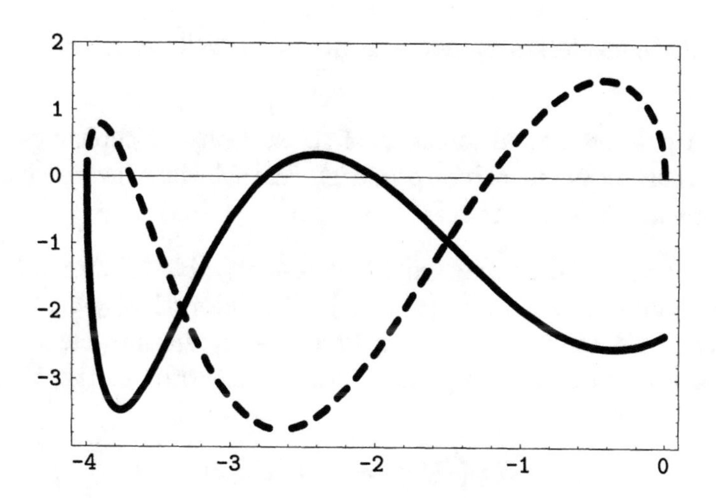


FIGURE 9.5. The condition (9.80) for a rectangular potential well with $R = 1$ and $V_0 = 4$. The dashed line represents the left-hand side of Eq. (9.80); the solid line shows the right-hand side. The condition is fulfilled where the lines interSection For this example, there are two possible energies which lead to a square-integrable solution of the Schrödinger equation: $E_0 = -3.33702$, and $E_1 = -1.50881$.

If we evaluate the right-hand side of this equation we obtain

$$\begin{pmatrix} c \\ 0 \end{pmatrix} = \frac{b\,e^{-2\kappa R}}{2\,\kappa\,k_0} \begin{pmatrix} e^{2\kappa R}\,(k_0^2+\kappa^2)\,\sin 2k_0 R \\ 2\,\kappa\,k_0\,\cos 2k_0 R - (k_0^2-\kappa^2)\,\sin 2k_0 R \end{pmatrix}. \tag{9.77}$$

with

$$\kappa = \sqrt{-2E}, \qquad k_0 = \sqrt{2(E+V_0)}, \qquad (-V_0 < E < 0). \tag{9.78}$$

The condition (9.77) can only be satisfied if

$$2\,\kappa\,k_0\,\cos 2k_0 R - (k_0^2-\kappa^2)\,\sin 2k_0 R = 0. \tag{9.79}$$

Inserting the definitions of $\kappa$ and $k_0$ in the equation above gives

$$2\sqrt{-E(V_0+E)}\,\cos(2\sqrt{2(V_0+E)}\,R) = (V_0+2E)\,\sin(2\sqrt{2(V_0+E)}\,R). \tag{9.80}$$

This is a transcendental equation for $E$ which is best solved numerically. Figure 9.5 shows an example with two values of $E$ satisfying the condition.

### 9.8.2. Energy spectrum for the potential well

As a second order equation the stationary Schrödinger equation in one dimension has always two linearly independent solutions. Which solutions can be accepted from the quantum-mechanical point of view depends on the energy under consideration. For the potential well we have the following situation:

(1) $E < -V_0$: There are no solutions of the stationary Schrödinger equation with a quantum-mechanical interpretation. All solutions have an exponential increase toward at least one side.

(2) $-V_0 < E < 0$: There may exist square-integrable solutions $\psi(E_k, x)$ at certain energies $E_1, E_2, \dots, E_n$ ($n$ finite) in the interval $-V_0 < E < 0$. These energies can be determined from Eq. (9.80). A square-integrable solution of the stationary Schrödinger equation is an eigenvector of the Hamiltonian operator,

$$H\,\psi(E_k, x) = E_k\,\psi(E_k, x). \tag{9.81}$$

Solutions of the time-dependent Schrödinger equation are

$$\psi(E_k, x, t) = \psi(E_k, x)\exp(-\mathrm{i}\,E_k\,t), \qquad k = 1, \dots, n, \tag{9.82}$$

and linear combinations (superpositions) of these functions. A particle described by a linear combination of the stationary solutions $\psi(E_k, x, t)$ remains near the region of the potential well for all times. These are bound states.

(3) $E > 0$: All energies $E > 0$ lead to two solutions, which outside the well are given by plane waves. One has to form wave packets to obtain solutions with a quantum-mechanical interpretation. These solutions are scattering states.

### 9.8.3. The scattering matrix

The scattering matrix connects the incoming waves with the outgoing waves, while the transition matrix

$$T = T(-V_0, 0, R)\,T(0, -V_0, -R) \tag{9.83}$$

connects the waves on the left-hand side with the waves on the right-hand side. From

$$\begin{pmatrix} c \\ d \end{pmatrix} = T \begin{pmatrix} a \\ b \end{pmatrix} = \begin{pmatrix} T_{11} & T_{12} \\ T_{21} & T_{22} \end{pmatrix} \begin{pmatrix} a \\ b \end{pmatrix}, \tag{9.84}$$

we can determine the relation between the coefficients of the incoming wave, $a$, $d$, and the coefficients of the outgoing wave, $b$, $c$. We obtain

$$\begin{pmatrix} b \\ c \end{pmatrix} = S \begin{pmatrix} a \\ d \end{pmatrix}, \tag{9.85}$$

where

$$S = \frac{1}{T_{22}} \begin{pmatrix} -T_{21} & 1 \\ \det T & T_{12} \end{pmatrix}. \tag{9.86}$$

The scattering matrix $S$ is unitary for energies $E > 0$. The definition of the transition matrix $T$ can also be used for other values of the energy. The explicit form of the matrix element $T_{22}$ for the potential well is

$$T_{22} = \frac{e^{-\kappa R}}{2\,\kappa\, k_0} \left(2\,\kappa\, k_0 \cos k_0 R - (k_0^2 - \kappa^2) \sin k_0 R\right). \tag{9.87}$$

We see that the elements of the $S$-matrix become singular when $E$ approaches the energy of a bound state because $T_{22} = 0$ is just the condition (9.79). Indeed, for a bound state, the coefficients $a$ and $d$ are both zero, while the coefficients $b$ and $c$ of the exponentially decaying parts are nonzero. Hence the relation (9.85) shows that the elements of $S$ must be infinite for any bound-state energy. Indeed, we have the following general fact: The analytical continuation of the $S$-matrix as a function of the energy $E > 0$ to arbitrary complex values of $E$ has poles of first order at the bound-state energies of the system.

## 9.9. Parity

The rectangular well discussed in Section 9.8 has the symmetry property

$$V(x) = V(-x). \tag{9.88}$$

Correspondingly, the eigenfunctions of the Schrödinger operator also have a symmetry with respect to reflections, see Fig. 9.6.

The observation of symmetry properties of the potential is always helpful for the solution of the Schrödinger equation. The following considerations are not limited to one space dimension. They are valid for $x \in \mathbb{R}^n$ as well.

### 9.9.1. The parity transformation

In order to investigate the reflection symmetry we introduce a linear operator $P$ by defining

$$(P\psi)(x) = \psi(-x). \tag{9.89}$$

The operator $P$ is called the *parity transformation*. Changing the sign of the argument of the wave function obviously has no influence on the norm of the wave function. Thus, the parity transformation is unitary. It is easy to see that $P$ is symmetric and hence self-adjoint (because it is defined everywhere),

$$P^\dagger = P = P^{(-1)}, \qquad P^2 = 1. \tag{9.90}$$

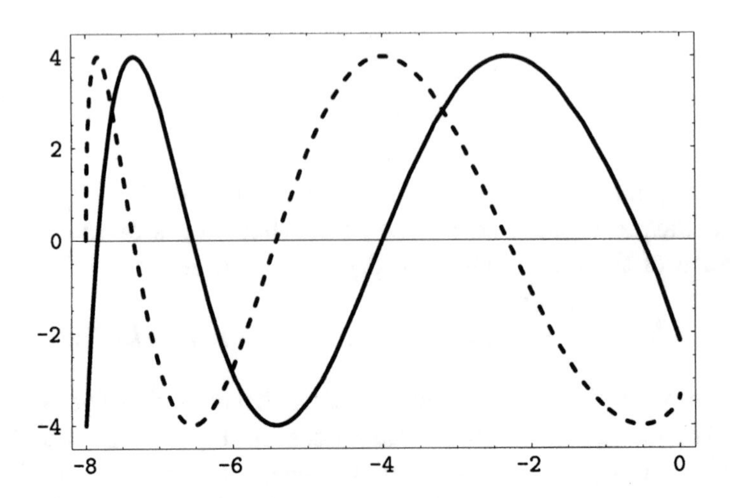


FIGURE 9.6. The energy eigenfunctions for a rectangular potential well are either even, $\psi(x) = \psi(-x)$, or odd, $\psi(x) = -\psi(-x)$.

These properties imply that the operator $P$ has at most two eigenvalues, $+1$ and $-1$. (Unitarity implies that the spectrum of $P$ is on the unit circle in $\mathbb{C}$, self-adjointness implies that the spectrum is real). The projection operators onto the (infinite-dimensional) eigenspaces are given by

$$P_{\text{even}} = \tfrac{1}{2}\,(1+P), \qquad P_{\text{odd}} = \tfrac{1}{2}\,(1-P). \tag{9.91}$$

If $\psi$ is an arbitrary square-integrable function, we may define

$$\psi_{\text{even}} = P_{\text{even}}\psi, \qquad \psi_{\text{odd}} = P_{\text{odd}}\psi. \tag{9.92}$$

The function $\psi_{\text{even}}$ is an *even function* in the sense

$$\psi_{\text{even}}(x) = \psi_{\text{even}}(-x) \quad \text{or} \quad P\psi_{\text{even}} = \psi_{\text{even}}. \tag{9.93}$$

(If $\psi_{\text{even}} \neq 0$, then it is an eigenfunction of $P$ belonging to the eigenvalue $+1$). Similarly, the *odd function* satisfies

$$\psi_{\text{odd}}(x) = -\psi_{\text{odd}}(-x) \quad \text{or} \quad P\psi_{\text{odd}} = -\psi_{\text{odd}}. \tag{9.94}$$

The reflection symmetry (9.88) of the potential means that the parity operator commutes with the operator of multiplication by $V$,

$$PV(x)\psi(x) = V(-x)\psi(-x) = V(x)(P\psi)(x), \tag{9.95}$$

for all $\psi$ in the domain of $V$. We also note that $P$ anticommutes with the position and momentum operators,

$$Px = -xP, \qquad Pp = -pP. \tag{9.96}$$

The latter property is an immediate consequence of the fact that

$$\frac{d}{dx}\,\psi(-x) = -\psi'(x). \tag{9.97}$$

We conclude from (9.96) that the parity transform commutes with the free Hamiltonian $H_0 = p^2/2$. If the potential has the reflection symmetry, then $P$ commutes with the Hamiltonian $H = H_0 + V$,

$$[H, P] = HP - PH = 0, \qquad \text{on the domain of } H. \tag{9.98}$$

This property has the important consequence that the eigenfunctions of the energy operator $H$ can always be chosen as eigenfunctions of the parity operator $P$. Let $\psi$ be an arbitrary eigenfunction of the energy, $H\psi = E\psi$. Then the commutation property (9.98) implies that

$$H\psi_{\text{even}} = H\,\tfrac{1}{2}\,(1+P)\,\psi = \tfrac{1}{2}\,(1+P)\,H\psi = E\psi_{\text{even}}$$

and similarly $H\psi_{\text{odd}} = E\psi_{\text{odd}}$. Because any function $\psi$ can be written as the orthogonal sum $\psi = \psi_{\text{even}} + \psi_{\text{odd}}$, we find that the eigenspace of $H$ corresponding to $E$ is spanned by even and odd functions.

Whenever an eigenvalue is nondegenerate, that is, if the eigenspace is one-dimensional, then the even and odd parts of the eigenfunction must be linearly independent. Because these parts are orthogonal, one of them must vanish. A nondegenerate eigenvalue of a Hamiltonian that is parity invariant has either an even or an odd eigenfunction.

It is a general property of the one-dimensional Schrödinger equation that all its eigenvalues are nondegenerate.

### 9.9.2. Example: The rectangular well

The condition (9.79) can be simplified by exploiting the symmetry of the system under a parity transformation. This symmetry implies that the eigenfunctions are either even or odd functions. For an even function, the coefficient $c$ of $\underleftarrow{\omega}$ to the left of the well must be equal to the coefficient $b$ of $\underrightarrow{\omega}$ on the right-hand side. The first line in Eq. (9.77) gives for this situation

$$2\kappa k_0 = (k_0^2 + \kappa^2)\,\sin 2k_0R.$$

Inserting this into (9.79) and simplifying the expressions involving doubled arguments of trigonometric functions, we obtain the condition

$$k_0\,\sin k_0R - \kappa\,\cos k_0R = 0 \qquad \text{(for an even eigenfunction).} \tag{9.99}$$

For odd eigenfunctions $\psi(x) = -\psi(-x)$ we have $c = -b$ in Eq. (9.77) and a similar consideration leads to

$$\kappa\,\sin k_0R + k_0\,\cos k_0R = 0 \qquad \text{(for an odd eigenfunction).} \tag{9.100}$$

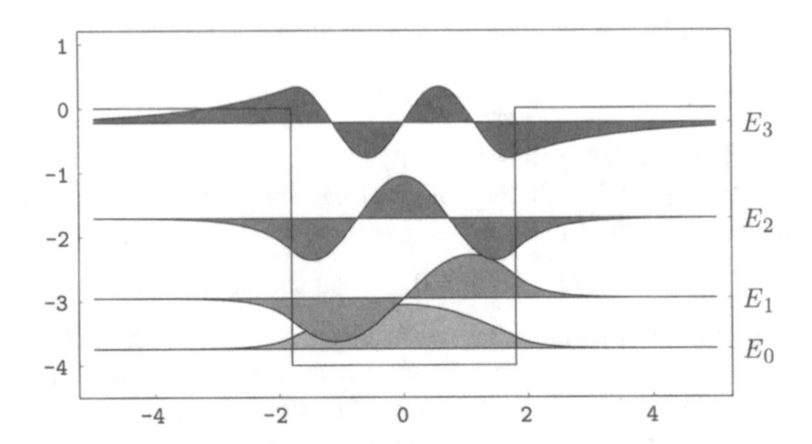


FIGURE 9.7. The energy eigenvalues for a rectangular potential well with $R = 2.5$ and $V_0 = 8$ are the zeros of these curves. The solid line shows the left-hand side of Eq. (9.99) as a function of the energy $E$, the dashed line represents the left-hand side of Eq. (9.100). The solid line determines the eigenvalues with even eigenfunctions, the dashed line corresponds to the odd eigenfunctions.

Figure 9.7 plots the right-hand sides of the new eigenvalue conditions showing that even and odd eigenfunctions alternate in the well, the eigenfunction with the lowest energy (the ground state) being even.

The wave function inside the well is a superpositon of $\underleftarrow{\omega}(E+V_0, x)$ and $\underrightarrow{\omega}(E+V_0, x)$. The only even superpositions are multiples of

$$\underrightarrow{\omega}(E+V_0, x) + \underleftarrow{\omega}(E+V_0, x) = 2 \cos k_0 x, \tag{9.101}$$

while the only odd superpositions that can be formed of plane waves are multiples of

$$\underrightarrow{\omega}(E+V_0, x) - \underleftarrow{\omega}(E+V_0, x) = 2\mathrm{i} \sin k_0 x. \tag{9.102}$$

EXERCISE 9.1. *Solve the stationary Schrödinger equation with a well potential using the ansatz*

$$\psi(E, x) = \begin{cases} N \cos k_0 x, & \text{for } 0 \leq x < R, \\ N (\cos k_0 R \exp \kappa R) \exp(-\kappa x), & \text{for } R < x \end{cases} \tag{9.103}$$

*for an even eigenfunction, and a similar ansatz for an odd eigenfunction. Show that the condition of continuity of $\psi'$ at $x = R$ is equivalent to (9.99) (resp. to (9.100) for the odd eigenfunction).*

EXERCISE 9.2. *With $\psi(E,x)$ defined as in Eq.* (9.101), *calculate*

$$\|\psi(E,\cdot)\|^2 = 2\int_0^\infty |\psi(E,x)|^2\,dx$$

*and show with the help of Eq.* (9.99) *that a normalized wave function is obtained for*

$$N = \frac{1}{\sqrt{R+\frac{1}{\kappa}}}. \tag{9.104}$$

*Prove the analogous result for the odd eigenfunction.*

Several movies on the CD show scattering phenomena in two dimensions. Obstacle scattering is shown in the chapter on boundary conditions. The chapter about scattering theory contains several movies of transparent obstacles, that is, potential barriers of finite height. Two-dimensional steps are shown in CD 7.8, barriers in CD 7.16. The scattering at smoothed spherical barriers is visualized in CD 7.20. Here the second movie shows a particle that partially penetrates into a half-crystal.

*Appendix A*

# Numerical Solution in One Dimension

In most cases, the Schrödinger equation with an external field cannot be solved analytically. Hence many of the movies had to be generated with the help of numerical methods. In order to solve the time-dependent Schrödinger equation we use a finite difference scheme, the Crank–Nicolson formula. We consider, as an example, the Schrödinger equation in one space dimension with an electrostatic potential $V(x, t)$ (a magnetic field in one-dimension can always be gauged away).

## A.1. Discretization of the Schrödinger Equation

We consider the equation

$$a \frac{\partial}{\partial t} \psi(x,t) = -b \frac{\partial^2}{\partial x^2} \psi(x,t) + V(x,t)\, \psi(x,t) \tag{A.1}$$

with a given initial value

$$\psi(x, t_0) = \psi_0(x). \tag{A.2}$$

For the numerical treatment we restrict the position space to some finite interval, say $[x_0, x_n]$. We have to prescribe the behavior of the solution at the borders of the interval. This can be done with suitable boundary conditions, for example

$$\psi(x_0, t) = \psi(x_n, t) = 0, \quad \text{for all } t. \tag{A.3}$$

The boundary conditions will certainly influence the behavior of the solution as discussed in Chapter 5. If there are no boundary conditions for the original equation (A.1), the boundary effects have to be treated as a numerical artifact. The best way to avoid unwanted boundary effects is to place the initial wave packet (for example, a well localized Gaussian function) sufficiently far away from the borders of the domain $[x_0, x_n]$. The domain has to be chosen large enough, so that the wave packet does not come close to its borders during the time-interval of interest.

The numerical solution of (A.1) is based on a discretization of space and time which allows us to approximate the partial differential equation by a matrix equation. The interval $[x_0, x_n]$ is divided into $n$ equal sized subintervals of size

$$\Delta x = \frac{x_n - x_0}{n}. \tag{A.4}$$

The points are

$$\{x_0, x_1, \dots, x_i, \dots, x_{n-1}, x_n\}. \tag{A.5}$$

Time is discretized into time-steps of length $\Delta t$. The sequence of times is

$$\{\dots, t_0, t_1, \dots, t_j, \dots, t_n, \dots\}. \tag{A.6}$$

For the values of the solution and of the potential at the space-time points $(x_i, t_j)$ we write

$$\psi(x_i, t_j) = \psi_i^j, \qquad V(x_i, t_j) = V_i^j. \tag{A.7}$$

The derivatives are replaced, as usual, by the difference quotients

$$\left.\frac{\partial}{\partial t}\psi\right|_{x_i,t_j} \approx \frac{\psi_i^{j+1} - \psi_i^j}{\Delta t}, \tag{A.8}$$

$$\left.\frac{\partial^2}{\partial x^2}\psi\right|_{x_i,t_j} \approx \frac{\psi_{i+1}^j - 2\psi_i^j + \psi_{i-1}^j}{(\Delta x)^2}. \tag{A.9}$$

We insert these expressions into the Schrödinger equation. On the right-hand side of Eq. (A.1) we take the average over the times $j$ and $j+1$ and obtain

$$\begin{aligned}\frac{2a}{\Delta t}\left(\psi_i^{j+1} - \psi_i^j\right) = &-\frac{b}{(\Delta x)^2}\left(\psi_{i+1}^{j+1} - 2\psi_i^{j+1} + \psi_{i-1}^{j+1}\right)\\ &-\frac{b}{(\Delta x)^2}\left(\psi_{i+1}^j - 2\psi_i^j + \psi_{i-1}^j\right)\\ &+ V_i^{j+1}\psi_i^{j+1} + V_i^j\psi_i^j,\end{aligned} \tag{A.10}$$

for all $j$ and $i = 1, 2, \dots, n-1$, and the boundary condition becomes

$$\psi_0^j = 0 = \psi_n^j, \quad \text{for all } j. \tag{A.11}$$

Using the abbreviations

$$u = b/(\Delta x)^2, \qquad v = a/\Delta t, \tag{A.12}$$

and

$$a_i = -u\,\psi_{i+1}^j + (2u + 2v + V_i^j)\,\psi_i^j - u\,\psi_{i-1}^j, \tag{A.13}$$

$$d_i = 2v - 2u - V_i^{j+1}, \tag{A.14}$$

we can write Eq. (A.10) as

$$u\,\psi_{i+1}^{j+1} + d_i\,\psi_i^{j+1} + u\,\psi_{i-1}^{j+1} = a_i, \tag{A.15}$$

where $i = 1, \ldots, n-1$. Setting $a_0 = a_n = 0$ we can write everything (including the boundary condition) in matrix form:

$$\begin{pmatrix} 1 & 0 & & & & \\ 0 & d_1 & u & & 0 & \\ & u & d_2 & u & & \\ & & \ddots & \ddots & \ddots & \\ & 0 & & u & d_{n-1} & 0 \\ & & & & 0 & 1 \end{pmatrix} \begin{pmatrix} \psi_0^{j+1} \\ \psi_1^{j+1} \\ \psi_2^{j+1} \\ \vdots \\ \psi_{n-1}^{j+1} \\ \psi_n^{j+1} \end{pmatrix} = \begin{pmatrix} a_0 \\ a_1 \\ a_2 \\ \vdots \\ a_{n-1} \\ a_n \end{pmatrix}. \tag{A.16}$$

This inhomogeneous linear equation gives $\psi$ at time $t_{j+1}$ in terms of the solution at time $t_j$. The boundary condition is preserved automatically.

## A.2. Solution of a Linear Equation with Tridiagonal Matrix

We have to solve Eq. (A.16) which is a linear inhomogeneous equation in $\mathbb{R}^n$ of the form $T\psi = a$ with a tridiagonal matrix T. Because of its importance we give the solution in the general case. We assume that $T$ is a matrix of the form

$$T = \begin{pmatrix} d_0 & u_0 & & & & \\ l_1 & d_1 & u_1 & & 0 & \\ & l_2 & d_2 & u_2 & & \\ & & \ddots & \ddots & \ddots & \\ & 0 & & l_{n-1} & d_{n-1} & u_{n-1} \\ & & & & l_n & d_n \end{pmatrix}. \tag{A.17}$$

It is possible to write $T = FG$ with triangular matrices $F$ and $G$ of a particularly simple type:

$$F = \begin{pmatrix} f_0 & 0 & & \cdots & 0 \\ l_1 & f_1 & 0 & & \vdots \\ 0 & \ddots & \ddots & \ddots & \\ \vdots & & l_{n-1} & f_{n-1} & 0 \\ 0 & \cdots & 0 & l_n & f_n \end{pmatrix}, \quad G = \begin{pmatrix} 1 & g_0 & 0 & \cdots & 0 \\ 0 & 1 & g_1 & & \vdots \\ & \ddots & \ddots & \ddots & 0 \\ \vdots & & 0 & 1 & g_{n-1} \\ 0 & \cdots & & 0 & 1 \end{pmatrix}.$$

It is easy to determine the matrix elements $f_i$ and $g_j$ from the equation $T = FG$. We can now solve the equation $T\psi = a$ by first determining a vector $b$ from $Fb = a$ and then solving $G\psi = b$ for $\psi$. For triangular matrices with only one off-diagonal row this can be done very easily.

## A.3. Crank–Nicolson Method for the Schrödinger Equation

The following algorithm summarizes the considerations above. Assume that the solution of Eq. (A.1) is given at a time $t_j$ on space points $x_i$, as a vector $(\psi_0^j, \psi_1^j, \dots, \psi_n^j)$. We determine $(\psi_0^{j+1}, \psi_1^{j+1}, \dots, \psi_n^{j+1})$, the solution at time $t_{j+1}$ by the following calculation.

1. Step:

$$\begin{aligned} a_i &= (2u + 2v + V_i^j)\psi_i^j - u(\psi_{i+1}^j + \psi_{i-1}^j), \quad i = 1, \dots, n-1, \\ d_i &= 2v - 2u - V_i^{j+1}, \qquad i = 1, \dots, n-1., \end{aligned}$$

2. Step:

$$\begin{aligned} f_1 &= d_1, \\ f_i &= d_i - u^2/f_{i-1} \quad \text{for } i = 2, \dots, n-1. \end{aligned}$$

3. Step:

$$\begin{aligned} b_0 &= 0, \\ b_i &= (a_i - ub_{i-1})/f_i \quad \text{for } i = 1, \dots, n-1. \end{aligned}$$

4. Step:

$$\begin{aligned} \psi_n^{j+1} &= 0, \\ \psi_i^{j+1} &= b_i - u\psi_{i+1}^{j+1}/f_i, \quad \text{for } i = n-1, \dots, 1. \\ \psi_0^{j+1} &= 0. \end{aligned}$$

## A.4. Discussion

The restriction to a finite interval and the discretization of space leads to an approximation of the partial differential equation (A.1) by a system of ordinary differential equations. The wave function is only calculated at the points $x_k$, $k = 0, \dots, n$ and is hence approximated by a vector in $\mathbb{C}^{n+1}$. With the approximation of the Hamiltonian by a difference operator according to Eq. (A.9) we obtain

$$a \frac{d}{dt} \vec{\psi}(t) = \hat{H}\vec{\psi}(t), \quad \vec{\psi} = (\psi_0, \psi_1, \dots, \psi_n) \in \mathbb{C}^{n+1}, \tag{A.18}$$

where $\hat{H}$ is the symmetric tridiagonal matrix

$$\hat{H} = \begin{pmatrix} 1 & 0 & & & & \\ 0 & 2u+V_1 & -u & & 0 & \\ & -u & 2u+V_2 & -u & & \\ & & \ddots & \ddots & \ddots & \\ & 0 & & -u & 2u+V_{n-1} & 0 \\ & & & & 0 & 1 \end{pmatrix}. \tag{A.19}$$

Let us assume that the potential and hence the matrix $\hat{H}$ does not depend on time. In this case the solution of the system of ordinary differential equations (A.18) belonging to an initial vector $\vec{\psi}(0)$ is given by

$$\vec{\psi}(t) = \exp\Big(\frac{1}{a}\,\hat{H}\,t\Big)\,\vec{\psi}(0), \tag{A.20}$$

where the exponential function of any matrix $A$ is defined by the power series

$$\exp A = \mathbf{1} + A + \frac{A^2}{2!} + \frac{A^3}{3!} + \cdots . \tag{A.21}$$

Advancing the solution by one time step $\Delta t$ is done by applying the exponential operator to the state vector at time $t_j$,

$$\vec{\psi}(t_{j+1}) = \exp\Big(\frac{1}{a}\,\hat{H}\,\Delta t\Big)\,\vec{\psi}(t_j). \tag{A.22}$$

For the numerical calculation of this time step, one could, for example, truncate the power series after a finite number of summands. For example, the first order *explicit Euler method* is obtained by writing

$$\exp\Big(\frac{1}{a}\,\hat{H}\,\Delta t\Big) \approx \mathbf{1} + \frac{1}{a}\,\hat{H}\,\Delta t. \tag{A.23}$$

The Crank–Nicolson method is obtained by the approximation

$$\exp\Big(\frac{1}{a}\,\hat{H}\,\Delta t\Big) \approx \Big(\mathbf{1} - \frac{1}{2a}\,\hat{H}\,\Delta t\Big)^{-1}\Big(\mathbf{1} + \frac{1}{2a}\,\hat{H}\,\Delta t\Big). \tag{A.24}$$

Writing

$$\Big(\mathbf{1} - \frac{1}{2a}\,\hat{H}\,\Delta t\Big)\,\vec{\psi}(t_{j+1}) = \Big(\mathbf{1} + \frac{1}{2a}\,\hat{H}\,\Delta t\Big)\,\vec{\psi}(t_j) \tag{A.25}$$

instead of Eq. (A.22) immediately leads to Eq. (A.10). The Crank–Nicolson method is of second order in $\Delta t$. Writing

$$(1-B)^{-1} = 1 + B + B^2 + B^3 + \cdots, \qquad \text{with} \quad B = \frac{1}{2a}\,\hat{H}\,\Delta t$$

(the power series converges for $\Delta t$ small enough), then the right-hand side of (A.24) becomes

$$\begin{aligned}(1-B)^{-1}(1+B) &= (1+B+B^2+\cdots)(1+B) = 1+2B+2B^2+\cdots \\ &= 1+\frac{1}{a}\hat{H}\,\Delta t+\frac{1}{2}\Big(\frac{1}{a}\hat{H}\,\Delta t\Big)^2+\cdots,\end{aligned}$$

which reproduces the power series of the exponential function up to order $(\Delta t)^2$. Thus, the Crank–Nicolson method is more accurate than the Euler method—at the expense of being implicit. The calculation of $\psi(t_{j+1})$ according to Crank–Nicolson involves an additional matrix-inversion (the solution of the linear system (A.16)). This is not necessary for the Euler method, which gives $\psi(t_{j+1})$ explicitly in terms of $\psi(t_j)$. But there is another advantage of the Crank–Nicolson method. For the Schrödinger equation, we have $a = \mathrm{i}$ in Eq. (A.18). Because $\hat{H}$ is a Hermitian matrix, the exponential operator $\exp(-\mathrm{i}\hat{H}\Delta t)$ is a unitary matrix. The approximation (A.24) is also unitary. Hence the Crank–Nicolson method (unlike the unstable explicit Euler method) automatically preserves the norm of the solution.

*Appendix B*

# Movie Index

## 1. Visualization

CD 1.1. **Complex numbers and the color map**

1. Complex plane
2. Color code
3. Color map

CD 1.2. **RGB color cube**

CD 1.3. **HSB cone**

CD 1.4. **HSB cylinder**

CD 1.5. **HLS double cone**

CD 1.6. **Color sphere**

CD 1.7. **Stereographic projection and the color map**

CD 1.8. **Visualization of complex functions in one dimension**

CD 1.9. **Visualization of complex functions in two dimensions**

CD 1.10. **Many examples**

CD 1.11. **Jacobi elliptic functions**

CD 1.12. **Plot of real part**

CD 1.13. **Plot of vector field**

CD 1.14. **Complex map**

CD 1.15. **Color density plot**

CD 1.16. **Surface plot**

CD 1.17. **Isosurface in three dimensions**

CD 1.18. **Plane wave in two dimensions**

## 2. Fourier Analysis

CD 2.1. **Trigonometric sums**

1. Approximation of a Gaussian
2. Sum of exp($ix$) and exp($-ix$)
3. Complex Gaussian function
4. The Fourier coefficients

CD 2.2. **Fourier spectrum**

1. Real-valued Gaussian function
2. Complex Gaussian function

CD 2.3. **Approximation of a characteristic function**

CD 2.4. **Saw-tooth function**

1. Real-valued function
2. Function with complex values

CD 2.5. **Fourier expansion and translation**

CD 2.6. **From Fourier series to Fourier integral**

CD 2.7. **Translations**

1. Translation in $k$-space
2. Translation in $x$-space

CD 2.8. **Noncommutativity**

1. $x$-space translation first
2. $k$-space translation first

CD 2.9. **Scaling transformation**

CD 2.10. **Example of the Fourier transform** $1/(1+x^2)$

CD 2.11. **Step function**

1. Fourier transform and scaling
2. Approximation by continuous functions

CD 2.12. **A discontinuous function**

CD 2.13. **Example of the Fourier transform**: $\cos(ax)\exp(-x^2)$

CD 2.14. **Example of the Fourier transform**: $\cos(x-b)\exp(-x^2)$

CD 2.15. **Phase factor in momentum space**

CD 2.16. **Example of the Fourier transform**

CD 2.17. **Example of the Fourier transform**

CD 2.18. **Generalized derivative**

CD 2.19. **Action of the resolvent on a Gaussian**

CD 2.20. **Resolvent acting on a step function**

# 3. Free Particles

CD 3.1. **Free plane waves**

1. Real and imaginary part
2. Plot as a space curve
3. Plot filled with colors
4. Periodic superposition
5. Quasiperiodic superposition

CD 3.2. **Building a wave packet**

1. Superposition of real parts
2. `ArgColorPlot` of the superposition

CD 3.3. **Gaussian wave packets**

1. Free particle at rest
2. Slowly moving free particle ($k = 2$)

CD 3.4. **Fast wave packets**

1. Fast moving free wave packet ($k = 8$)
2. Very fast moving free particle ($k = 100$)

CD 3.5. **Two-dimensional wave packet (Gaussian at rest)**

CD 3.6. **Slow wave packet in two dimensions**

CD 3.7. **Wave function in momentum space**

CD 3.8. **Motion in phase space**

1. Slowly moving Gaussian wave packet
2. Gaussian at rest in position space

CD 3.9. **Energy representation**

1. Gauss function in energy space
2. Time evolution in energy space

CD 3.10. **Direction of motion**

1. Particle at rest
2. Slowly moving particle

CD 3.11. **Free time evolution**

1. Flow from red to yellow
2. Flow from green to blue

CD 3.12. **Step function**

1. Time evolution
2. Close-up in slow motion
3. Approximation of a step function

CD 3.13. **Two separated Gaussian peaks (Schrödinger cat state)**

1. Non-Gaussian initial condition

2. Time evolution
3. Asymptotic time evolution

CD 3.14. **Superposition realizing a boundary condition**

1. Dirichlet boundary condition
2. Neumann boundary condition

CD 3.15. **Dirichlet boundary condition in two dimensions**

1. Two Gaussians
2. Time-symmetric motion

CD 3.16. **Neumann boundary condition in two dimensions**

1. Wave function in position space
2. Position probability density

CD 3.17. **Exotic initial state (circularly symmetric)**

CD 3.18. **State evolving into a characteristic function**

CD 3.19. **Motion in a constant field**

1. potential energy = kinetic energy
2. potential energy > kinetic energy
3. potential energy < kinetic energy
4. potential energy + kinetic energy = 0
5. potential energy = 0 (free motion for comparison)

CD 3.20. **Magnetic gauge field in one and two dimensions**

1. Motion of a Gaussian in a gauge field
2. Motion in another gauge field
3. Gauge field in two dimensions

# 4. Boundary Conditions

CD 4.1. **Particle hitting a wall**

1. Elastic reflection ($k = 10$)
2. Elastic reflection ($k = 2$)
3. Elastic reflection ($k = 5$)
4. Symmetric reflection

CD 4.2. **Momentum space**

1. Elastic reflection with $k = 5$
2. Symmetric reflection with $k = 5$

CD 4.3. **Method of mirror waves**

1. Gaussian with $k = 5$
2. Time-symmetric state
3. Particle at rest near the wall

CD 4.4. **Neumann condition**

1. Particle at rest near the wall
2. Comparison with Dirichlet
3. Reflection and phase shift

CD 4.5. **Between two walls**

1. Gaussian function with mirrors
2. Slower moving initial function

CD 4.6. **Particle in a box**

1. Slow Gaussian initial function
2. Fast, well-localized initial state
3. Initially at rest near one wall

CD 4.7. **Dirichlet box in 2D**

1. Gaussian initial state
2. Diagonal motion
3. Centered Gaussian at rest

CD 4.8. **Dirichlet box in 2D**

1. Centered initial state
2. Asymmetric initial state

CD 4.9. **Dirichlet box**

1. States with a sharp energy
2. Superposition of states 1 and 2
3. Superposition of states 1 and 3
4. Superposition of states 2 and 3

CD 4.10. **Neumann box**

1. States with a sharp energy
2. Superposition of states 0 and 1
3. Superposition of states 1 and 2
4. Gaussian initial state

CD 4.11. **Dirichlet box**

1. Strange behavior of the unit function

CD 4.12. **Particle on a circle**

1. Superposition of eigenfunctions
2. Self-interference of a Gaussian
3. Gaussian moving on a circle
4. Gaussian with periodic BC

CD 4.13. **Quantum waveguide**

1. Motion parallel to walls
2. Reflection between walls
3. Converging walls

CD 4.14. **Periodic walls**

1. Wall with periodic structure
2. Sine wall with short wavelength
3. Sine wall with long wavelength
4. Angle of incidence 45 degrees

CD 4.15. **Obstacle Scattering**

1. Scattering at a large sphere
2. Scattering at a large square
3. Scattering at a small sphere
4. Scattering at a small square

CD 4.16. **Wall with a hole**

1. Penetrating Gaussian state
2. Dependence on hole diameter
3. Scattering at the edge of a wall

CD 4.17. **Double-slit experiment**

1. Interference pattern
2. One slit closed—no interference
3. Superposition of spherical waves

CD 4.18. **Double slit**

1. Faster Gaussian wave packet
2. Dependence on slit distance

CD 4.19. **Multislit**

1. Scattering of Gaussian state
2. Parameter dependence

# 5. Harmonic Oscillator

CD 5.1. **Classical motion**

CD 5.2. **Energy eigenstates**

1. Introduction
2. Gallery of eigenfunctions
3. Comparison with classical density
4. Color plot of eigenfunctions

CD 5.3. **Oscillating state 0+1**

1. Wave function in position space
2. Position probability density
3. Wave function in momentum space
4. Wave function in phase space

CD 5.4. **Oscillating state 0+2**

1. Wave function in position space
2. Position probability density

3. Wave function in momentum space
4. Wave function in phase space

CD 5.5. **Oscillating state 1+2**

1. Wave function in position space
2. Position probability density
3. Wave function in momentum space
4. Wave function in phase space

CD 5.6. **Superposition of three eigenstates** $(0 + 1 + 2)$

1. Wave function in position space
2. Position probability density
3. Wave function in momentum space

CD 5.7. **Linear combination of three eigenstates** $(0 + 1 - 2)$

1. Wave function in position space
2. Position probability density

CD 5.8. **Superposition of ten eigenstates**

1. Sum of first ten eigenstates
2. Boltzmann distribution of energies
3. Random phases

CD 5.9. **Eigenfunction expansion**

1. Build a Gaussian function (interactive experiment)
2. Energy representation
3. Energy representation of a shifted state
4. Time evolution in energy space

CD 5.10. **Coherent state (shifted eigenstate)**

1. Wave function in position space
2. Position probability density
3. Wave function in momentum space
4. Energy representation
5. Time period and gauge

CD 5.11. **Squeezed state (widespread initial localization)**

1. Wave function in position space
2. Position probability density
3. Wave function in momentum space

CD 5.12. **Squeezed state (sharper initial localization)**

1. Wave function in position space
2. Position probability density
3. Momentum space

CD 5.13. **Step function**

1. Approximation in position space

2. Position probability density
3. Momentum space

CD 5.14. **Coherent state**

1. Phase space motion
2. Circular motion in two dimensions
3. Elliptic motion in two dimensions

CD 5.15. **Diagonal motion (surface plots)**

1. Coherent state in two dimensions
2. Squeezed state in two dimensions (flat)
3. Squeezed state in two dimensions (sharp)

CD 5.16. **Particle at rest**

1. Centered squeezed state
2. Asymmetric centered state
3. Squeezed state in phase space

CD 5.17. **Phase space motion (contour plots)**

1. Squeezed state in phase space
2. Closer to the center
3. Very flat Gaussian state
4. Sharp Gaussian peak

CD 5.18. **Circular motion (surface plots)**

1. Circular motion of Gaussian
2. Another initial condition
3. Yet another initial condition ("the swimmer")

CD 5.19. **Constant shapes (time evolution of shifted eigenstates)**

1. Translated first eigenstate
2. Second eigenstate
3. First+second eigenstate

CD 5.20. **Gallery of angular momentum eigenstates**

## 6. Special Systems

CD 6.1. **Free Fall**

1. Dropping a particle
2. Throwing a particle

CD 6.2. **Free fall**

1. Exact solution in a linear potential

CD 6.3. **Free fall in two dimensions**

1. Letting a particle drop vertically
2. Vertical throw of a particle

CD 6.4. **Quantum ballistics**

1. Horizontal initial velocity
2. Parabolic trajectory

CD 6.5. **Bouncing ball**

1. Eigenfunctions
2. Oscillating state "0+1"
3. Oscillating state "0+2"

CD 6.6. **Bouncing ball**

1. Oscillating state "1+2"
2. Oscillating state "0+1+2"

CD 6.7. **Bouncing ball**

1. Gaussian initial function
2. Approximate revival of the initial state
3. Better localized Gaussian

CD 6.8. **Box in two dimensions with gravity**

1. Gaussian initial state

CD 6.9. **Box in two dimensions with gravity**

1. Eigenfunctions

CD 6.10. **Box in two dimensions with gravity**

1. Oscillating state

CD 6.11. **Magnetic potential**

1. Three-dimensional view of the Poincaré gauge
2. Constant field strength in two dimensions
3. Arrows and colors

CD 6.12. **Gauge transformation**

1. Translations of the vector potential

CD 6.13. **Classical motion**

1. Particle moving on a circle
2. Small circle—low energy

CD 6.14. **Translations ($E$ =const)**

1. Horizontal direction
2. Vertical direction
3. Translations do not commute
4. Shifted ground state

CD 6.15. **Translations of second kind**

1. Horizontal direction
2. Vertical direction

3. Noncommutativity

CD 6.16. **Coherent states**

1. Gaussian starting at $x = 0$
2. Zero momentum and low energy
3. Zero momentum and high energy
4. Gaussian in another gauge

CD 6.17. **Squeezed states**

1. Well localized at $x = 0$
2. Flat state starting at $x = 0$
3. Zero momentum and low energy
4. Vector potential at center

CD 6.18. **Eigenfunctions**

1. Eigenfunction "0-1"
2. Eigenfunction "2-2"
3. Translated eigenfunction "2-2"

CD 6.19. **Oscillating states**

1. Simple linear combination
2. Centered state "(0+1)-(0+2)"
3. Shifted initial state

CD 6.20. **Aharonov–Bohm effect**

1. Interference due to gauge potential
2. No interference for quantized flux
3. Dependence on field strength

# 7. Scattering Theory

CD 7.1. **Scattering coefficients for a potential step**

1. $R(E)$ and $T(E)$ as functions of the energy
2. $R(E)$ and $T(E)$ as functions of the step size

CD 7.2. **Step potential: Plane wave solutions**

1. Incident wave from the left
2. Step size to infinity
3. Incident wave from the right

CD 7.3. **Wave packets at a step**

1. Parts of the wave packet
2. Energy higher than step size
3. Total reflection at low energies

CD 7.4. **Potential step**

1. Gaussian at a potential step

2. Narrow energy distribution
3. Dependence on the step size

CD 7.5. **Momentum space**

1. Energy/momentum representation
2. Dependence on average momentum
3. Dependence on the step size
4. Potential step down ($V$ negative)

CD 7.6. **Fourier transform**

1. Step size equals mean energy
2. Step size is half of the energy
3. Potential step down

CD 7.7. **Continuous potential steps**

1. Step size = $0.8\langle E\rangle$
2. Step size = $1.2\langle E\rangle$
3. Step size = $1.6\langle E\rangle$
4. Step size = $\langle E\rangle$ (wide)

CD 7.8. **Potential step in two dimensions**

1. Step up, scattering from the front
2. Step down, from the front
3. Step up, scattering at 45 degrees
4. Step down, at 45 degrees

CD 7.9. **Potential barrier**

1. Coefficients depending on width
2. Solution depending on width
3. Coefficients depending on height
4. Solution depending on height

CD 7.10. **Gaussian at barrier**

1. Energy around a zero of $R$
2. Energy around a maximum of $R$
3. Reflection at both edges
4. Multiple reflections

CD 7.11. **Tunnel effect**

1. Wave packet penetrating a barrier
2. Very thin rectangular barrier
3. Tunneling through a thicker barrier

CD 7.12. **Fourier transform**

1. Scattering at a rectangular barrier
2. Scattering at a thick barrier
3. Tunneling through a thin barrier
4. Thin barrier, higher energy

CD 7.13. **Momentum distribution**

1. Dependence on the height $V$
2. Dependence on the average energy
3. Dependence on the thickness $R$

CD 7.14. **Rectangular barrier**

1. Influence of thickness on final state
2. Much reflection at a thin barrier
3. Less reflection at a thicker barrier

CD 7.15. **Smoothed barrier**

1. Wave packet and Fourier transform
2. Wave packet at a thicker barrier
3. Very thick barrier

CD 7.16. **Barrier in two dimensions**

1. Jumping over a wall
2. Penetrating a thin wall
3. Oblique wall
4. Scattering at a potential corner

CD 7.17. **Barrier in two dimensions**

1. Scattering at a circular barrier
2. Scattering at a smaller circle
3. Scattering at a square barrier

CD 7.18. **Scanning tunneling microscope**

1. Tip close to a conducting surface
2. Larger tip–surface distance
3. Scanning a surface with the STM

CD 7.19. **Gallery of STM images**

CD 7.20. **Several obstacles**

1. Two smoothed circular barriers
2. Half-crystal

*Appendix C*

# Other Books on Quantum Mechanics

The huge number of excellent books about quantum mechanics indicates that it is very difficult to write anything new and original in that field. The Bibliography contains a certainly incomplete list and I apologize to all who have made significant contributions and do not appear here. Moreover, I did not include books on special topics such as quantum scattering theory or relativistic quantum mechanics, or books which are only available in languages other than English.

My decision to write *Visual Quantum Mechanics* (i.e., a "Movie Book of Quantum Mechanics") has certainly been influenced by *The Picture Book of Quantum Mechanics* [10] (see also [11]) and by some recent books concerning the application of computer algebra systems (such as *Mathematica*) to quantum mechanics, see the books [18], [36].

Sometimes, I look into one of the famous old classical texts, for example, [5], [7], [14], [24], [39], [49], [48], and [68]. Perhaps the most famous and most classic are [15] and [52], but I would not recommend to begin with those. If you prefer a more intuitive approach, you should read Feynman's lectures [19], which are still incredibly modern, and [40].

If you want to train your problem solving skills, you should have a look at [22], [31], [70], or [78]. On my desk I also found [4], [12], [62], and [76] which I recommend for the mathematically inclined reader, but not without some previous knowledge of functional analysis. The standard for books about the mathematical methods of quantum mechanics is set by [64].

If you want to read about quantum mechanics and its weird interpretation without being disturbed by mathematical formulas, I recommend [30]. For more recent developments in the quantum measurement problem, see the book of Omnes [54]. People interested in the history of quantum mechanics will like [38].

An alternative approach to quantum physics using Feynman's path integral formalism is explained in [20] and [66].

## Bibliography

[1] L. E. Ballentine. *Quantum Mechanics.* Prentice-Hall, Englewood Cliffs, NJ, 1990.
[2] J. E. Bayfield. *Quantum Evolution: An Introduction to Time-Dependent Quantum Mechanics.* Wiley, New York, 1999.
[3] G. Baym. *Lectures on Quantum Mechanics.* W. A. Benjamin, Reading, MA, 1969.
[4] F. A. Berezin, M. A. Shubin. *The Schrödinger Equation.* Kluwer Academic Publishers, Dordrecht, 1991.
[5] H. A. Bethe, E. E. Salpeter. *Quantum Mechanics of One- and Two-Electron Atoms.* Springer-Verlag, Berlin, 1957.
[6] H. A. Bethe, R. W. Jackiw. *Intermediate Quantum Mechanics.* 3rd edition (1986), Perseus Books, New York, 1997.
[7] D. I. Blokhintsev. *Principles of Quantum Mechanics.* Allyn and Bacon, Boston, 1964.
[8] A. Böhm. *Quantum Mechanics.* Texts and Monographs in Physics, Springer-Verlag, New York, 1979.
[9] D. Bohm. *Quantum Theory.* Dover, New York, 1989.
[10] S. Brandt, H. D. Dahmen. *The Picture Book of Quantum Mechanics.* 2nd edition. Springer-Verlag, New York, 1995.
[11] S. Brandt, H. D. Dahmen. *Quantum Mechanics on the Macintosh*, and *Quantum Mechanics on the Personal Computer.* 2nd edition. Springer-Verlag, New York, 1994.
[12] H. L. Cycon, R. G. Froese, W. Kirsch, B. Simon. *Schrödinger Operators.* Springer-Verlag, New York, 1987.
[13] C. Cohen-Tannoudji, B. Diu, F. Laloë. *Quantum Mechanics.* (Two volumes). Wiley, New York, 1978.
[14] A. S. Davydov. *Quantum Mechanics.* Pergamon Press, Oxford, 1965.
[15] P. A. M. Dirac. *The Principles of Quantum Mechanics.* Oxford University Press, Oxford, 1958.
[16] R. Eisberg, R. Resnick. *Quantum Physics of Atoms, Molecules, Solids, Nuclei, and Particles.* 2nd edition, Wiley, New York, 1985
[17] P. Exner, M. Havlicek, J. Blank. *Hilbert space operators in quantum physics.* Springer-Verlag, New York, 1997.
[18] J. M. Feagin. *Quantum Methods with Mathematica.* 2nd edition. Springer-Verlag, New York, 1998.
[19] R. P. Feynman, R. B. Leighton, M. Sands. *The Feynman Lectures on Physics, Vol. 3, Quantum Mechanics.* Addison-Wesley, Reading, MA, 1965.
[20] R. P. Feynman, A. R.. Hibbs. *Quantum Mechanics and Path Integrals.* McGraw-Hill, New York, 1965.
[21] D. D. Fitts. *Principles of Quantum Mechanics.* Cambridge University Press, Cambridge, 1999.
[22] S. Flügge. *Practical Quantum Mechanics.* (Two volumes). Springer-Verlag, New York, 1971.
[23] A. Galindo, P. Pascual. *Quantum Mechanics I, II.* Springer-Verlag, New York, 1990, 1991.
[24] S. Gasiorowicz. *Quantum Physics.* 2nd edition, Wiley, New York, 1996.
[25] D. T. Gillespie. *A Quantum Mechanics Primer.* Halsted Press, New York, 1970.
[26] K. Gottfried. *Quantum Mechanics.* Benjamin-Cummings Publ. Co., Reading, MA, 1994.
[27] H. S. Green. *Matrix Mechanics.* Noordhoff, Groningen, 1966.

[28] N. J. B. Green. *Quantum Mechanics 1. Foundations*, and *2. The Toolkit*. Oxford University Press, New York, 1996, 1998.

[29] W. Greiner. *Quantum Mechanics: An Introduction*. 3rd edition. Springer-Verlag, New York, 1994.

[30] J. Gribbin. *In Search of Schrödinger's Cat : Quantum Physics and Reality*. Bantam, New York, 1986.

[31] I. I. Goldman, V. D. Krivchenkov. *Problems in Quantum Mechanics*. Dover, New York, 1993.

[32] A. Goswami. *Quantum Mechanics*. McGraw-Hill, New York, 1997.

[33] K. Hannabuss. *An Introduction to Quantum Theory*. Clarendon Press, Oxford, 1997.

[34] W. Heisenberg. *Physical Principles of the Quantum Theory*. Dover, New York, 1967.

[35] J. R. Hiller, I. D. Johnston, D. F. Styer. *Quantum Mechanics Simulations: The Consortium for Upper-Level Physics Software*. Wiley, New York, 1995

[36] M. Horbatsch. *Quantum Mechanics Using Maple*. Springer-Verlag, New York, 1995.

[37] J. E. House. *Fundamentals of Quantum Mechanics*. Academic Press, New York, 1998.

[38] M. Jammer. *The Conceptual Development of Quantum Mechanics*. 2nd edition. Springer-Verlag, New York, 1998.

[39] L. D. Landau, E. M. Lifshitz. *Quantum Mechanics*. Addison-Wesley, Reading, MA, 1958.

[40] J.-M. Lévy-Leblond, F. Balibar. *Quantics. Rudiments of Quantum Physics*. North-Holland, Amsterdam, 1990.

[41] R. L. Liboff. *Introductory Quantum Mechanics*. 3rd edition, Addison-Wesley, Reading, MA, 1997.

[42] H. J. Lipkin. *Quantum Mechanics. New Approaches to Selected Topics*. North-Holland, Amsterdam, 1973.

[43] J. P. Lowe. *Quantum Chemistry*. Academic Press, New York, 1993.

[44] F. Mandl. *Quantum Mechanics*. Wiley, New York, 1992.

[45] J. L. Martin. *Basic Quantum Mechanics*. Oxford Physics Series, Clarendon Press, Oxford, 1981.

[46] J. D. McGervey. *Quantum Mechanics*. Academic Press, New York, 1995.

[47] D. A. McQuarrie. *Quantum Chemistry*. University Science Books, Mill Valley, CA, 1983.

[48] E. Merzbacher. *Quantum Mechanics*. 3rd edition. Wiley, New York, 1997.

[49] A. Messiah. *Quantum Mechanics*. (Two volumes). North-Holland, Amsterdam, 1970.

[50] A. Modinos. *Quantum Theory of Matter: A Novel Introduction*. Wiley, New York, 1996.

[51] M. A. Morrison. *Understanding Quantum Mechanics*. Prentice-Hall, Englewood Cliffs, NJ, 1990.

[52] J. v. Neumann. *Mathematical Foundations of Quantum Mechanics*. (Princeton Landmarks in Mathematics and Physics). Princeton University Press, 1996.

[53] R. Omnes. *The Interpretation of Quantum Mechanics*. Princeton Univ. Press, Princeton, NJ, 1994.

[54] R. Omnes. *Understanding Quantum Mechanics*. Princeton Univ. Press, Princeton, NJ, 1999.

[55] R. K. Osborn. *Applied Quantum Mechanics*. World Scientific, Singapore, 1988.

[56] W. Pauli. *General Principles of Quantum Mechanics*. Springer-Verlag, New York, 1980.

[57] D. A. Park. *Introduction to the Quantum Theory*. 3rd. edition, McGraw-Hill, New York, 1992.

[58] P. J. E. Peebles. *Quantum Mechanics.* Princeton University Press, Princeton, NJ, 1992.

[59] A. Peres. *Quantum Theory: Concepts and Methods.* Kluwer Academic Publishers, Dordrecht, 1995.

[60] J. L. Powell, B. Crasemann. *Quantum Mechanics.* Addison-Wesley, Reading, MA, 1961.

[61] I. Prigogine, Stuart A. Rice (Editors). *New Methods in Computational Quantum Mechanics* (Advances in Chemical Physics , Vol 93). Wiley, New York, 1997.

[62] E. Prugovečki. *Quantum Mechanics in Hilbert Space.* 2nd edition. Academic Press, New York, 1981.

[63] A. I. M. Rae. *Quantum Mechanics.* 3rd edition. IOP Pub/Institute of Physics, Bristol, 1992.

[64] M Reed, B. Simon. *Methods of Modern Mathematical Physics.* (Four volumes). Academic Press, New York, 1972–1978.

[65] R. W. Robinett. *Quantum Mechanics. Classical Results, Modern Systems, and Visualized Examples.* Oxford University Press, New York, 1996.

[66] G. Roepstorff. *Path Integral Approach to Quantum Physics : An Introduction.* Texts and Monographs in Physics. Springer-Verlag, New York, 1996.

[67] J. J. Sakurai, San F. Tuan. *Modern Quantum Mechanics.* Addison-Wesley, Reading, MA, 1994.

[68] L. I. Schiff. *Quantum Mechanics.* 3rd edition. McGraw-Hill, New York, 1968.

[69] R. Shankar. *Principles of Quantum Mechanics.* 2nd edition. Plenum, New York, 1994.

[70] G. L. Squires. *Problems in Quantum Mechanics. With solutions.* Cambridge University Press, Cambridge, 1995.

[71] L Sobrino. *Elements of Non-Relativistic Quantum Mechanics.* World Scientific, Singapore, 1996

[72] W.-H. Steeb. *Quantum Mechanics using Computer Algebra.* World Scientific, Singapore, 1994.

[73] D. F. Styer. *The Strange World of Quantum Mechanics.* Cambridge University Press, Cambridge, 2000.

[74] A. Sudbery. *Quantum Mechanics and the Particles of Nature.* Cambridge University Press, Cambridge, 1986.

[75] F. Schwabl. *Quantum Mechanics.* 2nd edition. Springer-Verlag, New York, 1995.

[76] W. Thirring. *A Course in Mathematical Physics.* Vol. 3. Springer-Verlag, New York, 1981.

[77] J. S. Townsend. *A Modern Approach to Quantum Mechanics.* McGraw-Hill, New York, 1992.

[78] E. Zaarur, P. Reuven. *Schaum's Outline of Quantum Mechanics.* McGraw-Hill, New York, 1998.

# Index

Printed in the United States
By Bookmasters